TO THE EUPHRATES AND BEYOND

TO THE EUPHRATES AND BEYOND

Archaeological studies in honour of Maurits N.van Loon

Edited by
O.M.C.HAEX, H.H.CURVERS & P.M.M.G.AKKERMANS

A.A.BALKEMA / ROTTERDAM / BROOKFIELD / 1989

Published by

A.A.Balkema, P.O.Box 1675, 3000 BR Rotterdam, Netherlands

A.A.Balkema Publishers, Old Post Road, Brookfield, VT 05036, USA

ISBN 90 6191 866 9

Printed in the Netherlands

Contents

Foreword

Planning this volume with studies in honour of Maurits N. Van Loon was a pleasant and challenging task. When we began to enumerate all the possible contributors for this celebratory volume, we realized that it would be impossible to include all his former and present colleagues, his students, and his many other friends. Forced by practical considerations to keep the size of the book within certain limits we have decided to restrict the number of contributors to his former students and collaborators in the field. We are painfully aware that this meant that some of Van Loon's very good friends could not be invited to participate, and we can only hope that they will forgive us.

The contributions in this volume are offered as a tribute to one of the most productive, versatile and stimulating Archaeologists today. So much is standard praise. Perhaps Maurits Van Loon's unique and overriding quality, however, is a determination to expand the archaeological discipline and to examine critically some of its most fundamental assumptions. Whether collaborating with biologists, pedologists or chemists he continues a vigorous, wide-ranging search for new methods of study and new areas of relevance. It is characteristic of the wide range of Van Loon's interests that the articles presented here cover such a variety of fields with a wide chronological and geographical range.

It is only fitting that the contributors to this volume all have been collaborators or former students of Dr. Van Loon and that the initiative in conceiving and editing it has remained entirely in their hands. Principal responsability for initiating it rested with Odette Haex, Peter M.M.G. Akkermans and Hans H.Curvers. We wish to express our gratitude to the committee *Vrienden-cadeau Maurits van Loon* for the financial support in the genesis of this celebratory volume. Special thanks are expressed to the *Stichting Nederlands Museum voor Anthropologie en Praehistorie* (Amsterdam), and the *Allard Pierson Stichting* (Amsterdam).

Finally, the editors would like to thank Remko Jas who has helped to correct the first prints.

Notes on the prehistoric environment of the Syrian Djezireh

SYTZE BOTTEMA

At that last stop we made, I asked one old fool who was leaning against a tree if he thought it was getting hotter, year by year. All he said was, 'It's always been hot skipper, since the day the world was made.' 'And what day might that be?', I asked him. 'In the Ice Age, as I heard tell.' That was his reply.

Bryan Aldiss: *Helliconia Summer*

When one travels through the Syrian Djezireh nowadays, one meets a barren landscape, sometimes ploughed all over up to the horizon. Where no plough marks are visible, a hard gypsum layer or solid rock kept the farmers from turning over the soil. Still this is a fairly late development. Some twenty-five years ago, when the road along the Euphrates from Meskeneh up to Abu Kemal was still a dirt track with Egyptian vultures resting on the telegraph poles, dense shrubby steppe vegetation on state property that had been fenced in for military purposes showed the potential plant cover. At that time random ploughing, a risky way of farming, was beginning to develop. However, plenty of undisturbed soil was still present, carrying a vegetation that, although heavily suffering from overgrazing, showed some natural aspect.

It is not surprising that human habitation in the Djezireh concentrated upon the rare water sources. Not all these water sources could be efficiently exploited for agriculture. The water regime of the Euphrates included the inundation of the wide valley in the spring thus frustrating any attempts at farming on the lower river-banks. The water of this river could only be used at the time when pumps were constructed that were able to transport water to levels that could be farmed and that were not threatened by the spring floods. Of the rivers of northern Syria the water of the Balikh seems to have been the most accessible, as is also suggested by early records. The Khabour has cut a channel that at least nowadays makes the water run up to 3 metres below the level of the Pleistocene bed and the upland. This situation must have prevailed at least from the Late Bronze Age onward, as can be concluded from the minimum date of prehistoric canals constructed on both sides along the Habur for irrigation purposes. The water could be diverted upstream and led to arable land downstream (Ergenzinger et al., in press).

What conditions did prehistoric farmers meet in the Djezireh and how did they exploit the area? To obtain an answer to such questions cooperation between archaeologists and natural scientists is of fundamental importance. The results of the study of past vegetation

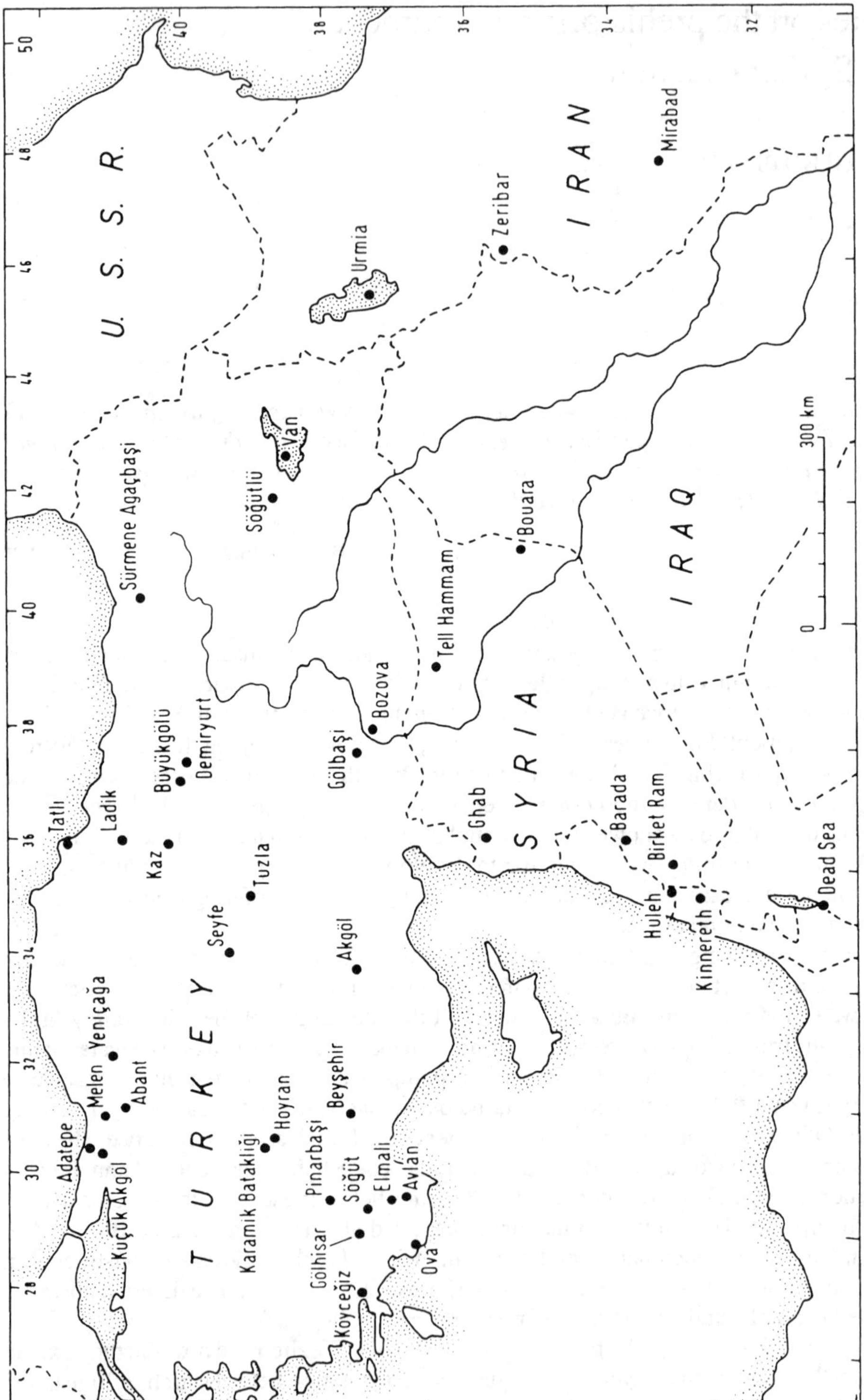

Fig. 1. Map of the Near East showing the pollen coring sites.

and climate have to be projected upon a background of archaeological information supplied by field surveys and excavations. Palaeobotanical research had already taken place in the area during the excavation of the prehistoric settlement of Mureybit, Selenkahiye, Hadidi, and Bouqras, which covered various time periods (van Zeist and Bakker-Heeres 1984 (1986); 1985 (1988); van Zeist and Waterbolk-van Rooijen 1985).

More recently attention has been paid to palynology in the area (Gremmen and Bottema, in press). In close cooperation with the University of Amsterdam's Archeological Mission to Syria, directed by Maurits N. van Loon, coring for pollen analysis formed part of extensive palaeobotanical sampling. At the invitation of Hartmuth Kühne (Freie Universität, Berlin), who directed archaeological excavations in the Habur area, a series of pollen cores were taken in that part. Some of the results of these recent palynological investigations in connection with human habitation also form part of this paper.

THE MODERN VEGETATION

A clear description of the modern vegetation of the eastern part of the area considered here is given by Frey and Kürschner (in press), while additional information on soil and climate of the Balikh area is supplied by Boerma (in press). Both studies are thought to be reasonably representative for the whole area.

The vegetation of the Djezireh can be divided into two main categories: one where trees are present and treeless vegetation. The first group includes steppe-forest on the Jebel Abdel-Aziz and riverine forest along the Euphrates, Balikh and Khabour. The steppe-forest on those slopes of the Jebel Abd el-Aziz which are less exposed to the sun still show pistachio (*Pistacia khinjuk*), almond (*Amydaglus orientalis*), hawthorn (*Crataegus azarolus*, *Crataegus aronia*) and cherry (*Cerasus microcarpa*). Of the herbs, mallow (*Alcaea rufescens*) and St. John's wort (*Hypericum triquefolium*) attract the attention of the traveller. Rejuvenation of the steppe-forest was and is very problematic because of grazing pressure, as can be witnessed on the Jebel Bichri, G. Mrâh and G. Abu Rezmeïne in central Syria, where *Pistacia atlantica* will gradually become extinct.

Very few remnants of riverine forest are found nowadays. Incidentally willow shrubs were seen along the Balikh and Khabour. Poplar (*Populus euphratica*) was often seen in plantations. Willem van Zeist and the present author witnessed a natural-looking riverine forest formed by *Populus euphratica* and some mulberry trees in 1965 on an island in the Euphrates, not far from Deir ez-Zor. The forest even housed a large rookery (*Corvus frugilegus*). Such islands were said to form part of large estates, and the forest was protected up to the time when the property of the landowners was re-distributed and given to smallholders, who were already cutting down the trees when we visited the island.

Due to human interference natural steppe vegetation is very rare or even absent. Already in the 1960s heavy overgrazing was seen. Grazing pressure was lower only in those parts where no water supply was found, although in 1965 water transport by car to the interior could be witnessed. Information on the actual, anthropogenic vegetation as well as on the potentially natural vegetation is given by Frey and Kürschner (in press). Frey and Kürschner postulate a wormwood (*Artemisia herba-alba*) shrub-steppe as the natural dominating vegetation in the Djezireh. Most of this vegetation cover has long

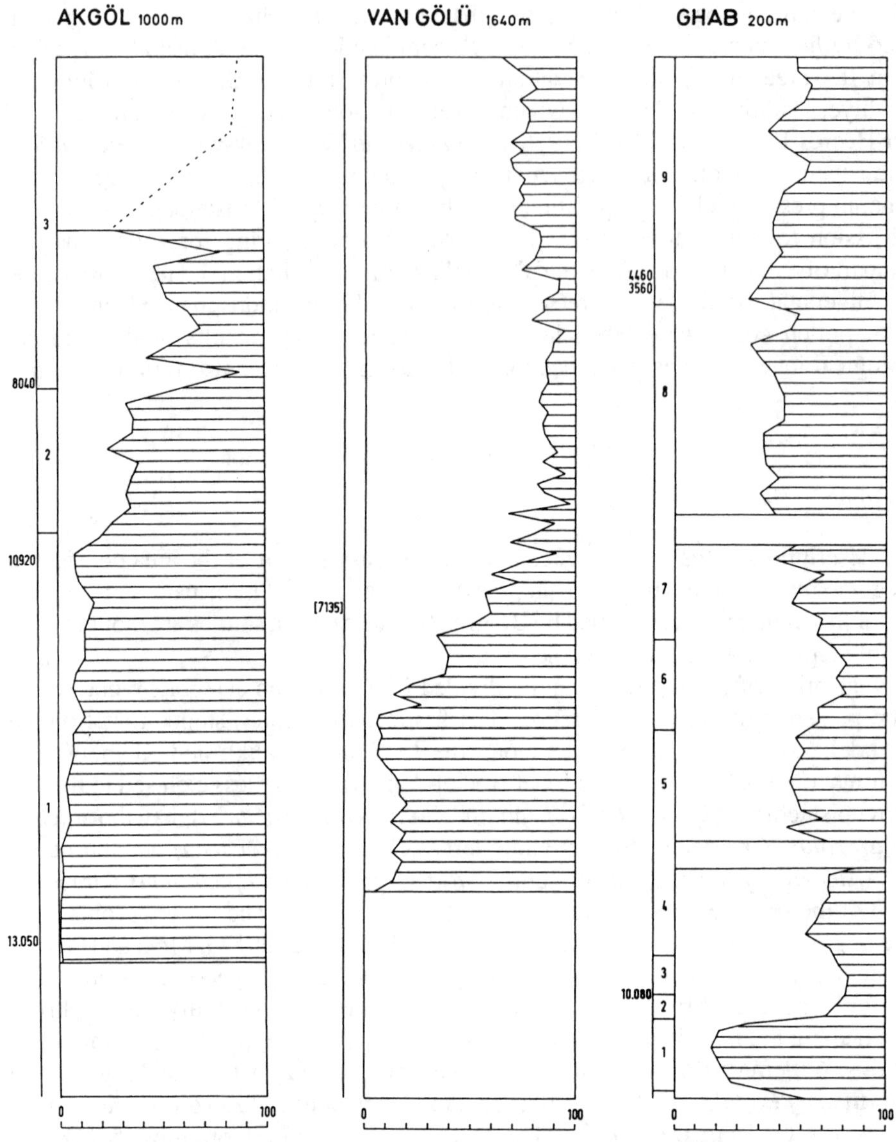

Fig. 2. The arboreal/non-arboreal pollen ratio for a selection of Near Eastern sites. White: the total amount of tree pollen. Hatched: pollen of Artemisia and Chenopodiaceae as most important steppe indicators.

since disappeared because of graizing, subsequent overgrazing, fuel gathering and in more recent times by extensive ploughing for hazardous rain-fed agriculture. These activities favoured unpalatable or prickly plant species avoided by the domestic animals, weeds of summer and winter crops, plough-resistant species, ruderals or plants of disturbed habitats. Irrigation involves typical weeds such as *Tribulis terrestris*, while abandoned saline fields may carry *Prosopis farcta* and *Alhagi maurorum*. Former

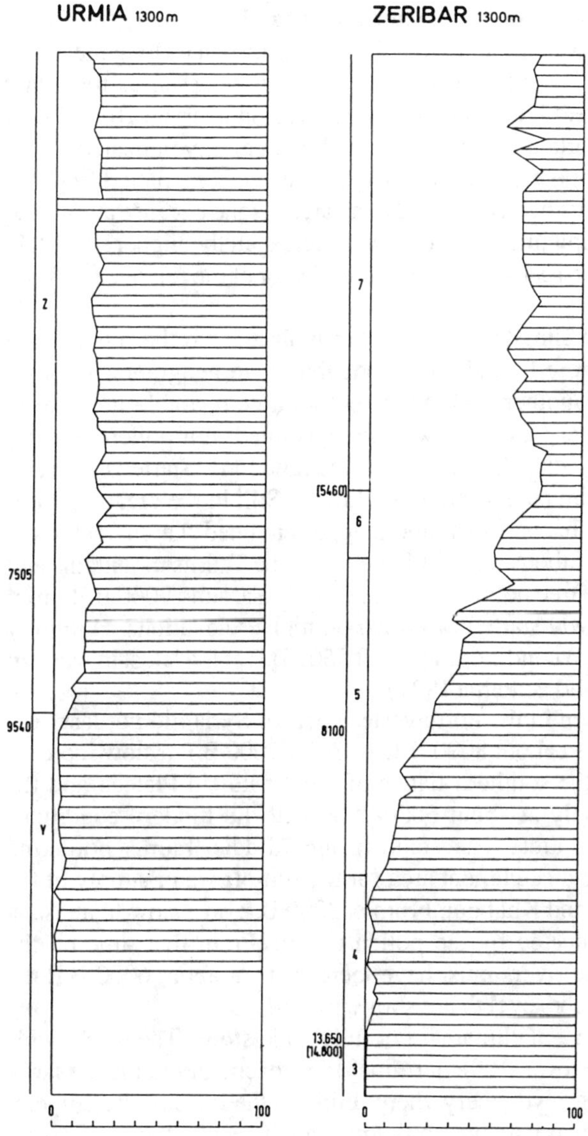

settlements or camp sites often show *Peganum harmala* or *Atriplex leucoclada*. On saline soils, halophytic vegetations are found rich in chenopods, such as *Haloxylon*, *Salsola*, *Noaea*, *Halocnemum*, *Aellenia*, *Seidlitzia*, *Suaeda* a.o., depending on local conditions.

PALAEO-VEGETATION AND CLIMATE

Some 15,000 years ago steppe or desert-steppe prevailed in the Eastern Mediterranean and the Near East, as can be concluded from the pollen record (fig. 1) (van Zeist and

Bottema 1982). Most of the area was devoid of any tree growth and tree refuge areas were only found where local conditions were more favourable, as for instance along the Black Sea coast or in open forest in the Levant. For that period it can be assumed that plant growth was limited by temperature. Global temperatures were on average 10° C lower than they are at present. More important however was the low humidity indicated by the nature of the steppe and desert-steppe vegetation which has been reconstructed from the pollen record. Tree growth must have been restricted to a narrow zone because at low and medium levels the amount of precipitaiton was too low, whereas in the higher mountain belt it was too cold. The AP/NAP ratio seems to be a good indicator of the relative moisture conditions (fig. 2).

At about 14,000 B.P. a change in climate must have taken place in northern Israel. An increase in oak pollen points to more humid conditions that were geographically of a fairly restricted character. It took another 3000 years before a comparable increase in precipitation was felt in northwestern Syria as was demonstrated in a pollen diagram from the Ghab (Niklewski and van Zeist 1970). At the same time forest started to expand in northwestern Turkey indicating an amelioration of climate. Still in the greater part of the Near East conditions did not change much and steppe remained the characteristic vegetation. On the southern edge of the Anatolian Plateau, on the Taurus mountains and on the Iranian Zagros an increase in tree pollen took place starting around 9000 B.P. From that time on oak and pistachio slowly started to spread on higher elevations. Humidity gradually increased, to reach modern conditions at about 5500 B.P. at the latest in western Iran in the Zeribar area (van Zeist and Bottema 1977).

The early increase in precipitation in the northwestern part of Syria did not last very long and dry conditions had returned about 8000 B.P. At 8000-7000 B.P. a slow increase in moisture became effective on the southern flanks of the Anatolian Plateau and the western side of the Zagros, but only at about 6000-7000 B.P. did modern conditions become established in the Taurus area. The south-facing Middle Taurus bordering northern Syria is very important as a catchment area for the run-off water supply of the Djezireh by the Euphrates, Balikh and Khabour. From c. 8000 B.P. on a slow increase in precipitation took place, as is indicated by the pollen record. From that time on the amount of water transported by the rivers must have increased. At about 6000 B.P. the river discharge established itself at the level that it shows nowadays.

The mechanism behind the change of climate is not fully understood. The world-wide rise in temperature that took place towards the terminal Pleistocene did not necessarily result in better conditions for tree growth everywhere. Initially the increase in temperature was not matched by an increase in precipitation, and it resulted in even drier conditions than before. Today the climate in large parts of the Near East is of the Mediterranean type and is generally effected by the westerly flow. (The reason that northern Israel ran so much out of line could be explained by the fact that during the terminal Pleistocene that part was not influenced by the westerly flow but by the monsoon).

In temperate Europe palynology has been of great help in reconstructing the past environment. This discipline depends upon the presence of pollen-bearing sediments or deposits. Only in some parts of the Near East are such conditions fulfilled, for instance at high elevations where a moisture regime prevailed that guaranteed peat growth, or the presence of perennial water bodies in which sediments such as clay were formed. Favourable conditions could occur in coastal areas that received enough precipitation to prevent sediments from drying up. It is obvious that humid conditions permitting peat growth or which are responsible for the development of perennial water bodies are absent or extremely rare in steppe regions. To reconstruct the past vegetation and climate of the steppe area, we have to rely upon the general picture of the Near East.

The development of the climate of the Djezireh has to be constructed mostly from the evidence obtained for adjacent areas (fig. l), and only to a lesser extent from the pollen record of the area itself. Besides, the latter evidence does not date back to before c. 7000 B.P. (Gremmen and Bottema, in press). So far no sediment covering the complete Holocene has been collected in which pollen has been preserved. The oldest sediment collected was obtained from the salt lake of Bouara, on the border of Syria and Iraq. In Bouara the beginning of the sedimentation is dated at c. 7000 B.P., at which time grey-blue clay was deposited on a brown/ochre coloured impenetrable clay, that may be of Late Pleistocene age. The 430-cm-long core was radiocarbon-dated 5730±120 B.P. (GrN-13376) at a depth of 340-350 cm. The formation of Lake Bouara is in phase with the development of the climate indicated by the pollen record of Lake Van and Söğütlü in East Anatolia and the lakes or Urmia, Zeribar and Mirabad in Northwest Iran (see also van Zeist and Bottema 1982).

It is questionable to what extent the Holocene moisture regime in the western part of the Syrian Djezireh resembled that of the eastern part. In this respect the reader is referred to the general description of the development of the climate in the Near East, where it is stated that for northern Israel the most humid conditions were found already from c. 14,000-11,000 B.P. on. More recent information on the Huleh basin, obtained by Uri Baruch (The Hebrew University, Jerusalem) and the present author confirm that at that time deciduous oak woodland predominated. Remarkably high values for Cerealia-type pollen draw attention. Cerealia-type pollen of the size that includes *Triticum dicoccoides* was regularly met with. It is not unlikely that such early favourable conditions predisposed the development of the Natufian culture. Still the distance between northern Israel and the Djezireh is rather large and the climatic development of the latter would more likely have been in phase with that or northwestern Syria, as has been studied for the Ghab valley (Niklewski and van Zeist 1970; van Zeist and Woldring 1980). In the Ghab valley the most favourable moisture conditions occurred from about 11,000-8000 B.P. More favourable conditions in the western Djezireh for the period under discussion are pointed at by the macrofossil record. In the Epi-Palaeolithic layers of the settlement of Mureybit, carbonized caryopses of wild einkorn were found during the excavation by van Loon (van Zeist and Casparie 1968; van Zeist and Bakker-Heeres 1984 (1986)). Nowadays wild einkorn is not recorded for the Mureybit area and the nearest stands of this wild cereal are found on the slopes of the Taurus, for instance the Karadağ above an elevation of 800 m. The habitation of Mureybit, the oldest prehistoric settlement recorded up to now, dates back to 10,500 B.P. Did the Mureybit people collect the wild einkorn as far away as up in Turkey, or did this cereal occur close to their settlement?

The Ghab record indicates that moisture conditions during part of the Late-Glacial and the first millennia of the Holocene were better in the Jebel Alaouite than afterwards. In this respect one should not forget that the global temperature for the period 11,000-8,000 B.P. was lower than at present. Especially at the beginning of the Holocene, the average annual temperature may have been a few degrees lower. Moister condition, present in the Alaouite area, may have extended somewhat further to the east, causing a shift of the belt of wild einkorn to the south, where it was exploited by the inhabitants of Mureybit.

Overgrazing must have been felt most at the margin of the distribution area of a species like wild einkorn. In this respect it is interesting to learn that at present parts of northern Iraq and Iran show an important increase in the amount of wild cereals in places where they were almost or completely absent. This is a result of the present war, which put an end to grazing because of the danger of mines. The rapid expansion of wild cereals in the Kurdistan area is of great interest for genetics in connection with commercial wheat-breeding, but also for research in ecological prehistory. At the moment field work is not possible and after future pacification natural scientists will have to do their field work faster than sheep and goats (International Meeting on Cereals, Eucarpia, Wageningen; February 1988). It seems, however, very unlikely that grazing or overgrazing led to the disappearance of wild einkorn in the Mureybit area in later periods. The geographical situation and especially the elevation do not seem suitable for the occurrence of this plant species in the Djezireh.

It is possible that the distribution pattern of prehistoric settlements can inform us on past climatic conditions. The presence of farming villages that relied on dry-farming is an indication of a certain level of precipitation. The limit for dry-farming coincides more or less with the 200-mm isohyet. According to TAVO map A VI 4 this isohyet runs c. 30 km south of Raqqa. However, measurements for the Raqqa station give 174 mm during the period 1951-1962. In this respect the mapping of Neolithic sites in the Balikh area by P.M.M.G. Akkermans is of interest. Neolithic people who settled about 8000 B.P. in the Balikh valley were one to two thousand years ahead of the distinct precipitation increase that is suggested by the pollen record of especially the Zagros Mountains. It is note likely that the vegetation showed a delay of a thousand years in reacting to the increase in precipitation.

Although an increase in humidity is concluded from the arboreal pollen values in the adjacent mountain ranges, trees did not expand to the Djezireh steppe. We are hardly or not at all informed on the nature of the upland steppe vegetation during the first half of the Holocene, and we do not know to what extent it differed from that of the upper part. In the first place no pollen diagrams from the area under discussion, which cover the complete Holocene, are available. In addition, there is evidence that many changes in plant cover of the last 6000 years were due to human activity.

The presence or absence, the appearance or disappearance of human cultures is explained too often by archaeologists as being related to climate. Certain forms of economy are directly linked with climate, for instance olive culture, dry-farming or date production, but the disappearance of farming is not necessarily caused by a change in climate. The appearance of Neolithic habitation in the Djezireh must be linked with the presence of water, as many settlements were founded for instance along the Balikh and the wadis running into this river and were not linked with favourable precipitation.

Akkermans (1988 and personal communication) noted that the early Neolithic sites in the Balikh valley, although linked with the drainage system, were not situated at some distance from the Balikh itself, but on the banks of branches of the Balikh or tributary wadis. Only where the stream had cut into the higher Pleistocene plateau, did the banks carry habitation as is, for instance, the case with Tell Assouad and Tell Damishliyye. Akkermans wonders whether the distribution of early Neolithic settlements points to more marshy conditions and a wetter climate around 8000 B.P. This idea is, however, not supported by the pollen evidence. For coastal Syria the early Holocene conditions became drier around 8000 B.P., whereas evidence from adjacent parts in eastern Anatolia and northwestern Iran points to an increase in precipitation becoming effective after about 7000 B.P.

How can we explain the preference of the Neolithic farmers for locations at some distance from the streams? Even today the traditional farmers of the Balikh area live in their beautiful multi-domed mudbrick farmhouses on the plateau on the west side of the Balikh one or more kilometres away from the water. When they were asked why they lived there, such farmers explained the choice of the location of their homesteads as a matter of drinking water. They said that they stayed away from the Balikh because if they settled any nearer to the river the bad taste of the river water could be traced in the ground water. Close to the water mosquitoes are a nuisance, but even the early Neolithic people of Nea Nikomedeia and Dikili Tash lived at the edge of the marsh, as do present-day Syrians who build along the Balikh.

The pollen and sedimentation record so far obtained for the Balikh or Khabour area hardly informs us about the climate of the Neolithic period. During coring in the shallow Wazgöl, north of Tell Hammam et-Turkman, a hard impenetrable layer was hit at a depth of 310 cm. In the Wazgöl sediment pollen was found down to 50 cm only. From the top down to 50 cm the pollen turned out to be progressively corroded; evidently the disappearance of the palynological contents in these upper layers is only a matter of time, on account of the repeated drying up of the lake. The fact that no pollen was preserved in the lower part of the sediment seems to prove that conditions were not wetter at that time than they are today. If significantly more humid conditions had occurred around 8000 B.P., one would expect relatively soft clay containing pollen to be present at deeper levels.

What is said above does not mean that the decision of early Neolithic people to settle at some distance from the water was not related to the climate that prevailed at that time. Palaeolimnological investigations of Lake Urmia in northwestern Iran (Kelts and Shahrabi 1986) in combination with palynological investigation of the same core taken from that lake (Bottema 1986) may be informative in this respect.

During the earlier part of the Holocene an influx of minerals from the adjacent mountains occurred in Lake Urmia. After that period the sediment changed into aragonite, carbonates and other precipitates that originated from the water itself. It seems obvious to explain the first phenomenon by an increase in precipitation resulting in a higher river discharge that washed in minerals. The latter phenomenon could result from dry conditions leading to increased evaporation. The pollen record of the older part, however, points to a very cold and dry open *Artemisia*/chenopod steppe, whereas at the time when various kinds of soluble salts started to precipitate, the pollen record indicates the development of vegetations that could have occurred only under a more humid regime. Very likely the inwash of minerals had been caused by torrential showers, which were indeed rare but which did fall occasionally and could easily have eroded the slopes, unhindered by any substantial plan cover. When a more stable and favourable moisture regime developed this resulted in a denser vegetation which in turn prevented the rapid run-off of the precipitation and melt water, a vegetation that even kept and evaporated an important part of the precipitation itself. Lake Urmia had in time turned from a freshwater body into a hypersaline lake, and a large part of the sediment was formed by local chemical precipitates.

A reason for keeping some distance from the stream channels in the Balikh area may have been the threat of flooding caused by torrential showers. It is not impossible that this regime changed during later periods, so that people could settle closer to the edge of the streams. In this respect Bar-Yosef's (1986) explanation of the function of the PPNA walls of Jericho is of interest. Bar-Yosef argues that those walls did not form part of a defence system against human intruders like the walls Joshua won his credits for much later, but that they were raised against the threatening floods of the Wadi el Mafjar. In the Balikh valley Neolithic habitation was present along the lateral channels. One wonders whether it was only the Balikh itself that flooded from time to time, floods that were fed by rain or melt-water from the mountains in Turkey. Such explanations are, however, rather speculative.

The preceding part shows that conclusions about the palaeoclimate based upon the settlement pattern have their limits. The pollen record as it is treated by van Zeist and Bottema (1982) is more suggestive, but a definite answer is not obtained from this discipline either as far as the Djezireh is concerned, as especially the Balikh valley lies in between areas with records that are not in phase.

Superimposed upon the natural development influencing the course of the climate were human activities. Anthropogenic pressure strongly influenced the more stable ecosystems where it very rapidly became the controlling factor. In vegetations that occurred under rather marginal conditions as regards precipitation, herding soon became destructive. In a more dynamic system like that of the Balikh valley and adjoining swamps, man was another dynamic factor that became dominant only after some time. If the actual and potential situation in the Balikh valley are considered, it looks as though a relatively short period of protection would restore the vegetation. Regeneration of the upland steppe would take much more time and the situation might even be irreversible to some extent.

It may be assumed that the exploitation of the upland (the area outside the valley) sooner or later resulted in over-cropping. The precise moment in time when such effects became felt is difficult to define. In this respect a case study of Ilse Köhler and Gary Rollefson on the effect of early Neolithic habitation, even leading to the destruction of the ecosystem and the subsequent disappearance of the instigators of the damage, drew attention (oral communication XIIth INQUA Congress, Ottawa 1987). Köhler and Rollefson produced evidence of the complete destruction of the woodland ecosystem in the Ain Ghazal area in northwestern Jordan. This resulted in erosion and may have been the reason for the disappearance of the early Neolithic habitation. According to Akkermans (personal communication) the southern Balikh valley was abandoned c. 8000 B.P., while about 500 years later the same can be said for most of the settlements in the north.

Compared to northwestern Jordan no oak woodland was present in the Djezireh. The find of small fragments of fruits of pistachio, almond and prune in Neolithic Tell Sabi Abyad (Akkermans 1987; 1988) is not sufficient proof of the occurrence of these trees in the neighbourhood. According to van Zeist (personal communication) these few remains may originate from a distant area. The rich fauna identified by H. Buitenhuis points to the presence of an extensive riverine forest as the only locality where trees were found.

The destruction of the vegetation of the upland decreased the carrying capacity for domestic herds and resulted in a gradual lowering of the biomass. This disappearance of plant cover caused a higher albedo, resulting in a large evaporation effect. The temperature of the soil, no longer protected by plants could be increased by maximally 5° C (Geddes 1983). The effect of the wind upon the soil would have increased if the plant cover had disappeared. The heating and drying effect of the wind causes greater loss in soil moisture than is the case if vegetation is present. In itself a protective plant cover evaporates considerable amounts of moisture, related to the total leaf surface. Still, apart from the important evapotranspiration of the leaves, the soil underneath would be protected to some extent. Especially the dark colour of the vegetation could effect higher precipitation because of the saltation effect. It is stated by Brown et al. (1982) that the destruction of forest can result in the loss of 60% or more of incident rainfall. Subsequent soil losses may be 50-200 times. As the Djezireh never carried any substantial forest, moisture loss will have been lower but the ever increasing anthropogenic pressure must have influenced the moisture balance in a negative way. Although anthropogenic activity may thus have changed the result of the climate, it is difficult to demonstrate the quantitative effect.

LOCAL EVIDENCE

In connection with van Loon's excavation in the Balikh valley and Kühne's investigations in the Khabour region palynological investigations in the Syrian Djezireh were carried out by Gremmen and Bottema (in press). Coring, mainly in oxbow lakes of the Balikh and Khabour or shallow basins (Waz Göl, Moi el Faidate, Chatounieh, Bouara), never exceeded 4 metres and generally came to a halt because of the presence of impenetrable layers, presumably old surfaces on which the formation of Younger Holocene sediment had started. The onset of sedimentation could be radiocarbon dated only in the Bouara depression, where organic material had been deposited. There sedimentation started about 7000 B.P. This is in line with the assumed increase in precipitation in the adjacent mountainous area. Still the run-off was never able to form perennial water bodies. Repeated to occasonal drying-up of lakes or marshes is assumed from the absence or corrosion of pollen and the red or yellow-brown colour of the sediment, both observations pointing to oxidation. Even Lake Chatounieh must have suffered from fluctuating levels as bordering deposits are devoid of pollen, indicating frequent drought. The Chatounieh sedimentological record very likely dates back to the second half of the Holocene, which was more humid than the first half.

The results of these investigations were informative, but not very promising. For a detailed report on the results the reader is referred to Gremmen and Bottema (in press). Some palynological results from Tell Hammam et-Turkman will be discussed here. The prehistoric site very likely started during or before the 'Ubaid period on an alluvial deposit of the Balikh. A short coring was performed to trace the virgin soil under the tell. At the east side of Tell Hammam a deep trench had been made through most of the accumulation of which the 'Ubaid layers had been excavated by Akkermans. The question remained whether the virgin soil could be traced accurately by coring. It is difficult to trace the level where the first strata of a settlement mound started on the virgin soil by means of a Dachnowsky sampler. At Tell Hammam attempts were made to trace the virgin soil by means of coring, but here like other sites in the Near East, no boundary could be seen. It was assumed that virgin soil represented by clay of the Balikh was found where the sediment showed a grey colour, a fine stratification and the typical striated fan-shape if the core sample was broken. From the part attributed to undisturbed clay under the 'Ubaid level, two samples A and B separated by 8 cm in depth have been analysed (Table 1).

Table 1. Pollen analysis of three samples from the Balikh valley in the Hammam et-Turkman area.

Spectra depth	Pre-'Ubaid A	Pre-'Ubaid B	Wazgöl 6 cm
Quercus calliprinos	.	.	0.6
Quercus cerris-type	.	.	1.8
Pinus	1.1	1.1	19.7
Humulus/Cannabis	.	.	0.3
Olea	.	.	1.6
Ostrya-type	.	.	0.2
Vitis	.	.	0.3
Juglans	.	.	0.3
Juniperus	.	.	0.2
Alnus	.	.	0.2
Abies	.	0.2	.
Cedrus	.	.	0.2
Fraxinus	.	.	0.3
Thymelaea	.	0.3	.
AP	1.1	1.6	25.5
Chenopodiaceae	1.9	1.0	11.1
Noaea-type	.	.	1.3
Aellenia-type	.	.	0.5
Artemisia	36.4	47.5	11.5
Centaurea solstitialis-type	6.3	7.4	2.2
Senecio-type	.	.	0.2
Matricaria-type	4.5	4.9	1.3
Xanthium	.	.	0.8
Carduus	.	0.2	.
Carthamus	0.7	1.8	.
Cirsium	0.7	.	.
Filago-type	.	0.4	0.2
Tubuliflorae	8.9	3.8	1.8
Liguliflorae	2.2	3.0	2.7
Plantago lanceolata-type	.	.	0.8
Plantago ovata-type	.	.	0.2
Plantago spec.	0.4	0.2	0.8
Cerealia-type	12.6	8.0	5.7
Gramineae	15.6	14.3	23.9
Caryophyllaceae	.	.	0.6
Pimpinella-type	.	0.2	.
Turgenia-type	.	1.8	.
Umbelliferae	.	1.8	2.4
Brassica-type	.	.	1.4
Capsella-type	.	.	0.6
Euphorbia	0.7	.	.
Crozophora	.	0.2	0.2
Helianthemum	.	0.2	0.2
Convolvulus	.	.	0.2
Calystegia	.	.	0.2
Cressa-type	0.4	1.0	.
Atraphaxis	.	0.2	.
Cuscuta	.	.	0.2
Solanum nigrum ssp. villosum	.	.	0.3

Table 1 *(continued)*

Spectra depth	Pre-'Ubaid A	Pre-'Ubaid B	Wazgöl 6 cm
Polygonum aviculare	.	.	2.1
Leguminosae	0.7	1.6	0.2
Leontice	0.4	.	.
Rumex cyprius-type	.	.	0.2
Valerianella	.	.	0.2
Aquilegia	.	1.1	.
Ranunculus sceleratus-type	0.4	.	0.5
Cephalaria syriaca-type	1.1	.	.
Prosopis	1.1	1.2	0.3
Sanguisorba minor-type	.	.	0.2
Galium-type	0.4	.	.
Linum austriacum-type	0.4	0.2	.
Ornithogalum-type	.	1.0	.
Muscari-type	.	0.2	.
Haplophyllum	.	0.3	.
Pollen sum	269	609	628
Cyperaceae	2.2	2.1	24.7
Sparganium-type	0.4	0.3	4.1
Potamogeton pectinatus-type	.	.	80.3
Dryopteris	.	.	0.2
Cheilanthes	.	1.8	.
Pediastrum integrum/boryanum brevicorne	.	.	48.4
Hystrix	.	15.0	
Indeterminata	16.4	3.3	5.3
Pollen preservation	±	+	++

The clay on which the 'Ubaid and younger periods have accumulated was firmly sealed and protected, comparable to the old surface under a burial mound. The construction of a burial mound over the soil surface protects the pollen in the surface layer from oxidation. The samples orginating from under the tell form a sharp contrast with those taken from a core obtained from an oxbow of the Balikh at a distance of about 100 m. In the oxbow sediment repeated drying-up was concluded from the ochre-yellow colour of the clay, and drying-up must have caused the oxidation of the less resistant pollen types and lowered the diversity in types considerably. The samples from underneath the tell are of a grey colour and pollen is preserved much better, as can also be concluded from the high diversity in types.

The age of samples A and B (Table 1) cannot be established with certainty. As the samples contain a large amount of Cerealia-type pollen grains, they cannot be expected to be older than early Neolithic times. The upper limit must be the Halaf period, as 'Ubaid layers are found on top of the clay. Roughly speaking, the age of the two samples lies between 6500-8000 B.P.

In Table 1 the pollen percentages are shown of samples A and B taken from under the

'Ubaid layers in Tell Hamman et-Turkman and a modern sample taken from the sediment in Wazgöl at a depth of 6 cm. Samples A and B contained charcoal, especially the lower sample A. Pollen preservation was better in sample B than in sample A, and this permitted the identification of a larger number of pollen types. The spectra are characterized by very low arboreal pollen values; even pine scores only 1.1%. The dominating type is that of *Artemisia*, partly *Artemisia herba-alba*-type, but it also includes other species of this genus. Grasses amount to c. 15%. This pollen may partly have been produced by plants growing in the Balikh valley.

In samples A and B Cerealia-type pollen measures 12.4 and 8.0%, respectively. Such high percentages point to cereal growing at the sampling spot or to threshing practices in the vicinity, because during threshing pollen is also released. Pollen of weeds that is also present favours the idea that crops were grown in the wet clay of the Balikh at the spot where later a settlement was formed. Nowadays some of these weeds, such as *Prosopis* and *Crozophora*, are to be found in irrigated fields. Other weeds met with are *Carthamus* and *Cephalaria* cf. *syriaca*. In later periods *Cephalaria syriaca* commonly occurred as an admixture in carbonized seed samples, as is evident from the investigations by van Zeist. The *Linum* pollen present in the assemblages does not belong to *Linum usitatissimum*, the species which was grown, but is from some wild species.

Somewhat puzzling is the presence of pollen that resembles *Cressa*. At present, *Cressa* occurs in wet and brackish fields (personal communication R. Neef), usually together with certain chenopod species. In the Tell Hammam samples, however, Chenopodiaceae pollen does not play any role.

In comparing the two spectra from pre-'Ubaid times with a modern spectrum from Wazgöl, conspicuous differences can be observed. Arboreal pollen amounts to 25.5% in Wazgöl, sharply contrasting with the 1.1 and 1.6% in samples A and B. Especially the absence of pine pollen in the pre-'Ubaid spectra points to an age of c. 7000 B.P. or older for the spectra A and B. That Chenopodiaceae values are much higher at present than at c. 7000 B.P. may be explained by the fact that salinization had not yet developed at that time. *Artemisia*, a pollen type that represents species which form an important part of the steppe vegetation, score four times higher in pre-'Ubaid time than in the modern sample.

The spectra A and B suggest that during Neolithic times it was drier than nowadays and that an *Artemisia*-steppe prevailed along the Balikh in the Tell Hammam area. Agriculture was practised in the Balikh valley, whereas outside this valley it was too dry for dry-farming. As some pollen of weed species typical of irrigated fields have been found for the pre-'Ubaid spectra, cereal farming in Neolithic times may have been based already on some form of irrigation, or was at least practised in moist alluvial soils.

ACKNOWLEDGEMENTS

I am much indebted to Maurits van Loon for the possibility to work in the Balikh area and I will not easily forget the pleasant times when I stayed in the Tell Hammam excavation camp. I received much help from Peter M.M.G Akkermans, Hans H. Curvers, Theo de Feyter, Diederik J.W. Meijer and Ferdinand Smit, who in turn put their weight upon the Dachnowski sampler during the field work. I am especially grateful to Peter M.M.G. Akkermans for his information and critical reading of the manuscript. The typing was done by Gertie Entjes-Nieborg and the English was corrected by Sheila van Gelder-Ottway.

REFERENCES

AKKERMANS P.M.M.G.

1987 'A late Neolithic and Early Halaf Village at Sabi Abyad, Northern Syria', Paléorient 13/1: 23-440.

1988 'The Neolithic of the Balikh valley, Northern Syria; A First Assessment. Symposium préhistoire du Levant no. 2, Lyon.

BAR-YOSEF O.

1986 'The Walls of Jericho: an Alternative Interpretation', *Current Anthropology* 27: 157-162.

BOERMA J.

this vol. 'Land evaluation in prehistory'.

BOTTEMA S.

1986 'A Late Quarternary Pollen Diagram from Lake Urmia (Northwestern Iran)', *Review of Palaeobotany and Palynology* 47: 241-261.

BROWN L.U. ET AL.

1982 *The Birds of Africa. II. Academic Press.*

ERGENZINGER P.J. ET AL.

in press 'The Reconstructions of Environment, Irrigation and Development of Settlement on the Habur in North-East Syria', in: *Conceptual Issues in Environmental Archaeology* (= Proceedings of the international symposium Oxford 1985, published by the University of Edinburgh).

FREY W. AND KÜRSCHNER

in press 'Die aktuelle und potentiell natürliche Vegetation im Bereich des unteren Habur (Nord-Ost Syrien). Mit einer Rekonstruktion der Vegetationsverhältnisse um Dur-Katlimmu/Tal Schech Hamad in mittel- und neuassyrischer Zeit', in: H. Kühne (Ed.), *Berichte der Ausgrabung Tell Schech Hamad/Dur-Katlimmu 1, Teil 2.*

GEDDES D.S.

1983 'Neolithic Transhumance in the Mediterranean Pyrenees', *World Archaeology* 15: 51-66.

GREMMEN W.H.E. AND BOTTEMA S.

in press 'Palynological Investigations in Northeast Syria', in: H. Kühne (Ed.), *Berichte der Ausgrabung Tell Schech Hamad/Dur-Katlimmu 1: Teil 2.*

KELTS K. AND SHAHRABI M.

1986 'Holocene Sedimentology of Hypersaline Lake Urmia, Northwestern Iran', *Palaeogeography, Palaeoclimatology, Palaeoecology* 54.

NIKLEWSKI J. AND VAN ZEIST W.

1970 'A Late Quarternary Pollen Diagram from Northwestern Syria', *Acta Botanica Neerlandica* 19: 737-754.

VAN ZEIST W. AND BAKKER-HEERES J.A.H.

1984(1986)'Archaeobotanical studies in the Levant. 3. Late Palaeolithic Mureybit', *Palaeohistoria* 26: 171-199.

1985(1988)'Archaeobotanical studies in the Levant. 4. Selenkahiye', *Palaeohistoria* 27.

VAN ZEIST W. AND BOTTEMA S.

1977 'Palynological Investigations in Western Iran', *Palaeohistoria* 19: 19-95.

1982 'Vegetation history of the Eastern Mediterranean and the Near East during the Last 20,000 Years', in: J.L. Bintliff and W. van Zeist (Eds.), *Palaeoclimates, Palaeoenvironments and Human ommunities in the Eastern Mediterranean Region in Later Prehistory.* (= BAR International Series 133). Oxford. Pp. 277-321.

VAN ZEIST W. AND CASPARIE W.A.

1968 'Wild Einkorn Wheat and Barley from Tell Mureybit in Northern Syria', *Acta Botanica Neerlandica* 17: 44-53.

VAN ZEIST W. AND WATERBOLK-VAN ROOIJEN W.,
1985 'The Palaeobotany of Tell Bouqras, Eastern Syria', *Paléorient* 11/2: 131-147.
VAN ZEIST W. VAN AND WOLDRING H.
1980 Holocene Vegetation and Climate of Northwestern Syria', *Palaeohistoria* 22: 111-125.

Landevaluation in prehistoric perspective: Some observations

J.A.K. BOERMA

From the beginning mankind has always exploited the natural resources of the earth. As Beek (1978) puts it: 'people have always been on the look out for land that suits their purposes: for building shelters, for providing food and fibre, for protection against wild animals' and 'land that was unsuitable was left idle as long as possible'. Soil has to be regarded as the main natural resource since it provides food, directly or indirectly. Consequently, the possibilities of soil for agricultural purposes have occupied mankind for a long time. For example 4000 years ago agricultural suitability of soil was used to establish a tax system in China. People were interested not in the soil itself but in its value for a certain use.

Also in modern times it has been the use that can be made of soil that interests people dealing with soils. Dokuchaiev, a pioneer of modern soil science, considered it necessary to express the value of soil for agricultural purposes. Soil genetic considerations, chemical and physical characteristics as well as repartition of types of soil were just a tool for his studies; not the ultimate goal.

Gradually observations on the use of soil for agricultural purposes were transformed into studies on the use of land for both agricultural and non-agricultural purposes. It must be emphasised that land in this respect is a much broader concept than soil. Land includes 'an area of the earth's surface which characteristics embrace all reasonably stable or predictably cyclic attributes of biosphere above and below this area, including those of the atmosphere, the soil and underlying geology, the hydrology, the plant and animal population and the results of past and present human activities' (Beek 1978). This resulted in the preparation of a framework for landevaluation (F.A.O. 1976) in order to provide a system to identify and execute the process of assessment of land performance when used for specific purposes.

Landevaluation should answer the following questions (F.A.O.1976):

- how is the land currently managed; what will happen if present practises remain unchanged?
- what improvements in management practises, within the present use, are possible?
- what other uses of land are physically possible and economically and socially relevant?
- which of these uses offer possibilities of sustained production or other benefits?
- what adverse effects, physical, economic or social, are associated with each use?
- what recurrent inputs are necessary to bring about the desired production and a minimum of adverse effects?
- what are the benefits of each use?

Introduction of a new form of landuse adds two questions to this list:

- what changes in the condition of the land are feasible and necessary and how can they be brought about?
- what non-recurrent inputs are necessary to implement these changes?

Summarizing one can say that management, landuse and its effects, socio-cultural aspects and benefits have to be considered. Although some of the questions listed here are not useful when dealing with landevaluation in a prehistoric context, it is very well possible to reconstruct the landuse potential for a given prehistoric site, environment and time, based on the same series of questions. In that case they have to be adapted to the conditions prevailing in the area and period considered. It is from this point of view that the relevance of the above-mentioned questions will be considered.

With regard to management questions, one must realize that in prehistoric time relatively few tools and virtually no machinery were known to man. Some aspects of management that can be considered are the following. In the first place, tools like 'primitive' hoes sometimes supplemented by men-driven ards. Another management aspect is also important today but in a different way: transportation (manor domesticated animal); in the prehistoric perspective only with respect to the distance to be covered, since it always had a low-velocity character; probably its availability and/or capacity played a minor role. A third aspect of management is its level; e.g. a distinction between rain-fed and irrigated arable landuse if appropriate. A more detailed distinction between the various forms of management does not seem useful considering the prehistoric conditions.

The question on improvement of management practices should not be considered with regard to the use of more sophisticated tools and the like, but mainly with regard to a change of system. For instance, if sufficient water is available in a semi-desert environment, one should consider the possibility - if culturally possible for the period in question - to change from rain-fed agriculture to irrigated agriculture.

Looking at the forms of landuse to be considered in a prehistoric context, the scope can be relatively wide. Taking into account the physical, economic and social terms of the site, the landuse may include purely nomadic landuse, several partly or episodic sedentary forms including gathering of wild fruits and 'wild crops', hunting, fishing and extensive grazing, as well as several sedentary forms including harvesting of agriculturally grown crops and fruits and more or less extensive herded grazing.

It is clear that the physical, economic and social terms cannot be looked at in the same way as in modern landevaluation. Physical terms can be derived directly from the land qualities, which does not represent an approach different from today's practice. Both economic and social terms are much more difficult and frequently, impossible to determine. No written sources are available to the investigator; the only source is the soil archive.

Economic terms can be derived from other factors like the physical environment itself, the size of a site, the density of settlements and the supposed number of inhabitants within a given area, the availability of qualified tracts of land within a certain distance as well as their accessability (obstacles, e.g a river to be crossed), *etc*. In this context one must realize that it's very likely that the number of known sites in a given prehistoric environment and time seldom represent the total of sites ever present in the area.

Social terms are of no relevance for a prehistoric form of land use since any form of reliable information is absent.

As far as the sustained production or other benefits resulting from a certain type of landuse are concerned, the relevance is clear. One must take two factors into account: the potential and the dimensions of the available land. A low potential of the land may lead to a quick wearing-out of the land, followed by abandonment and moving of the activity to another tract of land (i.e. shifting cultivation). In this context the dimensions, availability and distance of suitable land (determined among other things by the size of the settlement) may influence the period during which the land (soil) can restore itself and consequently influence the objective of sustained production. The same holds for the so-called other benefits.

The questions referring to adverse physical, economic or social effects are only relevant with respect to the physical effects, that is the environment. Neither the adverse economic nor the social effects can be taken into account since they concern the society (which we have no sound information about). Adverse physical effects may include erosion and deterioration of the natural vegetation.

The question of recurrent inputs necessary to realize the desired production can only be answered in a very limited sense, particularly since no real currency is involved. These particulary inputs have to be expressed in the form of required labour inputs which are difficult to estimate and must be derived from the land qualities. This factor can probably not be taken into account, since labour in the prehistoric perspective is a matter of availability and not of cost.

Finally the benefit of a certain form of landuse should not be expressed in the form of money. Benefits have to be seen as the possibilities offered by the environment for a given form of landuse. Here again, one is forced to consider the landuse potential instead of the benefit level.

The Framework of Landevaluation (F.A.O. 1976) presents two questions on the introduction of new forms of landuse. It is not necessary to take both questions into account because landevaluation in a prehistoric context is an evaluation of (the account possibilities of) the natural resources available to the inhabitants of a certain site in a certain period.

Landevaluation demands a distinction in several major types of landuse or, translated into prehistoric conditions, major types of use of the natural resources. The following forms have to be included:

- use of natural resources (environment) as they are without any specific input or output,
- use of natural resources for any kind of landuse of the collecting type,
- use of natural resources for cultivation purposes.

The most important form of use of natural resources without in- or output is simple and concerns mainly the site choice or part of it. It should be evaluated in terms of presence or absence of physical factors like flooding risk or topographical conditions and cultural factors like a stronghold position or the distance to water for domestic use. One can imagine several other factors.

The second major type of use of natural resources, i.e. exploitation of the natural resources in the form of collecting, includes a variety of landuse forms like the gathering of wild fruits, game and fish, of suitable rocks and clay for stoneware, tools and pottery respectively, or the collecting of building materials.

The third type, i.e. natural resources as a basis for landuse of the cultivation type, comprises forms like rain-fed arable landuse, irrigated landuse or grazing.

Other forms of exploitation of natural resources are not excluded. It is obvious that it is sometimes difficult to separate these major types of landuse.

The above mentioned observations make it clear that landevaluation in a prehistoric context must be limited to its physical aspects since economic and social aspects can be used in a derived form only. The physical aspects to be studied concern the climate, soil, rock, vegetation, topography, hydrology in their respective qualities or attributes. All of them concern physical conditions at the time the site was occupied, which means that one must consider palaeoconditions and reconstruct palaeoclimate, palaeovegetation, palaeo soil conditions, *etc*. These factors must be derived from the soil record mainly, usually in combination with the findings of comparable sites in order to establish the palaeo-environment.

In nearly all cases there is a complete absence of quantifiable assessments regarding output of the landuse form(s). This is mainly caused by the fact, that comparable sites within a given environment are scarce and/or far apart; a subjective fact, since unknown or vanished sites cannot be taken into account.

Only in relatively rare cases where the finds of a large number of sites in a relatively small area are available, may statistic manipulation of data provide some quantified assessments (cf. Kamermans et al. 1985). Thus, in most cases, prehistoric landevaluation must be limited to a qualitative one, based on the physical landuse potential. The physical landuse potential must take into account the known or supposed evidence of a certain level of development or management.

An example of landevalution according to this concept will be presented elsewhere (Boerma, in preparation).

REFERENCES

BEEK K.J.
1978 'Landevaluation for Agricultural Development', *I.L.R.I. Publications 23*.

BOERMA J.A.K.
in prep. 'Palaeoenvironment and Palaeolandevaluation Based on Actual Environmental Conditions of Tell Bouqras, East Syria'. F.A.O.
1976 'A Framework for Landevaluation', *F.A.O. Soils Bulletin* 32.

KAMERMANS H. ET AL.
1985 'Changing Patterns of the Prehistoric Landuse in the Agro Pontino', in: Caroline Mallone and Simon Stoddart (Eds.), *Papers in Italian Archaeology IV*, Oxford: BAR International Series 243. Pp. 53-68.

A microwear analysis of sampled borers from Netiv Hagdud, a PPNA settlement in the Jordan Valley

JOHANNES BUELLER

The tell of Netiv Hagdud is located about 10 km north of Jericho. The site is situated on a natural promontory, at an elevation of -180 m below sea level, overlooking the Salibyia depression. The depression, varying in elevation between -190 m and -240 m below sea level, contains several Natufian and preceramic Neolithic sites (Schuldenrein and Goldberg 1981). During the early Holocene, spring activity inside the depression resulted in a marshy environment in which the Natufians camped. Forced by increasing marshy conditions, the early Neolithic population avoided the bogs and situated their settlements on opposite margins of the depression: Netiv Hagdud on the western edge (Bar-Yosef et al. 1980) and Gilgal on a flint ridge on the eastern edge, nearer to the river Jordan (Noy et al. 1980).

Surface evidence indicates that Netiv Hagdud extended over an area of about 10.000 sqm. Sections exposed by modern construction work for a water reservoir revealed that the archaeological deposits have a depth of at least 3 m. At the western side of the tell an area of 100 sqm. has been excavated to a depth of 0.5 m. Here four oval to circular structures ranging in diameter from 4 to 7 m were brought to light. The walls, constructed by limestone blocks, were preserved to a height of 25 cm. In general, the excavated area showed a low density of artifacts. The flint assemblage consisted of core tools such as adzes, small tranchet axes, chisels and picks as well as polished limestone axes. Also present were notched arrowheads and borers. Two pieces of obsidian were recovered, both originating from Golü Dag in Anatolia (Perlman and Yellin 1980). The ground stone industry includes limestone blocks with cupmarks and limestone and basalt pestles. The low density of finds is not necessarily characteristic of the site as a whole since at nearby Gilgal, dating from the same period, the density of finds varies from zone to zone, possibly as a result of different activities. At Gilgal, retouched pieces, borers and arrowheads are common but axes are rare.

To date, no report on the faunal remains of Netiv Hagdud is available, but in neighbouring Gilgal remains of waterfowl like ducks and geese are plentiful. Among the mammals *G. gazella* is the dominant species but also woodland dwellers like aurochs, deer and boar are found (Tchernov in Noy et al. 1980).

On its western side, Netiv Hagdud was wholly covered by an accumulation of sand and gravel due to heavy slope wash. This is not surprising as the site is located on top of the alluvial fan of the Wadi al-Bagar. Bar-Yosef (1986) has pointed out that in fact many early Neolithic sites are located near the drainage outlets of streams. Such a location was chosen for two reasons: the convenience to have drinking and washing water in the

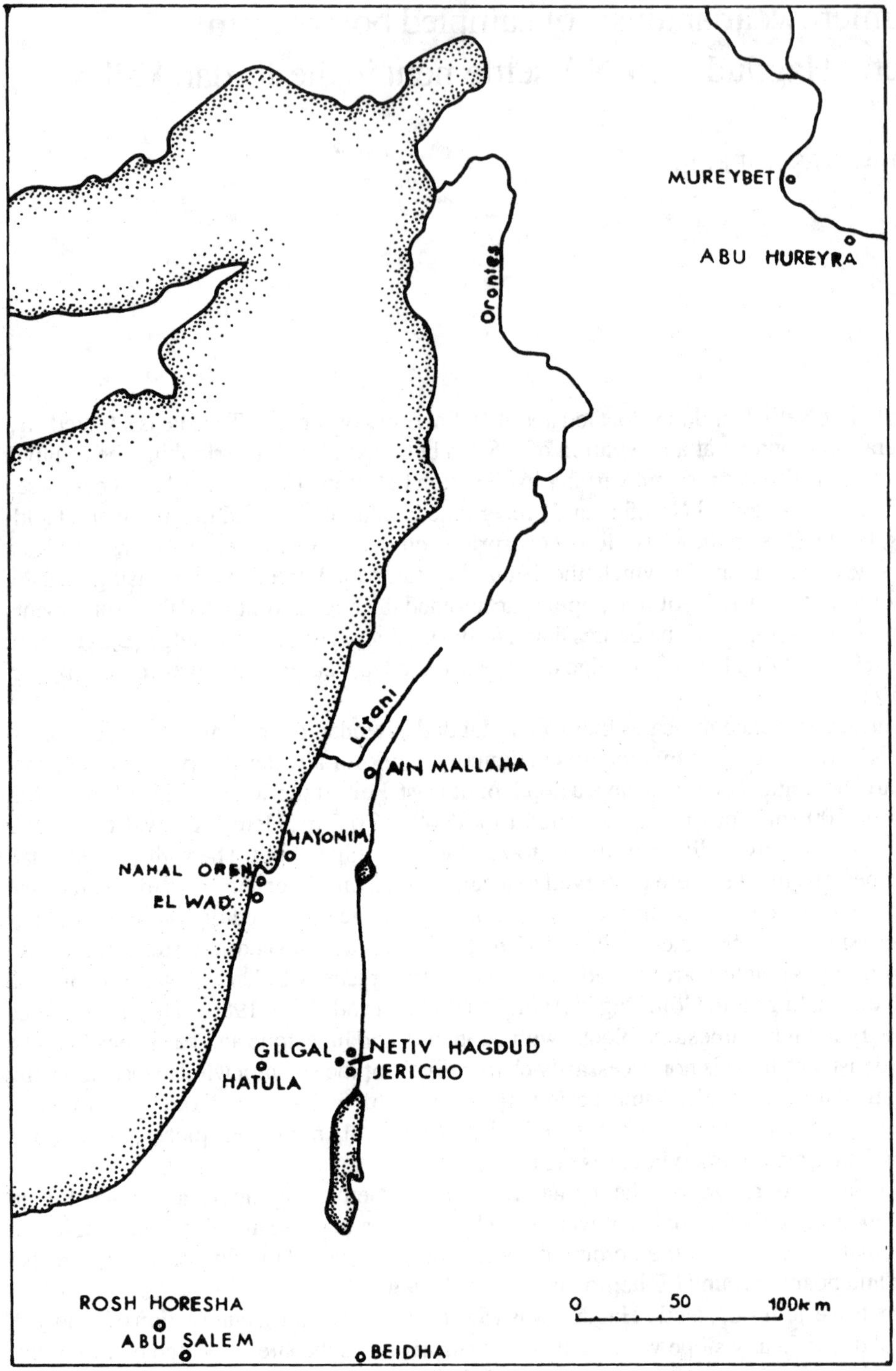

Fig. 1. Prehistoric sites mentioned in the text.

immediate vicinity, and, probably more important, the annual flooding brought a new veneer of soil which, in the absence of organic fertilizers (no animal husbandry), made a continuous cultivation possible. Settlements were located next to the fields and, depending on the local situation, needed protection against the annual or semi-annual mudflow. The 'city walls' of PPNA Jericho are considered to be such a protection by Bar-Yosef who also regards the terrace wall of Beidha (Kirkbride 1968) as a similar device. Early Neolithic settlement on alluvial fans resulted from two developments during the late Epipalaeolithic. These developments were sedentism, as evidenced by the faunal remains of the Late Natufian site of Hayonim (Davis 1983), and the cultivation of morphologically wild cereals, attested in the Late Natufian sites of Rosh Horesha (Horowitz 1977) and Abu Hureyra (Hillman in Moore et al. 1975).

Sites sharing the same cultural horizon as Netiv Hagdud include PPNA Jericho (Kenyon 1960), Nahal Oren II (Stekelis and Yizraeli 1963), Gilgal (Noy et al. 1980) and Mureybet II (Cauvin 1977), which is identical to Mureybet I-VIII (van Loon 1968).

Two radiocarbon dates are available from Netiv Hagdud: 8,230 ± 300 B.C. (RT 502 C) and 7,840 ± 380 B.C. (RT 502 A).

THE IMPORTANCE OF BORERS IN THE KHIAMIAN AND THE PPNA

From a very modest representation, around 2%, in the lithic assemblages of the Final Natufian, as defined by Valla (1984), the borer as a typological entity becomes increasingly important in the succeeding Khiamian. While lunates and other geometric microliths decrease and are replaced by notched arrowheads, the borers now easily become the second largest group or, as Ronen and Lechevallier (1985:162) put it: 'The Khiamian (of Hatula) may be defined as a Late Natufian with El-Khiam points and needle borers'. A few statistics will illustrate this point. In Hatula, where the Khiamian supersedes the Natufian, the borers increase in number from 0% to 20.8% of the lithic assemblage. At Salibiya IX, a Khiamian site near Netiv Hagdud, the borers and perforators make up 16.1% (Bar-Yosef 1980). In Mureybet IB 'les microperçoirs augmentent fortement' compared to the Natufian Phase IA (Cauvin 1977:23) and in Mureybet II the borers at some places constitute over 20% of the retouched tools (Cauvin 1977:28). Finally, in Gilgal, 'awls, borers and mèches de forêt form another large tool group next to the arrowheads' (Noy et al.1980:68).

From the above it will be clear that borers constitute an important element in the PPNA lithic toolkit.

THE MICROWEAR ANALYSIS

A sample of thirteen borers has been examined with the aid of an incident-light microscope at a uniform magnification of 200 X. Before examination, the archeological pieces were soaked in a 10% solution of HCl in order to remove any adhering mineral deposits. All relevant microwear occurrences were photographically recorded to enable an holistic analysis of each piece. An other advantage of an extensive photographic coverage is that the microwear analyst does not have to determine 'on the spot' the nature of the microwear phenomena. He or she can compare the wear traces with experimentally

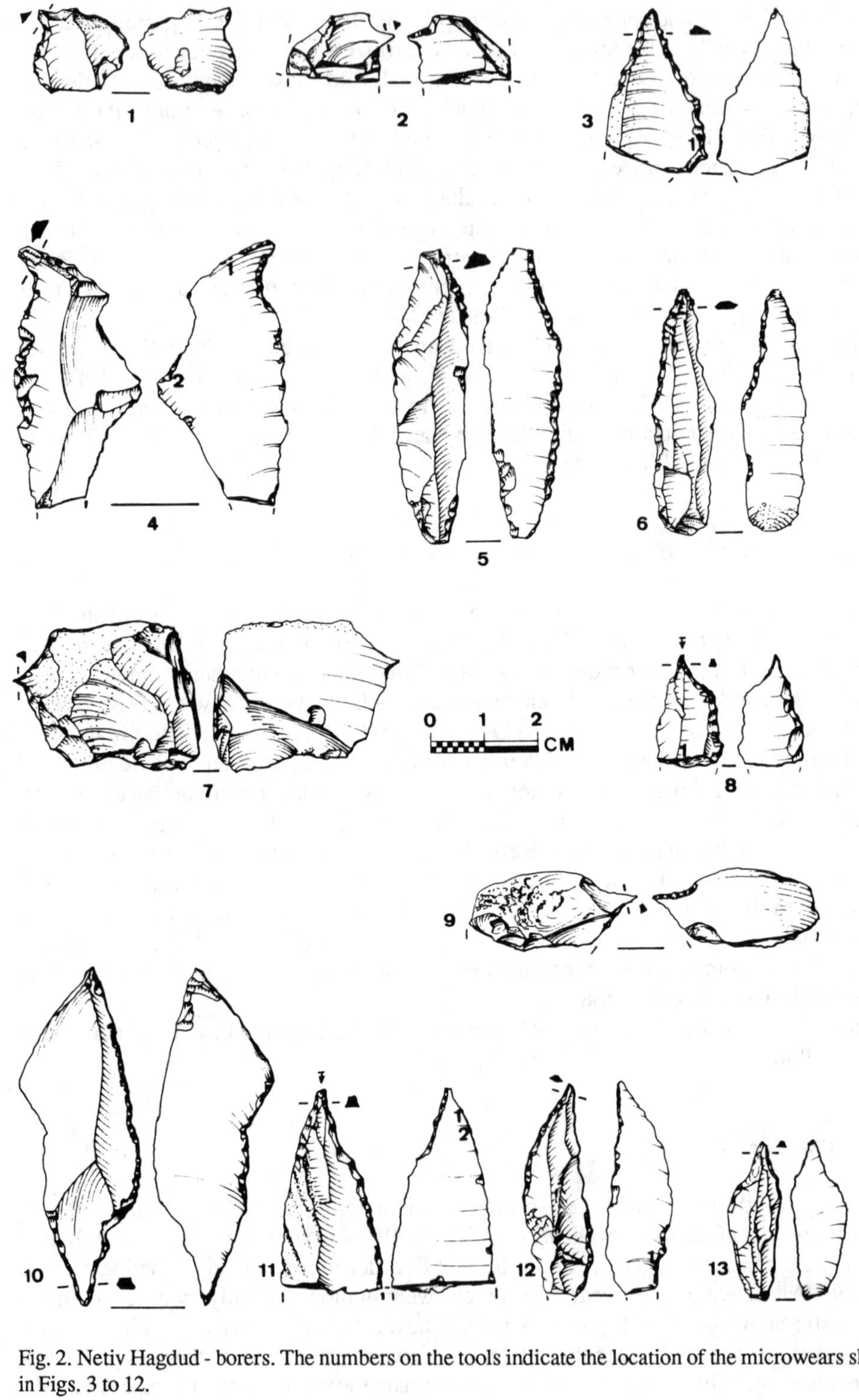

Fig. 2. Netiv Hagdud - borers. The numbers on the tools indicate the location of the microwears shown in Figs. 3 to 12.

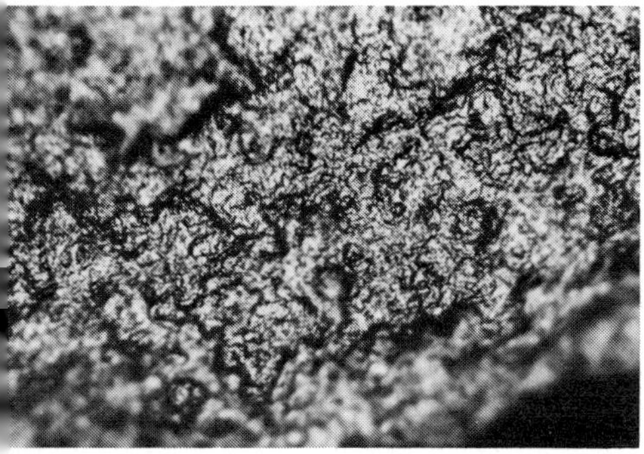

;. 3: Experimental piece HHE 12. Scraping dry ιe. Dull luster.

Fig. 4: Borer Fig. 2,, no. 11 (1). Mw (dry hide). Dull luster.

;. 5: Experimental piece TCE 5. Scraping dry .e. Narrow, deep, well-defined striation.

Fig. 6: Borer Fig. 2, no. 8 (1). Mw (dry hide). Narrow, deep, well-defined striation.

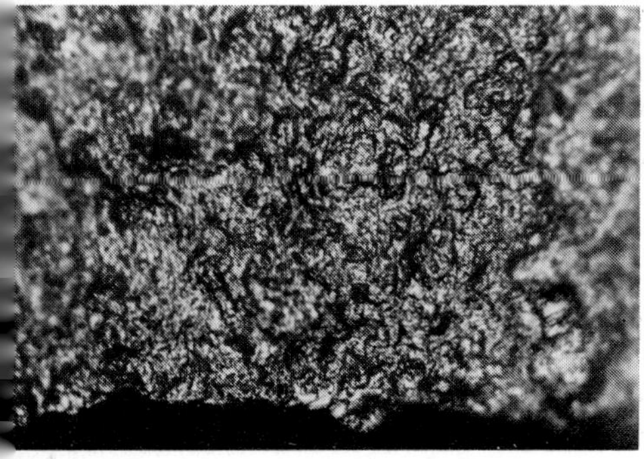

. 7: Experimental piece HHE 12. Scraping dry e. Linear features.

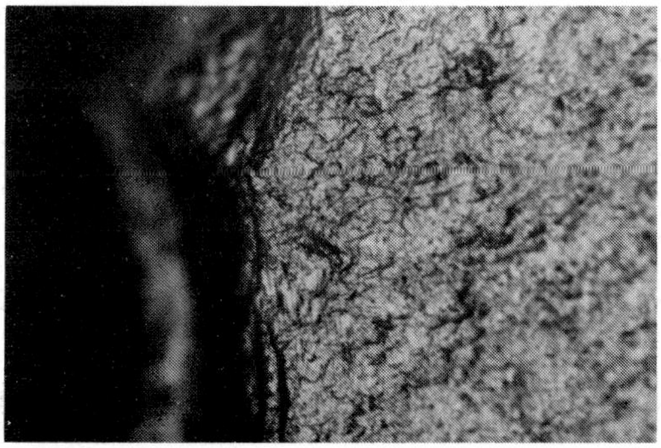

Fig. 8: Borer Fig. 2, no. 4 (1). Mw (dry hide). Linear features.

Fig. 9: Lunate Fig.2, no. 9 from Ain Mallaha. Mw (mastic). Bright luster and grooves.

Fig. 10: Borer Fig.2, no. 3 (1). Mw (mastic). Brig luster and grooves.

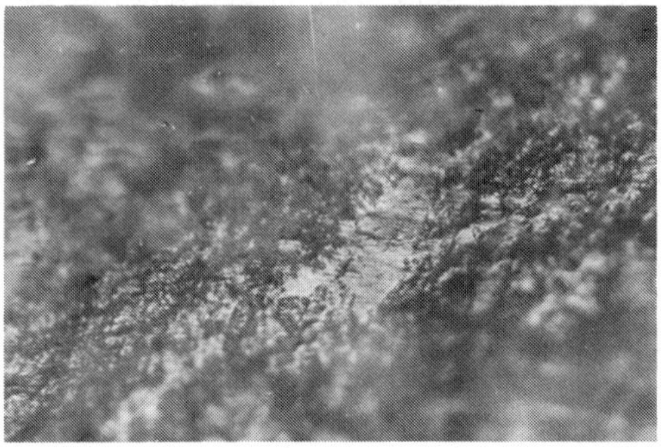

Fig. 11: Experimental piece NE 7. Cutting bone. Flat polish.

Fig. 12: Borer Fig.2, no. 4 (2). Mw (bone). F polish.

Fig. 13: Borer Fig.2, no. 12 (1). Mw (ocher). Capillar striations.

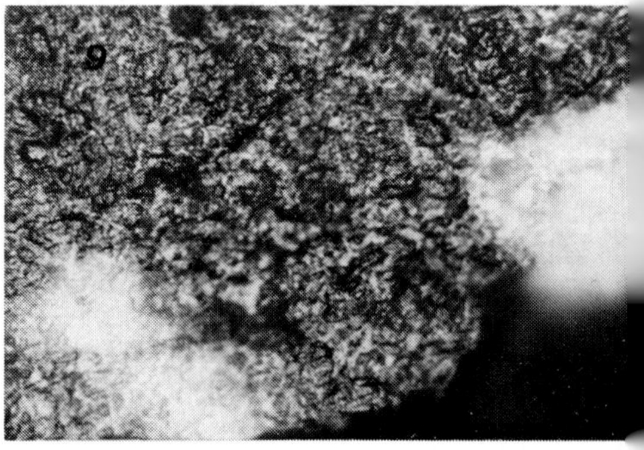

Fig. 14: Borer Fig. 2, no. 11 (2). Mw (?). Bri spots.

Table 1.

Microware anal.	Excavation	Functional microwear on the tip	Hafting; Fixing	Remark
1	NG83.L14.143	Dry hide	Bone; Dry hide	Capillar striations on hafted part
2	NG83.L14.176	Dry hide	Bone; Dry hide	Linear features on the tip
3	NG83.L40.3	Dry hide	Bone; Dry hide	Fixed in bone handle with aid of mastic
4	NG83.L40.13	Dry hide	Bone; Dry hide	Linear features on the tip
5	NG83.L40.85	Dry hide	Bone; Dry hide	No remarks
6	NG83.L40.91	Dry hide	– ; Dry hide	Bright spot on the tip
7	NG83.L40.96	Dry hide	– ; –	Hafting not analysed
8	NG84.L48.13	Dry hide	Bone; Dry hide	Narrow, deep, welldefined striation on hafted part
9	NG83.L14.77	Dry hide	Bone; Dry hide	No remarks
10	NG86.L1015.3	Dry hide	Bone; Dry hide	Band of discoloring on hafted part
11	NG86.L1015.6	Dry hide	Bone; Dry hide	No remarks
12	NG86.L1015.7	Dry hide	– ; Dry hide	Capillar striations on hafted part
13	NG83.L1015.11	Dry hide	– ; Dry hide	No remarks

obtained ones and will consequently be able to isolate any unknown phenomena. The most frequently occurring microwears have been described elsewhere (Keeley 1980:35-61; Bueller 1984:202-3). A quick look at Table 1 shows us that all functional microwears at Netiv Hagdud were caused by dry hide and therefore a description of this type of microwear will follow.

Most microwears develop from a luster into a polish. The term 'polish' is used here in the narrow sense of an alteration of the flint surface resulting in a white, smooth surface. The microwear of hide, however, never reaches that stage. It always shows a dull luster resembling, when the hide is still fresh, the luster caused by meat. The dull luster produced by hide frequently occurs on a knotty texture (see Figs. 3 and 4). When the hide becomes drier, the luster obtains a matter appearance. This change is quite gradual and makes it sometimes difficult to decide on the relative state of the worked hide (Unrath et al. 1986:163). Nevertheless, it was established that all hide microwears in the present sample of borers from Netiv Hagdud remained within the range of variation of dry hide. Helpful in the determination of the microwear caused by hide are the secondary phenomena like narrow, deep and well-defined striations (Figs. 5 and 6) and linear features (Figs. 7 and 8).

Nine of the examined borers showed the polish caused by bone on the robust part of the tool (Figs. 11 and 12). This polish resulted from hafting the borers in a bone handle. The tools were wrapped in a patch of hide and then jammed into the handle. The bone polish is located on those protruding parts which tore through the hide and came in contact with bone. The hide wrappings of small pieces like the borers illustrated by Fig. 2, nos. 6, 12 and 13 remained intact and consequently these tools do not show the bone polish but are completely covered by the dull luster caused by hide. No doubt, these borers had been hafted, too. The bulky borer shown in Fig. 2, no. 7, not examined for hafting traces, should not be considered an exception and was probably hafted as well. The prehistoric hideworker made use of the materials at hand and fixed his tools with little pieces of the hide which he was piercing. Hardly any difference could be discerned between the microwear caused by the worked hide and the one of the wrapping.

Sometimes a kind of mastic was used in addition to the hide to keep the tool from moving inside the handle. The microwear caused by the mastic shows an intense bright luster, as of wood, on a grooved surface (Figs. 9 and 10). Similar microwear has been found on the hafted parts of many lunates from the Natufian sites of el)Wad and Ain Mallaha and on the bottom of some Harif points from the Harifian site G VIII near Abu Salem (Bueller 1983).

All worked hides had been treated with ocher which left traces like capillar striations and a matt polish (Fig. 13). Ocher was rubbed into the hides in order to preserve them (Bueller 1988).

The 'bright spots' which occasionally occur amidst the hide microwear (Fig. 14) are an unknown phenomenon caused by an unidentified substance.

Summarizing the results of the microwear analysis, we can conclude that all examined borers were used on ocher)treated, dry hide only. There are strong indications that all borers were handled in an identical manner. They were used inside a bone handle, fixed with the aid of patches of the same hide that was being pierced.

DISCUSSION

What could be the purpose of this hide working? Evidently numerous tiny holes were made in cuttings of hide. Without stretching it too far one can hypothesize that these cuttings of hide were fashioned into clothing.

Ethnographic data show that the hide preferred for clothing is that of deer, because it can easily be worked soft, without proper tanning. The faunal remains recovered from early Neolithic sites in the Levant show no abundance of deer but gazelle is more than amply present. In Hatula, for instance, gazelle remains constitute 95% of the faunal assemblage in the Khiamian layer and also at Gilgal the gazelle remains are significantly more important than those of all other mammals. However, the presence of aurochs, boar and deer indicate that there must have been wooded areas in the vicinity of Gilgal and, consequently, of Netiv Hagdud. Tchernov (in Noy et al. 1980:80-1) wondered why people would roam great distances to obtain meat of the gazelle, whose habitat is the open grassland, whereas large mammals were available closer to the site. A possible explanation is that not the meat, but primarily the hide of the gazelle was the commodity pursued by early Neolithic man.

ACKNOWLEDGEMENTS

I would like to thank Danny Nadel, Hebrew University of Jerusalem, for kindly providing the archaeological material.

CATALOG (Figures placed between pages 24-25)

Fig. 3: Experimental piece HHE 12. Scraping dry hide. Dull luster.
Fig. 4: Borer Fig. 2,, no. 11 (1). Mw (dry hide). Dull luster.
Fig. 5: Experimental piece TCE 5. Scraping dry hide. Narrow, deep, well-defined striation.
Fig. 6: Borer Fig. 2, no. 8 (1). Mw (dry hide). Narrow, deep, well-defined striation.

Fig. 7: Experimental piece HHE 12. Scraping dry hide. Linear features.
Fig. 8: Borer Fig. 2, no. 4 (1). Mw (dry hide). Linear features.
Fig. 9: Lunate Fig.2, no. 9 from Ain Mallaha. Mw (mastic). Bright luster and grooves.
Fig. 10: Borer Fig.2, no. 3 (1). Mw (mastic). Bright luster and grooves.
Fig. 11: Experimental piece NE 7. Cutting bone. Flat polish.
Fig. 12: Borer Fig.2, no. 4 (2). Mw (bone). Flat polish.
Fig. 13: Borer Fig.2, no. 12 (1). Mw (ocher). Capillar striations.
Fig. 14: Borer Fig. 2, no. 11 (2). Mw (?). Bright spots.

REFERENCES

BAR-YOSEF O.

1980 'A Human Figurine from a Khiamian Site in the Lower Jordan Valley', *Paléorient* 6: 193-99.

1986 'The Walls of Jericho: An Alternative Interpretation', *Current Anthropology* 27: 157-62.

BAR-YOSEF O. ET AL.

1980 'Netiv Hagdud: A 'Sultanian' Mound in the Lower Jordan Valley', *Paléorient* 6: 201-6.

BUELLER J.

1983 'Methodological Problems in the Microwear Analysis of Tools Selected from the Natufian Sites of el Wad and Ain Mallaha', in: M.-C. Cauvin (Ed.), *Traces d'utilisation sur les outils néolithiques du Proche Orient*, Travaux de la Maison de l'Orient, no.5, Lyon. Pp. 107-127.

1984 'A Pilot Microwear Analysis of Sampled Tools of Layer 7 at Sefunim', in: A. Ronen (Ed.), *Sefunim Prehistoric Sites, Mount Carmel, Israel*, Oxford:BAR International Series 223. Pp. 201-214.

1988 'Handling, Hafting and Ocher Stains', in S. Beyries (Ed.), *Industries lithiques*, Oxford: BAR International Series 411.

CAUVIN J.

1977 'Les fouilles de Mureybet (1971-1974) et leur signification pour les origines de la sédentarisation du Proche-Orient', *Annual of the American Schools of Oriental Research* 44: 19-48.

DAVIS S.J.M.

1983 'The Age Profiles of Gazelles Predated by Ancient Man in Israel: Possible Evidence for a Shift from Seasonality to Sedentism in the Natufian', *Paléorient* 9: 55-62.

HOROWITZ A.

1977 'Pollen Spectra from Two Holocene Prehistoric Sites in the Har Harif (West Central Negev)', in E.S. Marks (Ed.), *Prehistory and Paleoenvironments in the Central Negev, Israel*, Vol. 2, Dallas: S.M.U. Press. Pp. 323-326.

KEELEY L.H.

1980 *Experimental Determination of Stone Tool Uses*. A Microwear Analysis, Chicago: University of Chicago Press.

KENYON K.M.

1960 'Excavations at Jericho 1957-1958', *Palestine Exploration Quarterly* 92: 1-21.

KIRKBRIDE D.

1968 'Beidha 1967: An Interim Report', *Palestine Exploration Quarterly* 100: 90-96.

VAN LOON M.

1968 'The Oriental Institute Excavations at Mureybit, Syria: Preliminary Report on the 1965 Campaign', *Journal of Near Eastern Studies* 27: 266-290.

MOORE A.M.T. ET AL.

1975 'The Excavation of Tell Abu Hureyra in Syria: A Preliminary Report', *Proceedings of the Prehistoric Society* 41: 50-77.

NOY T. ET AL.

1980 'Gilgal, A Pre-Pottery Neolithic A Site in the lower Jordan Valley', *Israel Exploration Journal* 30: 63-82.

PERLMAN I. AND YELLIN J.
1980 'The Provenience of Obsidian from Neolithic Sites in Israel', *Israel Exploration Journal* 30: 83-8.
RONEN A. AND LECHEVALLIER M.
1985 'The Natufian-Early Neolithic Site Hatula, near Latrun, Israel', *Quartaer* 35/36: 141-64.
SCHULDENREIN J. AND GOLDBERG P.
1981 'Late Quaternary Paleoenvironments and Prehistoric Site Distributions in the Lower Jordan Valley: A Preliminary Report', *Paléorient* 7: 57-71.
STEKELIS M. AND YISRAELI T.
1963 'Excavations at Nahal Oren. Preliminary Report', *Israel Exploration Journal* 13: 1-12.
UNRATH G. ET AL.
1986 'An Evaluation of Microwear Studies: A Multi-Analyst Approach', *Early Man News* 9/10/11: 117-176.
VALLA F.R.
1984 *Les industries de silex de Mallaha (Eynan) et du Natufian dans le Levant*, Mémoire et Travaux du Centre de Recherches Préhistoriques Français de Jerusalem 5. Paris: Association Paléorient.

Jebel Bishri and the Amorite homeland: The PPNB phase

JURIS ZARINS

Year in which Šar-kalli-šarri vanquished the Amorites in the Mountain of Ba-sa-ar (ḪUR.SAG Ba-sa-ar) (c. 2250 B.C.)

'I (Gudea) caused large stones to be brought to Lagash from Basalla (Ba-sal-la), the mountain of the Amorites.'(c. 2135 B.C.) (Buccellati 1966:236).

INTRODUCTION

Thanks to the research efforts of Maurits van Loon at Mureybit on the upper Euphrates, scholars began to see the developing Neolithic in a new light. Traditional assumptions about the sedentarization process and the impact it had on burgeoning populations were revised and new ideas began to emerge based on his excavations and others at such sites as Abu Hureyra, El Kowm, Bouqras, and Tell es Sinn. It has been almost axiomatic that in the popular image of the 'Fertile Crescent', a phrase coined in the very early twentieth century, little was left but to fill in some of the gaps concerning the range and distribution of the regions affected by the 'Agricultural Revolution' (Fig. 1). However, in examining the evidence more closely a number of interesting questions emerge. First, if the early sedentarization process affected human populations on the upper Euphrates (and sites such as Mureybit and Abu Hureyra were successful adaptations), why does Bouqras represent the southern manifestation of the PPNB expansion? Equally puzzling is the lack of early settlement along the southern Euphrates. Here the traditional explanation has suggested that Late Samarra/early Ubaid sites appeared to derive their impetus from a southern source, perhaps confirmed by the excavations at Eridu. Thus, in effect, we see no human settlements of the seventh millennium B.C. along the Mesopotamian plain or the Euphrates until we come to Bouqras. If we appeal to ecological considerations, we run into the main problem of this discussion, the appearance of seventh millennium B.C. populations in the desert regions west of the Euphrates from El Kowm/Taibe in the north to Jowf/Sakaka in the south. If we suggest that the development of pastoral nomadism in this region coincided with the appearance of the PPNB culture, we are left to ponder the symbiotic relationship between PPNB sites in the central and southern regions of the culture and their role in supporting sedentary settlements along the central and southern Euphrates. It may even be possible to suggest that this PPNB culture in the steppe regions was principally contributary and even necessary for the eventual successful occupation

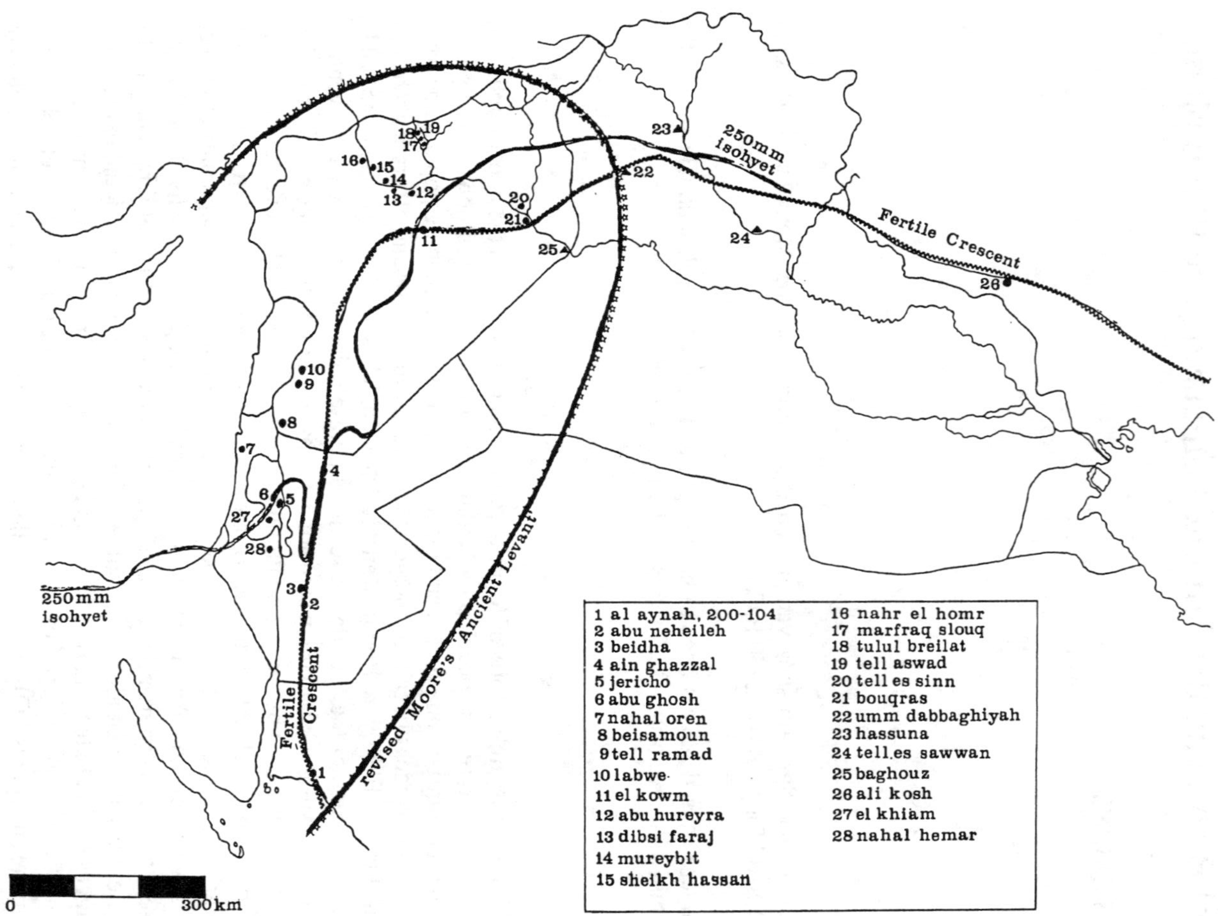

Fig. 1 The pertile crescent and related PPNB sites.

of the lower Mesopotamian plain. Finally, many scholars have repeated and affirmed the old hypothesis that 'Semitic' populations arose in periodic 'waves' from the desert and inundated existing sedentary populations in the Mesopotamian alluvium. If we again turn our attention to the nature and distribution of PPNB sites in the desert, we can see that occupation spread in a north to south direction and that the PPNB culture represented the apparent initial attempt at re-settling the desert regions after the end of the Upper Paleolithic or Epipaleolithic.

THE LEVANTINE PPNB

The PPNB cultural manifestation appears to solidify and expand on the earlier tenuous steps towards mixed agriculture perhaps begun in the Natufian and PPNA periods. The PPNB culture was first identified and defined in the southern Levant at Jericho and elsewhere (Kenyon 1960:47-57). Over the course of the last three decades, the sedentary aspect of the PPNB has come to be recognized at a series of sites in Saudi Arabia (Al Aynah:200-104, Hail: 206-36), Jordan (Beidha, Ain Ghazzal, Abu Neikheileh), Israel (Jericho, Nahal Oren, El Khiam), and Syria (Beisamoun, Tell Ramad, Labwe, Mureybit, Dibsi Faraj East, Sheikh Hassan, Nahr el Homr, Abu Hureyra). The most northerly sites are found on the Balikh and Khabur tributaries of the Euphrates (Copeland 1982; Cauvin 1972; Moore 1981). The southernmost Euphrates site to date appears to be Bouqras. Based on archaeologists' perception of this pattern, most refer to this site distribution as 'Levantine'. In fact, Moore suggests a concept labelled 'Ancient Levant' which includes substantially all of Syria, western Iraq, most of Jordan and a portion of Saudi Arabia (Moore 1979:64).

The details of the cultural characteristics of the period can be obtained from a convenient series of summaries (Singh 1974; Mellaart 1975:55-69; Bar-Yosef 1980:127-130; Moore 1981; 1982:9-16). For our purposes, we can emphasize aspects which will be pertinent to our presentation. The culture appears to span most of the seventh millennium B.C. with some debate over the initial beginnings and the transition to the later PN in the early sixth millennium B.C. Economically, according to Bar-Yosef, a radical change took place by the beginning of the period. However, while we have clear evidence of cereal and pulse domestication from the preceding PPNA period, the appearance of domesticated animals during the period remains a troublesome topic. Bar-Yosef (1980:128) suggests that most of the earlier (wild) game had given way to a dominance of goat-sheep and bovids by this period and that their domestication in later periods was not in doubt. Without committing himself to the issue of domesticity, he merely comments that a drastic faunal change was evident at Mureybit, Abu Hureyra, Beisamoun, Jericho and Abu Ghosh. Sheep remains at Jericho outside their natural habitat also suggest some initial manipulation. Mellaart (1975:65-6) suggests that a loose herding habit involving gazelle and other ungulates prevented intensive domestication during the PPNB and only the domestic goat is definitely attested. Clutton-Brock (1978; 1981:56-7, 60-1), on the basis of studying the faunal remains at Jericho, suggests that both sheep and goats were being domesticated by the eighth millennium B.C. The work of Legge at Abu Hureyra is echoed by Moore who suggests that sheep, goat and gazelle were herded by the PPNB period (Moore 1979:68). In a later review, Moore proposed that only goats and sheep were herded (1982:16) and that this strategy of herding was introduced during the PPNB

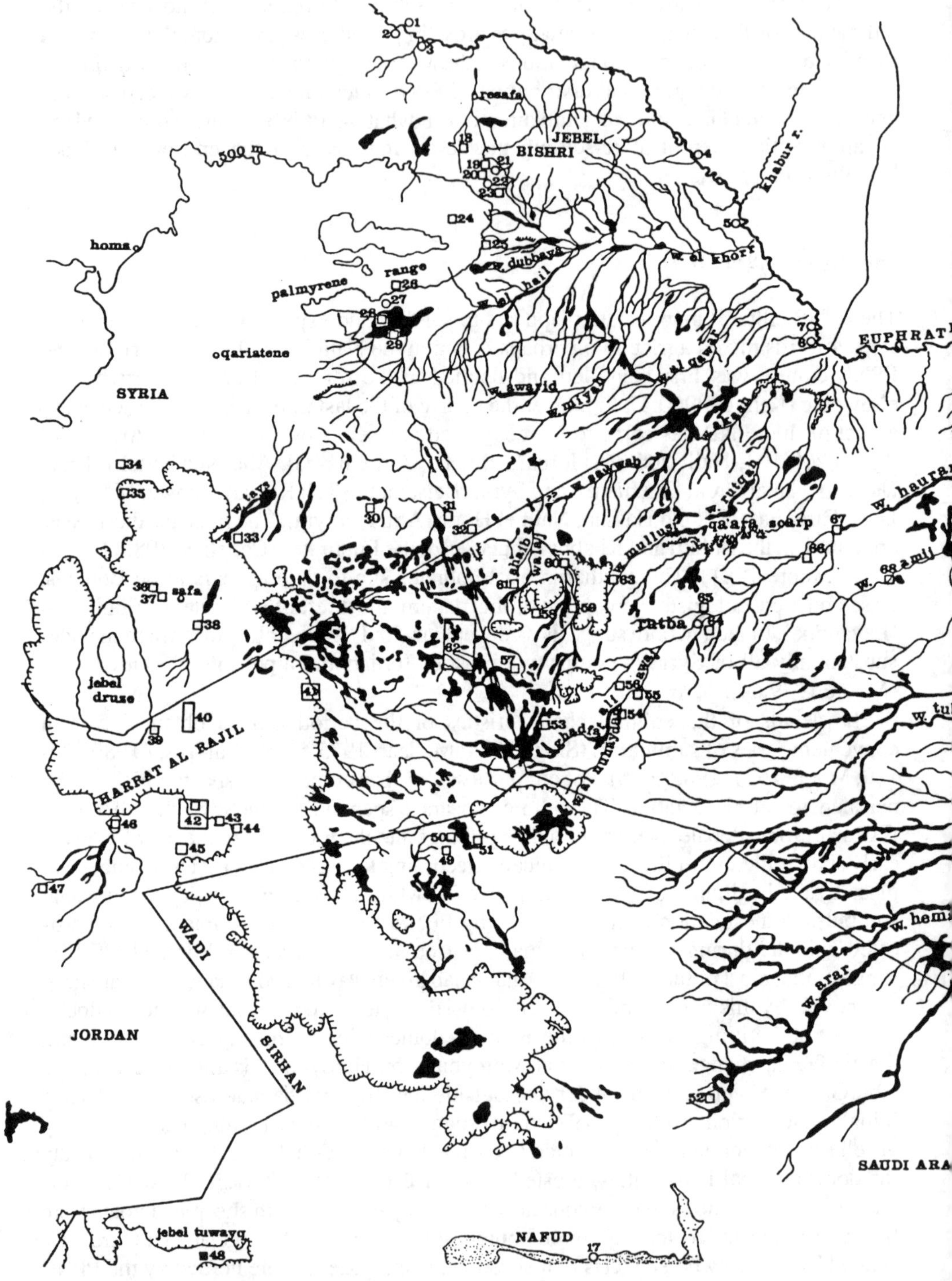

Fig. 2 The PPNB desert sites of South/East Syria, Western Iraq, Eastern Jordan and Northern Saudi Arabia.

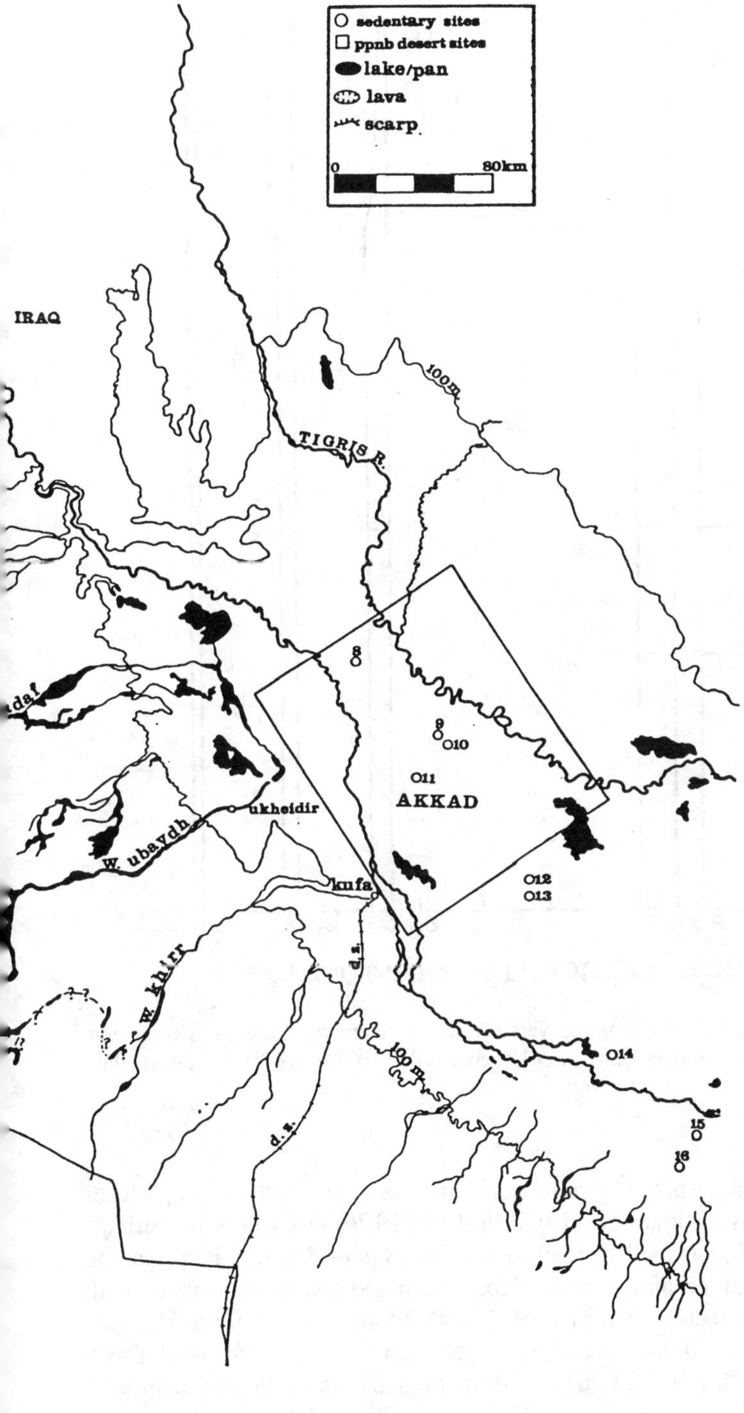

Site Identification numbers of Figure 2.

1. Mureybit
2. Dibsi Faraj
3. Abu Hureyra
4. Tell es-Sinn
5. Bouqras
6. Baghouz
7. Mari
8. Sippar
9. Uqair
10. Jemdet Nasr
11. Kish
12. Abu Salabikh
13. Nippur
14. Uruk
15. Ur
16. Eridu
17. Jowf
18. Moore 1981
19. Qdeir 1
20. Nad aouiyeh 4
21. Umm Tlel 2
22. Kowm
23. Tell Taibe
24. Watelin in: Dussaud 1929:150
25. Soukne (Rigm es-Saboun)
26. Douara Cave
27. Palmyra
28. Sabkhet Muh PPNB site 79
29. Sabkhet Muh PPNB site 122
30. Field Site 102 and 108 (1928)
31. Jebel Tenf (Field: 4, 12-13)
32. Field Site 16-17
33. Jebel Seys
34. (as-Sidd) Dmeyr
35. Deir es-Shemali
36. Umbachi
37. Khebariye
38. Nemara
39. Jawa
40. Shubeika (Betts 1983)
41. Qasr al-Burqu (Betts 1983)
42. Qa'a Mejalla (including Dhuweila 2202; Betts 1983)
43. Jebel aja 2321
44. Ibn al-Ghazzi 3133
45. Jebel Qurma (Betts 1983)
46. Azraq
47. Wadi Jilat (Dhobai)
48. Kilwa
49. Umm Wual
50. Bardawil
51. Dawqira
52. 201-42
53. Muhunsur (Field: 25,30)
54. Field Site 77
55. Field Site
56. El-Jidd Field: 21
57. Field Site 22
58. Field Site
59. Field Site 233
60. Field Site 15
61. Field Site
62. Tellul al-Basutin Field: 62, 67, 69, 70, 72, 81
63. Field Site
64. Field Site 36
65. Field Site 41, 51 Sugarloaf
66. Wright 1958
67. Qasr Muhaiwair (Tyracek and Amin 1981)
68. Watelin in: Dussaud 1929:150

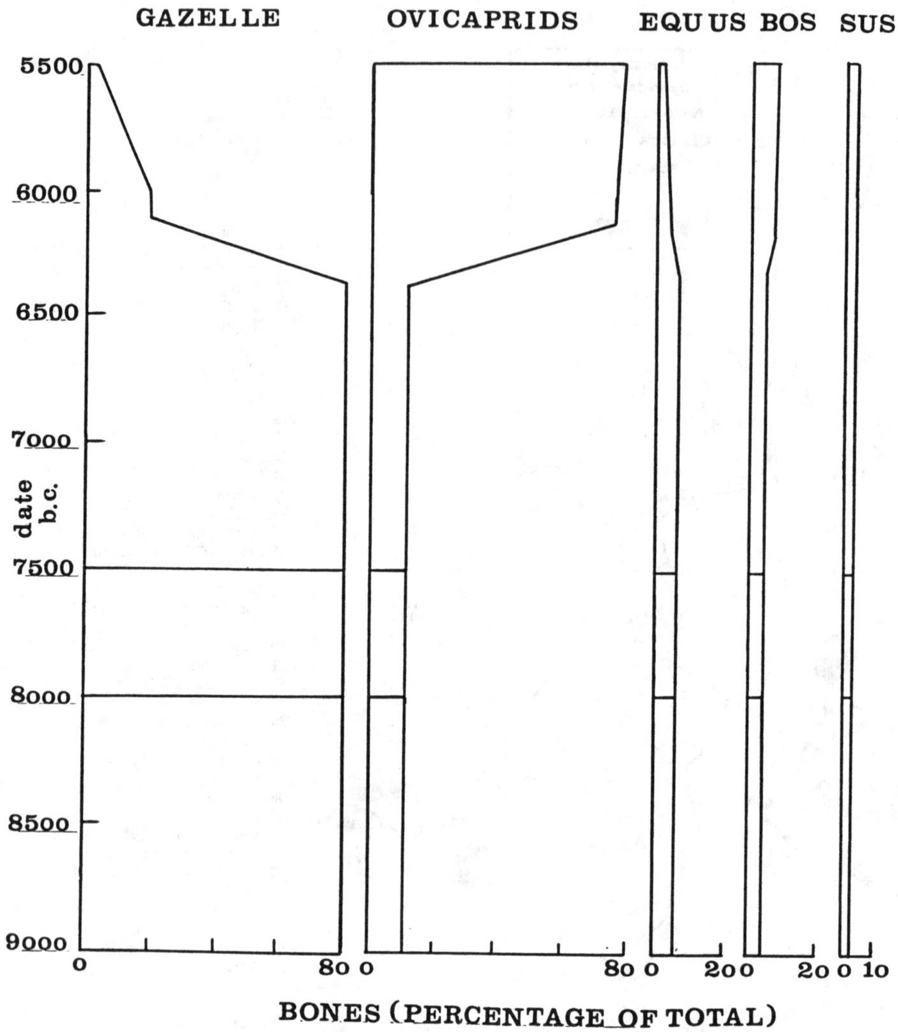

Fig. 3 Composite pattern of total animal bone recovered from Abu Hureyra, Mureybit, Beisamoun, Jericho, Abu Ghosh, Bouqras, and Tell es Sinn (after Legge and Rowley-Conwy 1987, Clason 1980, 1983 and Bar-Yosef 1980).

(1981:448, 456). Garrard suggests that gazelle continued to have a 'relationship closer than random predation 'during this period (Garrard 1984:126). At Bouqras, perhaps spanning the same time period as the latter half of the PPNB period at Abu Hureyra, the initial identification of faunal remains suggested that sheep and goat predominated with the goat being 'semi-domesticated' (Hooijer 1966:194). More recent work at Bouqras has revealed the presence of domesticated sheep and goats as well as wild sheep (Akkermans et al. 1982:57; Clason 1983:361). Clason noted that domestic cattle, initially found at Çatal Hüyük in Turkey ca. 6400 B.C. (Clutton-Brock 1981:66), have been identified at the PPNB site of Bouqras (Clason 1980; 1983:361; Buitenhuis 1984:213). In addition, domestic pig has also been identified at Bouqras (Clason and Buitenhuis 1978). At Abu Hureyra, the dramatic reversal of the relative percentages of gazelle vs. sheep/goat at ca. 6200 B.C. in favor of the latter suggests the introduction of sheep-goat animal husbandry (Fig. 3). Possibly, domestic cattle may have been introduced this time as well at Abu Hureyra (Legge and Rowley-Conwy 1987:95). Recent work at Ain Ghazal revealed that the majority of bones were from goats but some sheep were present. In both cases, 'the degree of domestication remains to be resolved' (Rollefson 1984:12). While white ware at Ain Ghazzal shows clear ties with northern Syria in the late seventh and early sixth millennia B.C., the presence of domestic cattle, sheep (?), and pig, may show that the impetus for these domesticates also came from the north. From this brief overview with its diverse opinions and comments, it appears that animal domestication was not far removed from its initial efforts at least in this area of the Near East.

There appears to have been a climatic optimum during this period as well. Mellaart speaks of a 'wetter phase' during the PPNB with the subsequent spread of the culture to the Syrian steppe sites such as Bouqras and Kowm (Mellaart 1975:65). Moore has postulated a major shift in vegetation zones between the beginning of the Natufian and the end of the PPNB (Moore1979:69). More recent evidence from southern Palestine suggests that the climate was 'wetter' during the seventh millennium B.C. (Horowitz 1979:248; Danin 1985:39-40 and Table 1; Rosen 1987:48-9). The work in the Palmyra basin and northern Saudi Arabia has also confirmed this opinion (Sakaguchi 1978:26, 109; 1987:14; Fujii et al. 1987:29; Garrard et al. 1981; Whitney 1983).

Lithic production was becoming specialized during the period and exotic or highly desired flint sources were exploited as part of an apparent long-range trade network involving the Mt. Carmel area, southern Syria/northern Arabia and central Syria (Mellaart 1975:65). Vulcanic obsidian and alabaster were traded through the Syrian desert as well (for a hitherto unidentified obsidian source in eastern Jordan, see Betts 1984a:34 and Fig. 3). Other long-distance trade items included jadeite/greenstone, turquoise and white ware. The 'markers' identified with lithic production include the distinctive bi-ended cores (often called naviform or boat-shaped) and crested blades as by-products which produced the desired long blades characteristic of the PPNB. These blades sometimes exhibited pressure retouch and were turned into reaping knives, distinctive tanged projectile points (e.g. Amuq and Byblos points), end-scrapers and commonly burins (Mortensen 1970; M.-C. Cauvin 1972:90-93 and Figs. 5-6; Mellaart 1975:63; Moore 1982:11 and Figs. 5-6).

THE DESERT PPNB

A number of researchers have viewed our study area as distinct *vis-à-vis* the settled zone. Geographically, we are describing a vast tableland sloping west to east draining into the Euphrates basin and the Arabian Gulf. This zone is characterized by intermittent wadi flow given to sheet and flash flooding and by local depressions such as the Wadi Sirhan, and the Azraq, Jafr and Jowf-Sakaka basins. A few sheer local elevations are present along the rim of the study zone such as Jebel Druze, the Palmyrene Range, Jebel Bishri and the Sinjar. The Harrat al Rajil lava field covers portions of Syria, Jordan and Saudi Arabia. Thus, the present-day landscape is dominated by features typical of such an arid zone: exposed rock formations, abrupt scarps, mesas, pinnacle rocks and yardangs.

Intervening flat-floored wadis filled with detritus, open plains, gravel flats, dune formations and pans are also highly typical (Zarins, in press). The relationship of the settled zone to our study area is more complex than usually defined. Many investigators have focussed on a basic dichotomy between the settled and desert zones. However, the zonal definitions are based on a complex interaction of climate, rainfall, fauna, flora and soil types. For example, Poidebard (1934) suggests a gradation based on rainfall (250 mm isohyet), separating 'climat syrien' from 'climat saharien' (sub-desertique vs. desertique). More recently, I suggested that the 250 mm isohyet was the critical line demarcing farming vs. pastoral zones (Zarins, in press). Geographers suggest a three-tier system between 1) the farming zone (greater than 350 mm), 2) steppe (350-200 mm) and 3) desert (less than 150 mm; Clawson et al. 1971:172). Moore also echoes this viewpoint describing an 'open forest, steppe and desert steppe' (1979:69). Zohary recognized two major phyto-geographical zones: the Irano-Turanian and Saharo-Arabian. The majority of the study area today lies within the latter zone (for a list of typical flora, see Zohary 1952:21-114; 1962; Allred 1968, Appendix 2). Recent re-evaluations of the fauna present suggest that a complex mix of Paleartic, Saharo-Arabian and Ethiopic elements occurred in the region (Harrison 1964-72; Tchernov 1982:127, Fig. 4.7).

The discovery of PPNB materials outside of the normal, expected sedentary zone actually preceded that of the Levant tell sites (Fig. 2). Already in the early twentieth century typical PPNB lithic material was being described from the Sinai but remained unrecognized as such (Currelly 1906:229-44, [Wadi Ahmar, Rothenberg's site 590]). Following World War I, RAF pilots began to overfly selected routes from Amman to Baghdad and identified certain enigmatic structures visible from the air which they called 'kites' (Maitland 1927; Rees 1929). In Syria, Poidebard in conjunction with other researchers compiled a similar impressive aerial portfolio describing large-scale 'enceintes fortifiées indigènes' (Poidebard 1934). Ground investigation of this region was first under-taken by Henry Field in 1926-1928, although the publication of the results was delayed until 1960 (for initial reports, see Field 1951:89, no. 9). The examination of the Kilwa region in the early 1930's and the excavations at Wadi Dhobai also failed to stir larger interest (Rhotert 1938; Waechter and Seton-Williams 1938). However, by the onset of World War II, it was known that three distinctive categories of artifactual remains could be found in this region consisting of rock art, lithics, and structural remains.

These early results, however, remained largely isolated and ignored as researchers were uncertain as to the cultural affiliation, date, and function of the material. More recently, intensive research began in three areas in the region in the 1960's and 1970's: 1) The Negev/Sinai, 2) eastern and southern Jordan, and 3) northern Saudi Arabia. It is neither within the aim or scope of this paper to detail the results of this work, (for a comprehensive bibliography, see Zarins, in press), since our focus is to examine the material found in southeastern Syria and western Iraq. However, a brief summary is necessary to place into proper focus the material which will be examined from our smaller study zone. A substantial number of sites have been found in the Negev. This, in turn, has allowed researchers to suggest a tripartite division of the PPNB based on knapping techniques, projectile point typology and radiocarbon dates (Bar-Yosef 1981b:221, 226; Noy and Cohen 1974; Zarins, in press, Table 1). Summary studies suggest site typology based on site size, architecture and lithic usage (Simmons 1981:32 ff; Goring-Morris and Gopher 1983; Baron 1981). In the Sinai, PPNB sites are known from both the North (Gilead and Goldberg 1976; Mintz and Ben-Ami 1977:242-3) and south (Copeland and Hours 1971:17; Goring-Morris and Mintz 1976:138; Bar-Yosef 1981b; 1981c; Goldberg and Bar-Yosef 1982). Based on the Negev sub-chronology, it appears that many sites belong to the late PPNB, a significant fact in parallel with Syrian sites (see below) (Bar-Yosef 1981c:45). In the Azraq region of Jordan a number of PPNB sites have been defined on the basis of stratigraphical association or surface concentrations (Garrard and Stanley-Price 1975-77; Garrard et al. 1985; 1987; Rollefson 1982:26-7). Extensive work by Betts in a three year campaign of survey and excavation in the eastern arm of Jordan has helped us considerably in defining the desert facies of the PPNB. Working principally in the lava field stretching from Syria to Saudi Arabia, the Harrat al Rajil, she noted a large number of sites dominated by blade burins on concave truncations (Betts 1982:27 ff). Other lithic material included the typical PPNB tanged projectile points. Structurally, Betts suggests her material falls into three classes and can be attributed to the PPNB (Betts 1984a:29-31; 1984b:78). The basic stone circle complex consisting of singular or interlocking stone rings appears to be the basic residence unit. Other circle complexes called 'jellyfish' have interlocking circles and outer enclosure walls (Betts 1982:25). Kites are very common here and in several cases can be attributed to the PPNB (Betts 1985:39 and Fig. 19; Betts 1983:9-10; Parr et al. 1978:36-7; Helms 1981:46-7 and Fig. 23). The so-called 'burin sites' based on sedentary site parallels mainly with the Euphrates region, should date to the second half of the seventh millennium B.C. (Betts 1984:29; Rollefson and Frohlich 1982). The southern extent of the PPNB associated with either structures or lithic scatters now appears to be north-central Saudi Arabia (Parr et al. 1978:39 and Pl.27). In 1977, a small site (201-42) was found on a limestone ridge in the Wadi Arar, northeast of Jowf-Sakaka. About 50 sqm.in size, the site contained the usual PPNB lithic forms including naviform cores, blade burins, and knives. The flint was identical to that found at Umm Wual and Dawqira, some 250 km. to the North (Parr et al 1978:36 and Pl. 39, 46-54). The typical burin PPNB assemblage also occurred at a similiar site (202-4) possibly associated with a small circle at Hadhilul, north of Linah (Gilmore et al. 1982:13). To the West, the southern PPNB line remains at Kilwa where a recent re-survey has confirmed Rhotert's earlier work.

In sum, we suggest that by ca. 6200 B.C. a recognizable PPNB culture had begun to penetrate the more marginal environments of the Near East both to the south in the Negev/Sinai and to the east of the Levant into Jordan, and Saudi Arabia (Fig. 2). This arrival of populations has been documented by lithic materials as well as by structural remains consisting of stone circles, circle complexes, and kites. In selected areas, rock art has also been attributed to the period. In contrast to the sedentary zones where the PPNB is seen as an early climax point succeeded by much site abandonment and area reduction, the PPNB in the desert zone is the harbinger of a later climax. Populations exploded throughout the Arabian peninsula in the suceeding PN and Chalcolithic periods.

This brief summary of the PPNB in the arid zones of Southwest Asia now allows us to examine a diverse data set obtained from southeast Syria and west Iraq. In neither region has much systematic work been undertaken since the initial efforts of Poidebard and the Field team. Beginning first with western Iraq, we have already noted that no seventh or early sixth millennium B.C. sites are known from either the southern Mesopotamian alluvium or along the southern and central Euphrates (Adams 1981:54). The geomorphology of the alluvium also continues to be puzzling, especially in relationship to the western margins of the central Euphrates (Buringh 1957; Voute 1957). Several recent studies have shown that the current Euphrates has existed in its present form since 8000

B.C. (Paepe 1971:19-20; Boerma 1983:362). Those few Samarra-like sites and the earliest Ubaid sites are to be found in the extreme southern alluvium associated with an early shoreline (Adams 1981:54; Wright 1981:300-1, 323) and no case can be made for any site in the alluvium contemporary with the PPNB period. Yet, we have a series of intriguing sites found in western Iraq (Fig. 2). At least twelve sites were reported by Henry Field in separate, rather random traverses of western Iraq between 1926 and 1928. They are clearly located on tributaries or main branches of wadis leading into the Euphrates. The westernmost sites are found on the braiding stream bed known in Iraq as Sha'ib al Walaj (sites 15, 22 and Tell Mahunsur sites 25 and 30). This watercourse actually flowed in a northerly direction until the Iraqi/Syrian border when it turned east and merged with a series of wadi/lakes flowing into the Euphrates. On two tributary arms of the Wadi Mullusi Field found site 242. Other sites are now known from this wadi area where it touches the Qaara escarpment and merges with the wadi Rutqah (Field 1951:89 and note 9 for additional references; Tyracek and Amin 1981:148, note 1). In the extreme southwestern corner of Iraq, sites are found on a series of large wadi/lake tributaries, Ghadfat Ali Awaj and An Nuhaydayn (site 77) both flowing into the Wadi Hauran. A number of sites have been reported on the Wadi Hauran itself including Field's site 36, Sugarloaf, 41, and 51 (Field 1960:49 ff). H.E. Wright, in passing, describes a stone circle village apparently also on the Wadi Hauran (1958:56). The most recently found site on the Wadi Hauran at Qasr Muhaiwir is approximately 125 km west of the Euphrates (Tyracek and Amin 1981). The authors suggest that Neolithic finds along the Wadi Hauran are fairly common (Tyracek and Amin 1981:148). To the south of Wadi Hauran, a parallel major tributary, Wadi Amiq also has reported material (Watelin in Dussaud 1929:150) approximately 150 km from the Euphrates (mention can also be made here of the area described by Field as Tellul al Basatin, now located in northeastern Jordan. But this area geomorphologically belongs to the internal drainage of the dessicated lava field, Harrat al Rajil and the Hamada, see Betts 1982-85; PPNB sites from this area labelled by Field include 62, 67, 69, 70, 72, and 81; Garrod 1960:139).

The material which was collected from these sites is now at the Field Museum in Chicago and was examined by the author in 1981. Garrod, in her initial study of the material in 1934, described it as belonging to the 'Wualian' after Umm Wual in Saudi Arabia (Garrod 1960:118-9). She ascribed the assemblage to the 'Tahunian' or early Neolithic of Jericho. Her analysis revealed that the assemblages often were dominated by an extraordinary high percentage of burins, especially angle burins (including simple, facetted, and multiple; for comments on the burin phenomenon, see Betts 1982:29-30). Other finds included end scrapers, hollow ended flakes, backed knives, steep scrapers, retouched flakes and naviform cores (Garrod 1960:120). Other finds such as the T-shaped implements, and tabular flint scrapers suggest a continuation of the tradition into the sixth millennium B.C. Architectural remains from the region have been noted (megaliths at El Jidd, stone circle complexes on the Wadi Hauran and Wadi Amiq), but these may belong to any period between the PPNB and the end of the Bronze Age. The nearest kites are reported from the Tellul al Basatin area just west of the Iraq/Jordan border. Rock art was described from the region for the first time in 1978. While little is known to date, reports suggest that material was seen around the rim of the Qa'ara depression and the adjacent wadis Mullusi and Aghari (Tyracek and Amin 1981:148, note 1). Rock art which has been actually seen and described comes from the Wadi Hauran at Qasr Muhaiwir. Two localities were discovered. In both cases, ibex and gazelles predominate but cattle are also present. Analysis of the style (Tyracek and Amin 1981, Fig. 3) shows close if not identical parallels to the more voluminous rock art known from north central Saudi Arabia. In particular, the rock art shows clear parallels to the material from Jubba lake in the Nafud. There the technique of pecking spots on the cattle (Adams et al. 1977, Pls. 12, 20; Parr et al. 1978, Pls. 28-31) is repeated in the Wadi Hauran (Fig. 3) as well as at Azraq in Jordan (Helms 1981:27, Fig. 9; for Negev/Sinai art, see Anati 1979:36 ff, style II, and plates of Wadi Huwara). The presence of the Jubba throw stick at Wadi Hauran is also paralleled at many northern Saudi Arabian sites. The depiction of these animals and associated human 'stickfigures' confirms our idea that ovicaprids and cattle were known to the inhabitants of the western desert of Iraq as well as to those of the sedentary sites of the Euphrates and the Levant (see above). The dating of the rock art remains somewhat problematic but the general classification is clear. The earliest style is represented by rare depictions in the Saudi Arabian repertoire and is best known from Kilwa where an early association to the PPNB can be entertained. (Rhotert 1938; for a summary, Anati 1979:29-35). The style depicted at Jubba and in the Wadi Hauran, as well as at Azraq, Hanakiya and elsewhere is somewhat later in date and should properly belong to the sixth millennium B.C. However, the assocation of lithic material at Wadi Hauran (Qasr Muhaiwir) may tie this rock art style with the later PPNB (Tyracek and Amin 1981:147).

As can be readily seen, the investigation by Field and others since 1926 has confirmed that PPNB sites existed in the wadi tributaries of the Euphrates. Unfortunately, nothing has been reported from a vast unexplored region of western Iraq and we have noted that PPNB materials have been found as far south as Linah in Saudi Arabia. A brief glance at major wadi and lake systems in the western desert promises to yield numerous sites along such wadis as Wadi Ghadaf (and its western tributaries), Wadi Tubal (flowing into a tremendous lake), Wadi al Ubbayid (a known historical route into the Euphrates now terminating at the well-known Qasr Ukhaidir. To the south Wadi Arar (with a PPNB site) merges into Wadi Hamir. The traditional Darb Zubaydah pilgrim route in the Abbasid period follows the Wadi al Khirr (among others) and it is on this tributary to the Euphrates that the PPNB site was found at Linah.

From Syria, the results are very uneven. We will concentrate principally on the region west of the Euphrates and south of the Palmyrene range ending at Jebel Bishri. The rough rectangle between Palmyra and the Iraqi border to the south then stretching to the Euphrates and up to Jebel Bishri contains no site information of any kind with two exceptions, both of which were explored as the result of a natural route taken by Field coming from western Iraq (PPNB sites 7 and 16 on the Wadi ash-Sha'lan and sites 4, 12-13 in the Jebel Tenf area; Garrod 1960:137). The eastern half of this rectangle is very similar to the topography described for western Iraq. A series of wadi systems (Al Wudyan) flowing into low-lying basins dominate the landscape, e.g. al Qaw'ar, Miyah, Al Uwayrid, al Hayl, and Dubbayah. Southern Syria again forms a large rectangle with the Palmyrene range on the north paralleled by the Jordanian border on to the south. The southern end of the rectangle is marked by Jebel Druze. Within this area, partially dominated by the northern extension of the vast Harrat al Rajil, (Helms 1981, Fig. 27), we have a number of sites reported from the 1920's both by the RAF, Field, and Poidebard. The only areas which have had more recent intensive archaeological work center around two basins: 1) Sabkhat Muh and Palmyra south of the Palmyrene massif and 2) the El Kowm basin between Jebel Bishri and the Palmyrene massif (or other brief references to this region by researchers, see the references in Buccellati 1966:237, notes 13-14; Buccellati and Buccellati 1967).

Artifactual information from the PPNB period in Syria again is restricted to three categories of data: 1) kites, 2) stone circle constructions and 3) lithic concentrations. Kites in this region of Syria constitute perhaps one of the only reliable guide markers, having been spotted from an aerial perspective initially (Maitland 1927; Rees 1929; Poidebard 1934). We can briefly describe their distribution along a southwest to northeast arc. The southernmost kites can be found around Jebel Druze and south of Safa (Poidebard 1934:194 and Pl. XCI/2; note the kite cut by wadi as Sham at Nemara, Pl. LI; Rees 1929:399) and possibly at Jebel Seys. The earliest kite mentioned in the literature is the one at Dmeyr called 'as Sidd', having been described by Musil in 1908 (1928:3, 4, 7). It was photographed by Poidebard (1934:43, 192, 194 and Pl. XIV; republished by Legge and Rowley-Conwy 1987:89). According to Watelin, kites may have been also located in the Al-Hamada near Jebel Tinf (Poidebard 1934:191, 192). Following the lava scarp to the south of Dmeyr, Poidebard describes another series of kites including ones at Deir Semali (Poidebard 1934:194). On the Wudian plain he reports over ten kites from the Souhne area on the Jebel Rigm as Saboun (Poidebard 1934:77-78, Pls. LXVIII-LXIX). Watelin saw kites to the north of Soukne and south of Gdeihm wells (Dussaud 1929:150; for other kites in the Palmyrene massif, see Helms 1981, Fig. 27 and between Homs and Al Qaryatein, see Legge and Rowley-Conwy 1987:91). To the west of Jebel Bishri, a series of kites were found between Resafa and Taibe (Poidebard 1934:79, 191 and 192, note 1; Legge and Rowley-Conwy 1987:91; Helms 1981, Fig. 27). The last area which has yielded an extensive number of kites is centered west of Jebel Sinjar, surprisingly across the Euphrates and east of the Khabur. Poidebard describes over sixteen kites and they concentrate in a few areas (1934, Pl. CXL). First, seven are seen around Jebel Cembe (Poidebard 1934:153, Pl. CXLV), one at Jebel Jeribe and one south of Goseibi. Two are found just north of Khatuniya on the wadi Frati. Three are reported on the slopes of Jebel Kokab just north of the Euphrates and south of Tell Aswad. The westernmost kite was seen near Jebel Gug along the Thallaba-Tell Barti road (Poidebard 1934:150) and finally, the northernmost kite reported to date is found at Tell Awan just north of Tell Brak (Poidebard 1934, Pl. CXL).

Other structures associated with the PPNB remain very problematical since no ground control has been exerted. Betts' work in Jordan has implied that a number of 'forts' or jellyfish complexes may belong to the PPNB period and, based on this assumption, we can note that similar structures have been identified in the Jebel Tinf area (Field 1960:52-54, 93; Poidebard 1934:126). Watelin observed circle complexes in the mountain passes between Jebel Bishri and Abu Rigmen (Dussaud 1929:150) and Buccellati noted a number of such structures in the vicinity of Jebel Bishri (pers. com.). This is confirmed by Helms who placed 'forts' in the vicinity of Jebel Bishri, Jebel Tinf, and the Sinjar (1981, Fig. 27). Unfortunately, many or most of these structures may belong to subsequent periods. For example, the structures at Khirbet el Umbachi and Hebariye excavated in the 1950's and were identified with the EB III/IV period (Dubertret and Dunand 1954).

Lithic collections from Syria also are rather rare. Again from the initial surveys of the 1920s by Field, a small portion of southern Syria in the Hamada was traversed. On the route from Rutba to Damascus, Field reports lithic material of the PPNB burin type from Wadi ash Shalan just north of the Jordan-Iraq line (sites 7, 16) and at Jebel Tinf itself

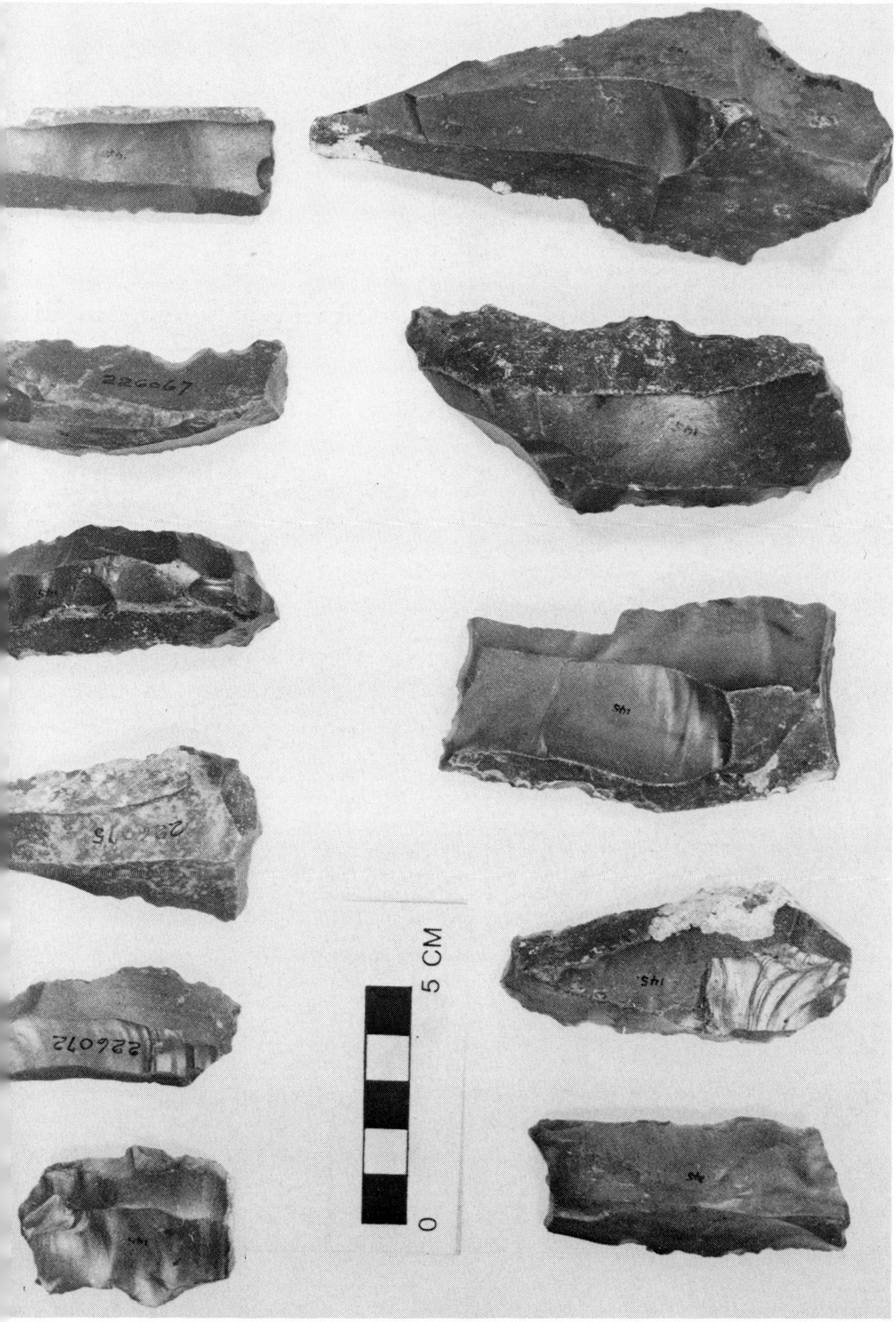

Plate 1.

Plate 2.

(sites 4, 12-13). These may be associated with the structural remains from Jebel Tinf mentioned above. Further along the road near Maqha ash-Shami, sites 102-103 may also belong to the period associated with a 'circle village' (Field 1960:139). More recently, the work in the Palmyra basin has suggested that a rather dense PPNB occupation is to be found around now fossilized lakes and stream beds (Sagaguchi 1978, Fig. 2.1). Moore lists at least thirteen PPNB lithic sites around the Palmyra basin or Sabkha Mouh (Moore 1981:446, Fig. 1). The initial investigations at the Douara cave failed to yield PPNB material *in situ* although some had been mixed with the Upper Paleolithic of the cave and the talus slope (Endo 1978:80; Sakaguchi 1987:9-10). In investigating the Wadi fans in front of the cave, the majority of lithic materials turned out to belong to the PPNB period (naviforms cores, crested blades, core tablets etc.; Endo et al. 1978:96-97). Surficial investigations around Sabkhet Mouh have turned up a number of PPNB sites clearly indicating an active lake phase in the region (Koizumi 1978:47 and Fig. 3.11; Sakaguchi 1987:14). These sites are found most often on terrace remnants and consist of simple inventories dominated by naviform core production, crested blades and truncation burins (e.g. site 79: Akazawa 1975, 1979; Abe and Akazawa 1977; Suzuki and Kobori 1970; Suzuki and Akazawa 1971; site 122: Fujii et al.1987:29-39).

Similarly, along the El Kowm basin, a number of sites have been located. Astride the route leading from the Middle Euphrates to the Wudian region and the Al Hamada (including the Palmyra region), the Kowm area lies between Jebel Bishri and the northern Palmyrene massif. Three tells in the region (El Kowm I/II), Qdeir I, and Umm Tlel 2 exhibit the classic characteristics of a settled PPNB community (e.g. white ware, structures and lithic industry; Cauvin 1981:471-477). Another site, Nadaouiyeh 4, north of the Tell el Kowm (Cauvin 1981, Fig.1), resembles the lithic scatters reported above from western Iraq and the Palmyra basin. Here, covering small hillocks of 3 hectares, is material dominated by burins and scrapers (Cauvin 1981:477-481 and Fig. 7).

In sum, we have seen that PPNB populations began to expand into the areas of the Near East often considered marginal to the 'Fertile Crescent'. This expansion has been well documented in eastern Jordan where the surveys and limited excavations have established some chronological reliablity and special context. Structures such as kites and forts can be seen as contemporary to lithic scatters and areas where rock art is exhibited. Sites in southern and eastern Syria as well as western Iraq have been less well investigated and their distribuition remains a matter of conjecture. It would appear that particularly the one class of material culture which can be compared to the settled sites, the lithics, show an astonishing similarity and contemporaneity. It remains now to examine this relationship and infer the connection to later historical groups in the region.

ORIGINS OF PASTORALISM

Anthropologists have long conducted a debate on defining pastoral nomadism. Several recent syntheses (Dyson-Hudson and Dyson-Hudson 1980; Khazinov 1984:16 ff; Goldschmidt 1979; Johnson 969) have suggested that all pastoral nomads share certain characteristics such as strong, localized kinship and lineage bonds, impermanent dwellings, movement to procure pasturage and water for herds and specific relationships to formal states. Herd animals are viewed as the main form of subsistence although alternate strategies are also known (Lancaster 1981). In sum, pastoral nomadism seems to stress animal husbandry a longpatterned migration routes exploiting marginal environments.

Developmental models studying the origins of pastoral nomadism have largely focussed on the historical record (Khazanov 1984:10-14; Ephal 1982; Briant 1982; Lees and Bates 1974) largely because researchers have concluded that a close symbiotic relationship exists between the urban/settled and nomadic peoples (Rowton 1973a-b, 1974; Rosenfeld 1965). In addition, the lack of theoretical underpinnings has seriously hampered progress in defining the early stages of pastoral nomadism (Khazinov 1984:89-90. Sahlins (1968:33), Krader (1981:499) and Spooner (1971:201) suggest that this adaptation was late in time and entirely secondary to agriculture. This way of life only developed, they conclude, when extensive population pressure in sedentary zones began, together with dramatic crop strategies. Alternately, Sherratt argues that nomadism developed only with the 'secondary products' revolution (exploitation of wool and milk) in the fourth millennium B.C. (Sherratt 1980, 1981, 1983; echoed more recently by Garrard 1985:45). More specifically, many have associated pastoral nomadism with the advent of the urban state (e.g. Lees and Bates 1974; Adams 1981; Rowton 1973a-b).

The evidence presented above, however, points the way towards a new interpretation. Pastoral nomadism began by the end of the seventh millennium B.C. as populations began to utilize the new discovery of animal husbandry in a new and dramatic way (for Iran, see Flannery 1965:1254-5). Our first line of reasoning has been summarized above. Based on faunal material found at Mureybat, Abu Hureyra, Bouqras, Tell es-Sinn (as well as at Jericho, Beisamoun, Ain Ghazzal, Beidha), it would appear that domestication of ovicaprids, bovids, and perhaps pigs began in the seventh millennium B.C. An emphasis on herding at Bouqras and Tell es Sinn is suggested based on the location of the site in a very low-fall rainfall area (150 mm currently) and the apparent domination of animals over plants in the economy (Clason 1979/80:38; such a conclusion may also be important if confirmed from the sites in the El Kowm basin).

Secondly, if domesticated animals were known to the inhabitants of the Fertile Crescent and the sites in the steppe/desert are chronologically contemporaneous and/or loosely inter-related, we may suggest that the sites in the steppe represent either those of pastoral nomads or hunters/gatherers (Betts 1982:30). This evidence is difficult to resolve. First, in terms of faunal materials, we are limited by our understanding of early morphological development in incipient pastoral nomadism, and at most sites we have no collected material. From sites where faunal material is present, the opinions are mixed. From the Negev/Sinai region, Simmons, on analyzing the data, concluded that seasonal transhumance was present, but there was no reliance on animal husbandry or food production (Simmons 1981:47; supported by Servello 1976:369). At Nahal Issaron, however, the excavators suggest that the recovered remains of animals, dominated by *Capra* sp. suggest the development of animal husbandry (Goring-Morris and Gopher 1983:162; supported by Baron 1981:61). In the southern Sinai, sites such as Wadi Tbeik, Igrat el Mihad and Rothenberg 590 have a preponderance of *Capra ibex* reported, suggesting only the presence of wild animals. However, based on domestic *Capra aegagrus* found at Beidha and Ain Ghazzal and the difficulty in identifying wild vs. domestic forms in initial phases, one should be cautious about interpretation. It may be that by the mid-seventh millennium, domestic ovicaprids had been introduced into the Sinai (Moore 1982:16; Baron 1981:65; Bar-Yosef 1981a:403; Tchernov and Bar-Yosef 1982; Clutton-Brock 1982:131; Uerpmann 1982). From southern Jordan, the complex at Wadi Jilat 7 (Waechter's site C) yielded a moderate quantity of animal bones and carbonized plant remains: mostly gazelle with smaller numbers of other wild species which included cattle. No ovicaprids were found (Garrard et al. 1986:23), supporting the evidence already drawn by Bate (1938). However, the two radiocarbon dates from the complex suggest a date of the early to mid-seventh millennium B.C. and the faunal remains from the site may correspond to the early Abu Hureyra pattern reported on the Euphrates for a similar time period. Similarly, from Dhuweila (2202) in the eastern desert of Jordan, Garrard identifies over 100 faunal remains as *Gazella* sp. But from Ibn el Ghazzi (3133) and Jebel Naja (2321) *Ovis/Capra* sp. are present. Garrard concludes that at Jebel Naja (2321) both *Capra* and *Ovis* are present and based on morphology it is impossible to make a judgement on domesticity. If they were domesticated he points out, 'the site is likely to be PPNB or later in date.' (Garrard in Betts 1985:45). Helms, in his analysis of the Jordanian sites, concludes that by the late PPNB, extensive relationships had developed between permanent agricultural settlements and *semi-nomadic pastoralists* and hunters (Helms 1987:51).

In our specific area of west Iraq and southern/eastern Syria the evidence is less clear, principally since the steppe/desert sites have not been analyzed or excavated in detail. Nevertheless, we can draw certain conclusions and ideas based on the sedentary zone development itself. First, along the Euphrates it is clear that animal husbandry began by the late PPNB in a startling reversal from the reliance on earlier wild fauna (Legge and Rowley-Conwy 1987) (Fig. 3). Legge and Rowley-Conwy conclude that domestication of the ovicaprids within the Abu Hureyra area was stimulated by the wide-spread killing of gazelles in the steppe/desert (using the kites of the early-middle PPNB). This created a demand for a steady supply of meat protein which could best be met by domesticating ovicaprids. This new innovation replaced the earlier reliance on gazelle which now verged on extinction and could not migrate northward in the summer season to complete their annual cycle (Legge and Rowley-Conwey 1987:95 and Figure on p. 91). In turn, this new reliance on domesticated animals continued the interaction with hunting/gathering groups in the desert who now turned towards the domesticated ovicaprids as an alternative means of survival. Cauvin interprets the PPNB material of Nadauoiyeh 4 then as the remains of groups 'more or less nomadic moving around the villages' (Cauvin 1981:482). Moore is less cautious. He states that the introduction of domesticated ovicaprids made possible for the first time the exploitation of the steppe zone by pastoral groups. They were semi-sedentary (?) and this way of life flourished as an alternative to increasing populations within the sedentary zone (Moore 1981:448).

No faunal analyses have been carried out on sites from western Iraq or south/eastern Syria and the suggested profile from sites such as Abu Hureyra and the dominance of *Ovis-Capra* at Bouqras and Tell es Sinn cannot yet be duplicated (Fig. 3). Nevertheless, two other arguments can conclude our suggestions here. It appears that substantial construction and the blossoming of sites in the PPNB can only be related to complex interactions with the settled zones which boomed when domestic animals were introduced. Second, the appearance of numerous sites in western Iraq along now dry watercourses argues for both climatic amelioration during this period (Moore 1979:69; Sakaguchi 1987:14; Fujii et al. 1987:38) and the ability of a new way of life to sustain human existence in the region. While much of western Iraq and southern Syria has yet to be adequately surveyed, it does appear that Late Upper and Epipaleolithic populations largely abandoned the steppe/desert (e.g., El Kowm, Cauvin 1981:481), particularly in the south. The PPNB settlements represent the initial re-settlement which was made largely possible by a new way of life: pastoral nomadism. It is also intriguing to note that the succeeding period within the Levantine sedentary zone represented a decline in fortunes but that most of the Arabian peninsula was now occupied due to the exceedingly rapid spread of pastoral nomadism.

JEBEL BISHRI

The presence of PPNB period populations in western Iraq and the region around Jebel Bishri and the Palmyrene has been demonstrated above. We have suggested that the driving force behind this expansion in the late seventh millennium B.C. was the adoption of pastoral nomadism, a way of life based on a symbiotic relationship to the sedentary zone. The 'sudden' appearance of the populations without any previous occupational history in the region (or at least a hiatus of over three millennia) strongly argues for a brief examination of ethnic/linguistic identity. We have known for some time that the two major ethno-linguistic components of the alluvium during the early historic period were the Sumerians and Akkadians. Traditionally, the southern plain was recognized by the early inhabitants themselves as being divided into southern Sumer and northern Akkad (Jacobsen 1939; Gelb 1961a:5). Linguistically, Akkadian is recognized as belonging to the Semitic family of languages and analysis of cuneiform texts suggests that people speaking Akkadian were well established in the alluvium by the EDII-III periods (2700 B.C.; Biggs 1966:77-78; 1967; 1974:27 and note 24; Cooper 1973:240-243). The center of political control in Akkad was often considered to be Kish based on our interpretation of the Sumerian King List (Jacobsen 1939a). Gelb has suggested that the civilizational aspect of the period between 2700 B.C. and 2400 B.C. in Akkad should be labelled 'Kish Civilization' (Gelb 1981:52 ff). Historians have generally considered the Akkadians to have been nomads coming from the west and settling in the alluvium sometime in the fourth millennium B.C. (Kienast 1981:83- 4) or in more remoter antiquity (Gelb 1961:27; Kramer 1948:160-1 and note 10; Gibson 1972:1-2). If the Akkadians are to be viewed as 'settled' and an integral part of the alluvium, their incursions into the plain (principally around Akkad) would have followed the later historical pattern attested for the Amorites (Gelb 1961:29 ff). These later Semitic populations identified as MAR.TU came from the western desert of Iraq and southeastern Syria and appear in the light of the history from ca. 2500 B.C. Historical mention of the Amorite populations occasionally ties them to Jebel Bishri, the mountain massif just west of the Middle Euphrates. It is sometimes referred to as their homeland (Buccellati 1966:236-242) However, MAR.TU is merely a convenient term for populations which had existed in the area since the end of the seventh millennium B.C. We have postulated that the Middle Euphrates in this region (including Jebel Bishri) saw the first development of pastoral nomadism based on animal husbandry, and the subsequent expansion was southward into southern Syria and western Iraq. Therefore, we would conclude that Jebel Bishri indeed represented the 'homeland' of the early Semitic populations who interacted principally with the settled populations of the southern alluvium and the Middle Euphrates. Therefore, the layering of Semitic influence on the alluvium is considerably more complex than previously thought. The Akkadians can certainly be seen as the MAR.TU predecessors in the fifth and fourth millennia B.C. and perhaps Gelb's early Kish rulers such as 'Scorpion', 'Fish', and 'Dog' may represent the earliest stratum from the sixth millennium B.C. (Gelb 1981:54-55, 67). This earliest infiltration and occupation of the alluvium coincided with the expanding

Sumerian (?) populations from the south and the Proto-Elamites from the east (Kramer 1948:160-1; Gelb 1961a:5). Since we have PPNB populations in western Iraq as early as the late seventh millennium B.C., and the earliest Ubaid settlements of Akkad (and Kish) do not antedate ca. 5500 B.C. (Gibson 1972:66, 111, 123, and note 43; map 2), it would be safe to conclude that Semitic nomadic pastoralists began to settle the northern alluvium from the western desert and interact with other populations as early as the seventh millennium B.C. and that early Semitic populations came from regions to the west of the central Euphrates Valley and mountain massifs such as Jebel Bishri.

ACKNOWLEDGEMENTS

It is indeed a great pleasure to dedicate this small contribution to Professor Maurits van Loon, the person most responsible for introducing me to the basics of tell archaeology. His unflagging efforts in stimulating education both in the classroom and the field will always be remembered with special fondness. In addition, I would like to thank Dr. Glen Cole at the Field Museum of Natural History in Chicago who permitted examination of the artifactual materials and notebooks from the Field Expedition to western Iraq and southern Syria in 1926-1928. These included the original study materials by Dorothy Garrod. In addition, my thanks goes to Andrew Garrard and Alison Betts at the British Institute of Archaeology in Amman for updating the latest developments in eastern Jordan.

REFERENCES

ABE Y. AND AKAZAWA T.
1977 'Burin Factory Site, Palmyra, Syria', *Bulletin of the National Science Museum*, Series D (Anthropology) 3: 1-22.

ADAMS R.MC.
1981 *Heartland of Cities*, Chicago: The University of Chicago Press.

ADAMS R.MC. ET AL.
1977 'Saudi Arabian Archaeological Reconnaissance 1976', *Atlal* 1: 21-40.

AKAZAWA T.
1977 'Douara Basin Sites: Prehistoric Workshop Station near Palmyra, Syria', *Bulletin of the National Science Museum*, Series D (Anthropology) 1: 11-24.
1979 'Flint Factory Sites in Palmyra Basin', in: K. Hanihara and T. Akazawa (Eds.), *Palaeolithic Site of Douara Cave and Palaeogeography of Palmyra Basin in Syria*, Part II, Tokyo: The University of Tokyo Press. Pp. 159-200.

AKKERMANS P.A. ET AL.
1982 'The 1976-1977 Excavations at Tell Bouqras', *Annales Archéologiques Arabes Syriennes* 32: 45-57.

ALLRED B.W.
1968 *Range Management Training Handbook for Saudi Arabia*, Rome: F.A.O.

ANATI E.
1979 *L'Arte Rupestre del Negev e del Sinai*, Milan: Jaca Books.

BARON A.G.
1981 'Adaptive Strategies in the Archaeology of the Negev', *Bulletin of the American Schools of Oriental Research* 242: 51-81.

BAR-YOSEF O.
1980 'Prehistory of the Levant', *Annual Review of Anthropology* 9: 101-33.

1981a 'The 'Pre-pottery Neolithic' Period in the Southern Levant', in: J. Cauvin and P. Sanlaville (Eds.), *Préhistoire du Levant*, Paris: CNRS. Pp. 555-69.

1981b 'Neolithic Sites in Sinai', in: W. Frey and H.P. Uerpmann (Eds.), *Beiträge zur Umweltgeschichte des Vorderen Orients*, Wiesbaden: Ludwig Reichert. Pp. 217-35.

1981c 'Paleolithic and Neolithic Desertic Adaptations in the Sinai Peninsula', in: *Union Internacional de Ciencias prehistoricas y protohistoricas X Congreso*. Comision IX, Cultura y Medio ambiente del Hombre fosil en Asia. Mexico City. Pp. 35-61.

BETTS A.

1982 'Prehistoric Sites at Qa'a Mejalla, Eastern Jordan', *Levant* 14: 1-34.

1983 'Black Desert Survey, Jordan: First Preliminary Report', *Levant* 15: 1-10.

1984a 'Black Desert Survey, Jordan: Second Preliminary Report', *Levant* 16: 25-34.

1984b 'Black Desert Survey, 1983', *Palestine Exploration Quarterly* 116:78.

1985 'Black Desert Survey, Jordan: Third Preliminary Report', *Levant* 17: 29-52.

BIGGS R.D.

1966 'The Abu Salabikh Tablets', *Journal of Cuneiform Studies* 20: 73-88.

1967 'Semitic Names in the Fara Period', *Orientalia: NS* 36: 55-66.

1974 *Inscriptions from Tell Abu Salabikh*, Chicago: The University of Chicago Press.

BOERMA J.A.K.

1983 'Environmental Conditions', in: P.A. Akkermans et al. 'Bouqras Revisited: Preliminary Report on a Project in Eastern Syria', *Proceedings of the Prehistoric Society* 49: 362-5.

BRIANT P.

1982 *État et pasteurs au Moyen-Orient ancien*, Cambridge: Cambridge University Press.

BUCCELLATI G.

1966 *The Amorites of the Ur III Period*, Naples: Studi Semitici.

BUCCELLATI G. AND KELLY-BUCCELLATI M.

1967 'Archaeological Survey of the Palmyrene and the Jebel Bishri', *Archaeology* 20: 305-6.

BUITENHUIS H.

1984 'Some Remarks on the Possible Development of *Bos Taurus* Breeds in the Near East', in: J. Clutton-Brock and C. Grigson (Eds.), *Animals and Archaeology, 3, Early Herders and Their Flocks*, Oxford: BAR International Series 202. Pp. 213-21.

BURINGH P.

1957 'Living Conditions in the Lower Mesopotamian Plain in Ancient Times', *Sumer* 13: 30-46.

CAUVIN J.

1972 'Sondage à Tell Assouad (Djézireh, Syrie)', *Annales Archéologiques Arabes Syriennes* 22: 85-103.

1981 'L'Occupation Néolithique de la Région d'el Kowm: Résultats 1978-1979', in: J. Cauvin and P. Sanlaville (Eds.), *Préhistoire du Levant*, Paris: CNRS. Pp. 471-483.

CAUVIN M.-C.

1972 'Note préliminaire sur l'outillage lithique de Tell Assouad (Djézireh)', *Annales Archéologiques Arabes Syriennes* 22: 90-93.

CLASON A.T.

1980 'The Animal Remains from Tell es Sinn Compared with Those from Bouqras', *Anatolica* 7: 35-54.

1983 'Faunal Remains' in: P.A. Akkermans et al., 'Bouqras Revisited: Preliminary Report on a Project in Eastern Syria', *Proceedings of the Prehistoric Society* 49: 359-362.

CLASON A.T. AND BUITENHUIS H.

1978 'Archaeozoologisch onderzoek in het Midden-Oosten', *Spiegel Historiael* 13: 677-87.

CLAWSON M. ET AL.

1971 *The Agricultural Potential of the Middle East*, New York: Elsevier.

CLUTTON-BROCK J.

1978 'Early Domestication and the Ungulate Fauna of the Levant during the Prepottery Neolithic Period', in: W.C. Brice (Ed.), *The Environmental History of the Near and Middle East Since the Last Ice Age*, New York: Academic Press. Pp. 29-40.

1981 *Domesticated Animals from Early Times*, London: British Museum of Natural History.

DE CONTENSON H. AND VAN LIERE W.J.
1966 'Premier Sondage à Bouqras en 1965, Rapport Préliminaire', *Annales Archéologiques Arabes Syriennes* 16: 181-192.

COOPER J.
1973 'Sumerian and Akkadian in Sumer and Akkad', *Orientalia* NS 42: 239-246.

COPELAND L.
1982 'Prehistoric Tells in the Lower Balikh Valley, Syria: A Report on the Survey of 1978', *Annales Archéologiques Arabes Syriennes* 32: 251-271.

COPELAND L. AND HOURS F.
1971 'A Microlithic Flint Site in the Wadi Rum, Jordan,and a Review of the Epipaleolithic of Northern Arabia', *PSAS* 2: 7-21.

CURRELLY C.T.
1906 'Gebel Musa and the Nawamis', in: F. Petrie, *Researches in Sinai*, New York: Dutton, Chapters 15-18.

DANIN A.
1985 'Palaeoclimates in Israel: Evidence from Weathering Patterns of Stones In and Near Archaeological Sites.'*Bulletin of the American Schools of Oriental Research* 259: 33-43.

DUBERTRET L. AND DUNAND M.
1954 'Les gisements ossiferes de Khirbet El-Umbachi et de Hebariye (Safa)', *Annales Archéologiques Arabes Syriennes* 4-5: 59-76.

DUSSAUD R.
1929 'Les releves du Capitaine Rees dans le Desert de Syrie', *Syria* 10: 144-163.

DYSON-HUDSON R. AND DYSON-HUDSON N.
1980 'Nomadic Pastoralism', *Annual Review of Anthropology* 9: 15-61.

ENDO K.
1978 'Stratigraphy and Paleoenvironments of the Deposits in and around the Douara Cave Site', in: K. Hanihara and Y. Sakaguchi (Eds.), *Paleolithic Site of the Douara Cave and Paleogeography of Palmyra Basin in Syria*, Tokyo: University of Tokyo Press. Pp. 53-81.

EPH'AL I.
1982 *The Ancient Arabs*, Leiden: E.J. Brill.

FIELD H.
1951 'Reconnaissance in Southwestern Asia', *Southwest Journal of Anthropology* 7: 86-102.
1960 *North Arabian Desert Archaeological Survey 1925-50*, Cambridge, Mass.: Harvard University, Peabody Museum.

FLANNERY K.
1965 'The Ecology of Early Food Production in Mesopotamia', *Science* 147: 1247-56.

FUJII S. ET AL.
1987 'Thaniyyet Wuker: A Pre-Pottery Neolithic B Site on the Lacustrine Terrace of Paleo-Palmyra Lake', in: T.Akazawa and Y. Sakaguchi (Eds.), *Paleolithic Site of the Douara Cave and Paleogeography of Palmyra Basin in Syria*, Tokyo: University of Tokyo Press, Pp. 29-39.

GARRARD A.N.
1984 'Azraq Survey 1982', *Levant* 16:v.
1985 'Appendix I: Faunal Remains', in: A. Betts, 'Black Desert Survey, Jordan: Third Preliminary Report', *Levant* 17: 39-49.

GARRARD A.N. AND STANLEY-PRICE N.P.
1975 'A Survey of Prehistoric Sites in the Azraq Basin, Eastern Jordan', *Paléorient* 3: 109-126.

GARRARD A.N. ET AL.
1981 'Environment and Settlement during the Upper Pleistocene and Holocene at Jubba in the Great Nefud, Northern Arabia', *Atlal* 5: 137-148.
1985 'Prehistoric Environment and Settlement in the Azraq Basin. A Report on the 1982 Survey Season', *Levant* 17: 1-28.
1986 'Prehistoric Environment and Settlement in the Azraq Basin: An Interim Report on the 1984 Excavation Season', *Levant* 18: 5-24.

1987 'Prehistoric Environment and Settlement in the Azraq Basin: An Interim Report on the 1985 Excavation Season', *Levant* 19: 5-25.

GARROD D.
1960 'The Flint Implements', in: H. Field, *North Arabian Desert Archaeological Survey 1925-50*, Cambridge, Mass.: Harvard University, Peabody Museum. Pp. 111-123.

GELB I.J.
1961 'The Early History of the West Semitic Peoples', *Journal of Cuneiform Studies* 15: 27-47.
1961a *Old Akkadian Writing and Grammar*, Chicago: The University of Chicago Press.
1981 'Ebla and the Kish Civilization', in: L. Cagni (Ed.), *La Lingua di Ebla*, Naples: Istituto Universitario Orientale. Pp. 9-73.

GIBSON Mc G.
1972 *The City and Area of Kish*, Miami, Florida: Field Research Projects.

GILEAD I. AND GOLDBERG P.
1976 'Prehistoric Survey in the Kadesh Barnea Area', *Israel Exploration Journal* 26: 137.

GILMORE M. ET AL.
1982 'Preliminary Report on the Northwestern and Northern Regions Survey, 1981 (1401)', *Atlal* 6: 9-24.

GOLDBERG P. AND BAR-YOSEF O.
1982 'Environmental and Archaeological Evidence for Climatic Change in the Southern Levant', in: J.L. Bintliff and W.V. Zeist (Eds.), *Paleoclimates, Palaeoenvironments and Human Communities in the Eastern Mediterranean Region in Later Prehistory*, Oxford: BAR International Series 133. Pp. 399-414.

GOLDSCHMIDT W.
1979 'A General Model for Pastoral Social Systems', *Pastoral Production and Society*, Cambridge: Cambridge University Press. Pp. 15-27.

GORING-MORRIS A.N. AND GOPHER A.
1983 'Nahal Issaron: A Neolithic Settlement in the Southern Negev', *Israel Exploration Journal* 33: 149-162.

GORING-MORRIS A.N. AND MINTZ E.
1976 'Surveys in Southern Sinai', *Israel Exploration Journal* 26: 137-138.

HALDAR, A.
1971 *Who Were the Amorites?*, Leiden: E.J. Brill.

HARRISON D.L.
1964 *The Mammals of Saudi Arabia*, 3 vols, London: E. Benn.

HELMS S.W.
1981 *Jawa: Lost City of the Black Desert*, Ithaca, N.Y.: Cornell University Press.
1982 'Paleo-Beduin and Transmigrant Urbanism', in: A.Hadidi (Ed.), *Studies in the History and Archaeology of Jordan*, I, Amman: Department of Antiquities. Pp. 97-113.
1987 'Jawa, Tell Um Hammad and the EB I/Late Chalcolithic Landscape', *Levant* 19: 49-81.

HOOIJER D.A.
1966 'Preliminary Notes on the Animal Remains found at Bouqras and Ramad in 1965', *Annales Archéologiques Arabes Syriennes* 16: 193-195.

HOROWITZ A.
1979 *The Quaternary of Israel*, New York: Academic Press.

JACOBSEN T.
1939 'The Assumed Conflict Between the Sumerians and Semites in Early Mesopotamian History', *Journal of the American Oriental Society* 59: 485-495.
1939a *The Sumerian King List*, Chicago: The University of Chicago Press.

JOHNSON D.L.
1969 *The Nature of Nomadism*, Chicago: The University of Chicago Press.

KENYON K.
1960 *Archaeology in the Holy Land*, New York: Praeger.

KHAZANOV A.M.
1984 *Nomads and the Outside World*, Cambridge: Cambridge University.

KIENAST B.
1981 'Die Sprache von Ebla und das Altsemitische', in: L.Cagni (Ed.), *La Lingua di Ebla*, Naples: Istituto Universitario Orientale. Pp. 83-98.

KOIZUMI T.
1978 'Climato-genetic Landforms around Jabal ad Douara and Its Surroundings', in: K. Hanihara and Y. Sakaguchi (Eds.), *Paleolithic Site of the Douara Cave and Paleogeography of Palmyra Basin in Syria*, Tokyo: The University of Tokyo Press. Pp. 29-51.

KRADER L.
1981 'The Ecology of Nomadic Pastoralism', in: J.G. Galaty and P.E. Salzmann (Eds.), *Change and Development in Nomadic and Pastoral Society*, Leiden: E.J. Brill. Pp. 499-510.

KRAMER S.N.
1848 'New Light on the Early History of the Ancient Near East', *American Journal of Archaeology* 52: 156-164.

LANCASTER W.
1981 *The Rwala Beduin Today*, Cambridge: The Cambridge University Press.

LEES S. AND BATES D.G.
1974 'The Origins of Specialized Nomadic Pastoralism: A Systemic Model', *American Antiquity* 39: 187-193.

LEGGE A.J. AND ROWLEY-CONWY P.
1987 'Gazelle Killing in Stone Age Syria', *Scientific American* 248: 88-95.

MAITLAND R.A.
1927 'The Works of the "Old Men" in Arabia', *Antiquity* 1: 197-203.

MELLAART J.
1975 *The Neolithic of the Near East*, London: Thames and Hudson.

MINTZ E. AND BEN AMI D.
1977 'Neolithic Occurrences', in: O. Bar-Yosef and J.L. Phillips (Eds.), *Prehistoric Investigations in Gebel Maghara, Northern Sinai*, Jerusalem: Hebrew University. Pp. 219-244.

MOORE A.M.T.
1979 'A Pre-Neolithic Farmer's Village on the Euphrates', *Scientific American* 240: 62-70.
1981 'North Syria in Neolithic 2' in: J. Cauvin and P. Sanlaville (Eds.), *Préhistoire du Levant*, Paris: CNRS. Pp. 445-456.
1982 'A Four Stage Sequence for the Levantine Neolithic, ca. 8500-3750 B.C', *Bulletin of the American Schools of Oriental Research* 246: 1-34.

MORTENSEN P.
1970 'A Preliminary Study of the Chipped Stone Industry from Beidha', *Acta Archaeologica* 41: 1-54.

MUSIL A.
1928 *Palmyrena*, New York: The American Geographical Society.

NOY T. AND COHEN R.
1974 'Nahal Boqer: An Early Pre-Pottery Neolithic B Site', Mitequfat Ha'even 12: 15-25.

PAEPE R.
1971 'Geological Approach of the Tell ed-Der Area', in: H. Gasche and R. Paepe (Eds.), *Tell ed-Der*, I. Rapport préliminaire sur la Première Campagne, Fevrier 1970, Leuven: Editions Peeters. Pp. 9-27.

PARR P.J. ET AL.
1978 'Preliminary Report on the Second Phase of the Northern Province Survey 1397/1977', *Atlal* 2: 29-51.

POIDEBARD A.
1934 *La Trace de Rome dans le Desert de Syrie*, Paris: Paul Geuthner.

REES V.C.
1929 'The Transjordan Desert', *Antiquity* 3: 389-407.

RHOTERT H.
1938 *Transjordanien, Vorgeschichtliche Forschungen*, Stuttgart: Verlag Strecher und Schroeder.

ROLLEFSON G.O.
1982 'Preliminary Report on the 1980 Excavations at Ain el-Assad', *ADAJ* 26: 5-35.
1984 'Ain Ghazal : An Early Neolithic Community in Highland Jordan near Amman', *Bulletin of the American Schools of Oriental Research* 255: 3-14.
ROLLEFSON G.O. AND FROHLICH B.
1982 'A PPNB Burin Site on Jebel Uweinid, Eastern Jordan', *ADAJ* 27: 189-198.
ROLLEFSON G.O. AND KAFAFI Z.
1985 'Khirbet Hammam: A PPNB Village in the Wadi el-Hasa, Southern Jordan', *Bulletin of the American Schools of Oriental Reserach* 258: 63-69.
ROSEN S.A.
1987 'Demographic Trends in the Negev Highlands: Preliminary Results from the Emergency Survey', *Bulletin of the American Schools of Oriental Research* 266: 45-58.
ROSENFELD H.
1965 'The Social Composition of the Military in the Process of State Formation in the Arabian Desert', *JRAI* 95: 75-96, 174-194.
ROWTON M.B.
1973a 'Autonomy and Nomadism in Western Asia', *Orientalia: NS* 42: 247-258.
1973b 'Urban Autonomy in a Nomadic Environment', *Journal of Near Eastern Studies* 32: 201-215.
1974 'Enclosed Nomadism', *JESHO* 17: 1-30.
SAHLINS M.
1968 *Tribesmen*, Englewood Cliffs, N.J.: Prentice-Hall Co.
SAKAGUCHI Y.
1978 'Palmyra Pluvial Lake', in: K. Hanihara and Y. Sakaguchi (Eds.), *Paleolithic Site of the Douara Cave and Paleogeography of Palmyra Basin in Syria*, Tokyo: Tokyo University Press. Pp. 5-28.
1987 'Paleoenvironments in Palmyra District during the Late Quaternary', in: T. Akazawa and Y. Sakaguchi (Eds.), *Paleolithic Site of the Doura Cave and Paleogeography of Palmyra Basin in Syria*, Tokyo: Tokyo University Press. Pp. 1-27.
SERVELLO A.F.
1976 'Nahal Divshon: A Pre-pottery Neolithic B Hunting Camp', in: A.E. Marks (Ed.), *Prehistory and Palaeoenvironments in the Central Negev, Israel*, I, Dallas: SMU Press. Pp. 349-370.
SHERRATT A.
1980 'Water, Soil and Seasonality in Early Cereal Cultivation', *World Archaeology* 11: 313-330.
1981 'Plough and Pastoralism: Aspects of the Secondary Products Revolution', in: I. Hodder et al. (Eds.), *Pattern of the Past: Studies in Honour of David Clarke*, Cambridge: Cambridge University Press. Pp. 261-305.
1983 'The Secondary Exploitation of Animals in the Old World', *World Archaeology* 15: 90-104.
SIMMONS A.H.
1981 'A Paleosubsistence Model for Early Neolithic Occupation of the Western Negev Desert', *Bulletin of the American Oriental Society* 242: 31-50.
SINGH P.
1974 *Neolithic Cultures of Western Asia*, New York: Seminar Press.
SPOONER B.
1971 'Towards a Generative Model of Nomadism', *Anthropological Quarterly* 44: 198-210.
SUZUKI C. AND AKAZAWA T.
1971 'Manufacturing Technique of the Stone Artifacts from Palmyra, Syria', *Journal of the Anthropological Society of Nippon* 79: 105-127.
SUZUKI C. AND KOBORI I. (EDS.)
1970 *Report of the Reconnaissance Survey on Paleolithic Sites in Lebanon and Syria*, Tokyo: University of Tokyo Press.
TCHERNOV E.
1982 'Faunal Responses to Environmental Changes in the Eastern Mediterranean during the Last 20,000 Years', in: J.L. Bintliff and W. V. Zeist (Eds.), *Palaeoclimates, Palaeoenvironments*

and Human Communities in the Eastern Mediterranean Region in Later Prehistory, Oxford: BAR International Series 133. Pp. 105-127.

TCHERNOV E. AND BAR-YOSEF O.

1982 'Animal Exploitation in the Pre-Pottery Neolithic B Period at Wadi Tbeik, Southern Sinai', *Paléorient* 8: 17-37.

TYRACEK J. AND AMIN R.M.

1981 'Rock Pictures (Petroglyphs) near Qasr Muhaiwair, Iraqi Western Desert', *Sumer* 37: 145-148.

UERPMANN H.P.

1982 'Comments' in: J.L. Bintliff and W. van Zeist (Eds.), *Paleoclimates, Palaeoenvironments and Human Communities in the Eastern Mediterranean Region in Later Prehistory*, Oxford: BAR International Series 133. Pp. 131, 451.

VOUTE C.

1957 'A Prehistoric Find Near Razzaza (Karbala Liwa)', *Sumer* 13: 135-148.

WAECHTER J. AND SETON-WILLIAMS V.M.

1938 'The Excavations at Wadi Dhobai, 1937-1938 and the Dhobaian Industry', *JPOS* 18: 172-186, 292-298.

WHITNEY J.W.

1983 *Erosional History and Surficial Geology of Western Saudi Arabia*, Jeddah: Ministry of Petroleum and Mineral Resources.

WRIGHT H.E. JR.

1958 'An Extinct Wadi System in the Syrian Desert', *Bulletin of the Research Council of Israel* 7: 53-60.

WRIGHT H.T.

1981 'The Southern Margins of Sumer' in: R. Mc. Adams, *Heartland of Cities*, Chicago: The University of Chicago Press. Pp. 295-345.

ZARINS J.

in press 'Archaeological and Chronological Problems within the Greater Southwest Asian Arid Zone, 8500-1850 B.C', in: R.W. Ehrich (Ed.), *Absolute Chronologies in Old World Archaeology*, Chicago: The University of Chicago Press. Chapter 3.

ZOHARY M.

1952 'Ecological Studies in the Vegetation of the Near Eastern Deserts', *Israel Exploration Journal* 2: 201-215.

1962 *Plant Life of Palestine, Israel, and Jordan*, New York: Ronald Press.

Les débuts de la céramique sur le Moyen-Euphrate (6500-5500) B.C.[1]

MARIE LE MIÈRE

La région considérée ici s'étend de part et d'autre de l'Euphrate et de ses affluents depuis les steppes du Sud-Est de la Turquie jusqu'à la frontière syro-irakienne (Fig. 1). Si la présence du fleuve donne une certaine unité à cette région, sa vaste extension du Nord au Sud entraîne évidemment des conditions climatiques un peu différentes suivant que l'on se trouve encore en bordure des montagnes du Taurus oriental ou déjà en zone désertique en aval du confluent du Balikh. Et si des installations sédentaires ont apparu au Nord de l'isohyète de 200 mm dès le IXè millénaire, c'est seulement au milieu du VIIè millénaire que toute la région est occupée avec l'apparition des premières installations au Sud.

Cette région n'a pas encore été explorée systématiquement, néanmoins les travaux des dix dernières années -fouilles de sauvetage effectuées en Turquie, prospection de la vallée du Sajour, nouvelles prospections et fouilles de la vallée du Balikh, fouille du site de Bouqras- ont apporté de nombreuses données nouvelles. Même s'il s'agit dans la plupart des cas de prospections et de sondages restreints, ces travaux permettent de commencer à entrevoir la place de la vallée de l'Euphrate pendant la période qui va de l'apparition de la céramique à la maîtrise de cette technique, de 6500 à 5500 B.C.

En effet, alors que dans la première moitié du VIIè millénaire les sites connus sont tous acéramiques, aux environs de 6000 B.C. la céramique est répandue à peu près dans tout le Proche-Orient. L'élaboration de la technique céramique se serait donc effectuée dans la deuxième moitié du VIIè millénaire. Et, d'après les quelques sites céramiques de cette période, comme Tepe Guran S (Zagros) (Mortensen 1963) ou Çatal Hüyük XII-IX (Anatolie) (Mellaart 1964 et 1966), les premières céramiques semblent se caractériser par un matériel peu abondant, des catégories de poterie peu variées, limitées à une, parfois deux séries, la présence d'inclusions végétales, le traitement des surfaces par lissage ou polissage uniquement,de très rares décors de peinture, des formes très simples et très monotones (fermées ou droites, convexes, sans col ni carène, à fonds plats ou convexes et bords très peu élaborés), une facture qualifiée fréquemment de grossière, en tous cas une épaisseur importante relativement aux séries postérieures (Le Mière 1986). La simplicité et la monotonie des premières céramiques sont peut-être dues à une maîtrise encore limitée de la technique. En effet, d'après les quelques données dont nous disposons, peu de particularismes régionaux se montrent dans ce matériel. Dès les environs de 6000 B.C., alors que la céramique se généralise et apparaît sur tous les sites du Proche-Orient, sauf ceux du Levant Sud (Israel, Jordanie et désert syrien), celle-ci va évoluer très vite et se différencier de façon nette selon les régions, au niveau des pâtes comme au niveau des formes ou des décors.

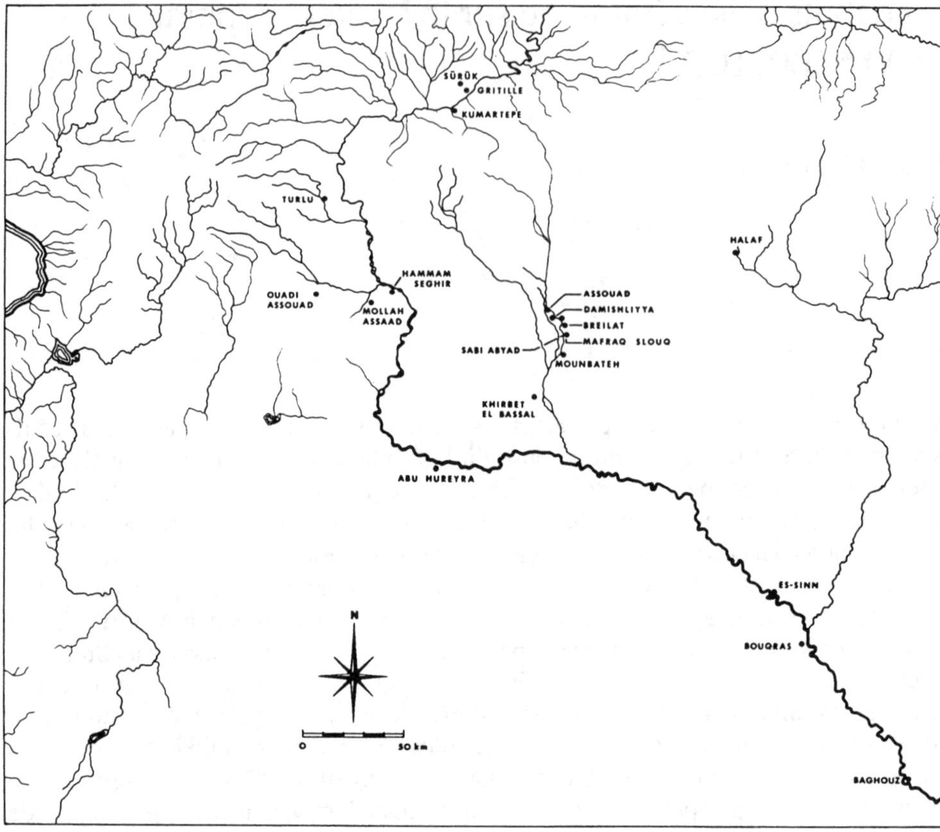

Fig. 1. Les premières céramiques du Moyen-Euphrate.

Les régions céramiques qui entourent le Moyen-Euphrate au début du VIè millénaire sont à l'Est la Mésopotamie du Nord, au Nord-Ouest l'Anatolie et à l'Ouest la Syrie du Nord - Cilicie. Grossièrement la céramique *proto-Hassuna* (Bader 1975) de Mésopotamie du Nord [Tell Sotto, Kül Tepe, Yarim Tepe 1 niv. XII-IX (Merpert et al. 1977; 1978; 1981); Umm Dabaghiyah (Kirkbride 1972; 1973); Telul eth-Thalathat II niv. XVI-XV (Fukai et Matsutani 1981); Hassuna Ia (Lloyd et Safar 1945)] peut se définir par la présence de deux séries, l'une à inclusions végétales, de couleur claire, polie ou non et une série sans inclusions végétales, engobée rouge ou non et parfois polie. Les formes sont principalement carénées et concaves, avec en particulier des formes fermées à carène (double-ogee); les cols sont rares, les moyens de préhension absents, les décors variés (décors en relief et décors peints surtout).

En Anatolie la série claire à inclusions végétales des niveaux les plus anciens disparaît au profit d'une série foncée sans inclusions végétales, polie, aux formes convexes, généralement fermées, sans col, avec des tenons souvent percés, très rarement décorée d'incisions et exceptionnellement peinte.

La céramique de Syrie du Nord - Cilicie vers 6000 B.C.[(Amuq A (Braidwood et Braidwood 1960); Mersin XXXII-XXVII (Garstang 1953); Ras Shamra VB (de Contenson 1977); vallée du Queiq (Mellaart 1981)] se compose en général d'une série grossière à inclusions végétales, plus ou moins claire, parfois polie, et d'une série plus fine sans inclusions végétales, foncée et souvent polie (*DFBW*). Les formes sont, pour les deux séries, convexes, très rarement carénées, éventuellement avec col, à tenons; les décors d'impression sont courants.

Cette description sommaire mériterait d'être nuancée, mais elle fournit néanmoins quelques repères, qui permettront de situer les débuts de la céramique du Moyen-Euphrate par rapport aux régions qui l'entourent.

Nos connaissances sur le Moyen Euphrate étant, malgré les récents travaux, relativement dispersées chronologiquement et géographiquement, et d'importance très variable, une présentation synthétique de la céramique de cette région paraît difficile; c'est pourquoi nous la présenterons site par site, en suivant le cours du fleuve, afin d'essayer de relever tous les indices significatifs.

Gritille (Voigt 1985; 1986a) (Fig. 1). Ce site a livré seulement 3 tessons tout à fait à la fin de la séquence préhistorique (phases B et A), dont un seul est dans un contexte stratigraphique sûr (phase A). Il est décrit comme peu compact, de couleur gris foncé, poli et appartenant à un vase sans col. Les deux autres tessons sont très petits et ont le même aspect, avec des inclusions minérales très fines; ils se trouvaient dans un contexte moins sûr dans la phase B. Celle-ci serait datée du début du VIè millénaire.

Sürük (Voigt 1986a; comm. pers.) (Fig. 1). La céramique de ce site de prospection n'a pas encore été publiée mais seulement mentionnée. Elle serait comparable à celle de Kumartepe (cf. *infra*).

Kumartepe (Roodenberg et al. 1984) (Fig. 1). Présente dès la base du sondage effectué sur ce site, la céramique est abondante et caractérisée par deux séries principales; une série grossière, la plus fréquente, généralement à inclusions végétales, de couleur claire, rarement polie, dont les formes sont convexes ou rectilignes, le plus souvent fermées ou droites, avec de rares cols peu marqués et de nombreux tenons horizontaux ou verticaux percés; une série plus fine, aux dimensions plus réduites, sans inclusions végétales, de couleurs plus variées (grise, beige, rouge ou marron), polie ou non, dont les formes sont assez semblables à celles de la série grossière, hormis la présence de formes concaves et de bords plus élaborés, et peut-être aussi de carènes. Il faut également signaler l'existence de deux autres séries de céramique, sans inclusions végétales, qui sont peu abondantes: l'une à engobe rouge et l'autre unie non polie, à pâte fine, de même que l'existence de quelques tessons peints de lignes verticales rouges.

L'ensemble de ces caractères permet de comparer ce matériel de façon précise avec celui de Syrie du Nord- Cilicie, mises à part les couleurs moins foncées de la série polie sans inclusions végétales qui, par ailleurs, correspondrait au *DFBW*. On notera que la série de la vallée du Queiq que J. Mellaart compare à la série *DFBW* est elle aussi plus claire (Mellaart 1981). Une date vers le milieu du VIè millénaire, telle qu'elle est suggérée dans le rapport préliminaire, est peut-être un peu récente, car la plupart des traits sont ceux de la céramique Amuq A, bien que quelques éléments Amuq B commencent à apparaître. On notera que, si le site a été occupé à d'autres périodes, il n'est pas fait mention de matériel Halaf ou Obeid.

Turlu (Mellink 1964; Moore 1979) (Fig. 1). La céramique des niveaux de base (1-2) du sondage effectué par J. Perrot en 1962 comporte une série grossière, la plus importante, à inclusions végétales, de couleur beige ou marron, parfois polie. Quelques exemplaires de cette série sont peints de lignes rouges ou noires. La deuxième série, que A. Moore attribue aux niveaux de base mais dont J. Perrot signalait l'association avec la céramique Halaf (niv. 3-4), est très dure, de couleur grise, noire ou parfois rouge, et polie. Les formes des deux séries sont les mêmes: vases fermés sans col ou avec col, vases ouverts, fonds plats.

Les informations sont un peu succinctes, notamment sur le nombre de tessons, qui permettrait de juger de la signification de l'absence d'autres séries. L'absence de série foncée dans les niveaux de base rapprocherait plutôt ce matériel de celui des sites du Balikh (cf. *infra*), alors que si sa présence était vérifiée, la parenté avec la céramique de Syrie du Nord - Cilicie serait évidente, même s'il est difficile de le situer précisément dans la chronologie.

Tell Ouadi Assouad (Qara Dere) (Copeland 1985; de Contenson 1985) (Fig. 1). La céramique ramassée sur ce site n'est pas décrite, mais néanmoins la présence dans le matériel lithique de pointes de Byblos et d'une pointe d'Amuq a fait penser que la céramique, faite à la main, pourrait peut-être dater du VIè millénaire; ce n'est qu'une hypothèse. On notera qu'il n'y avait pas en surface de céramique Halaf.

Tell Mollah Assaad (Copeland 1985; de Contenson 1985) (Fig. 1). Quelques tessons grossiers ont été trouvés en surface, qui pourraient contenir des inclusions végétales. Ils sont de couleur beige; la seule forme décrite est une jarre sans col avec cordon, qui rappelle une forme courante à Tell Assouad (Le Mière 1979), mais aussi dans l'Amuq (Braidwood et Braidwood 1960, fig. 21). Une coupe a livré du matériel qui ne comportait pas de céramique et a été attribué à une occupation acéramique de la fin du VIIè millénaire. Il faut néanmoins remarquer que le matériel lithique trouvé dans le même secteur du tell que la céramique est tout à fait semblable à celui de la coupe et que l'ensemble, y compris les enduits blancs et la vaisselle blanche, pourrait correspondre à une seule occupation. Il n'y a pas trace d'une occupation Halaf.

Hammam Seghir (Moore 1985) (Fig. 1). La céramique de ce site de prospection est noire ou brune et polie, peut-être de type *DFBW* puisqu'elle est attribuée au VIè millénaire. Le site n'a pas révélé en surface de matériel préhistorique postérieur.

Tell Abu Hureyra (Moore 1975) (Fig. 1). La céramique n'apparaît que dans le niveau supérieur du site, ce dernier étant en grande partie acéramique. L'installation du niveau supérieur présente un certain nombre de caractères qui la différencient des installations précédentes. Elle est datée de la première moitiée du VIè millénaire. Le matériel céramique se limite à une très petite quantité de tessons, soit moins de 100. Les observations sur la pâte concernent 30 tessons que j'ai pu examiner. Une partie seulement des tessons contient des inclusions végétales sans que l'on puisse vraiment distinguer deux séries différentes sur ce critère. L'ensemble est de couleur claire. Un seul tesson se distingue: il est gris et plus mince que les autres et pourrait être rapproché des céramiques *DFBW*. Ces tessons ne donnent pas d'indication précise sur les formes hormis qu'ils sont convexes. Les formes signalées dans le rapport préliminaire sont droites, ouvertes ou fermées avec col, convexes. On note la présence de ce qui semble être un double cordon (Moore 1975, fig. 8:5), d'une carène (Moore, comm. pers.) ainsi que d'un ou deux tessons peints. Ces données sont trop succinctes pour permettre d'en tirer de véritables conclusions, les rares indices rappelant plutôt la céramique de Syrie du Nord - Cilicie.

Tell Assouad (Cauvin 1972; Le Mière 1979) (Fig. 1). Le sondage en escalier effectué sur ce tell, dont M. Mallowan avait précédemment sondé les niveaux Halaf du sommet (Mallowan 1946), n'a fourni de céramique que dans les deux marches inférieures. Cette situation singulière de la céramique peut trouver plusieurs explications, que nous ne discuterons pas ici; signalons seulement que les résultats obtenus à Tell Damishliyya (cf. *infra*) montrent une succession chronologique plus 'attendue' de niveaux précéramiques puis céramiques, qui reste à prouver à Tell Assouad. Quoiqu'il en soit la céramique de Tell Assouad, trouvée dans des niveaux datés de la deuxième moitié du VIIè millénaire, se caractérise par une grande simplicité: deux séries seulement, toutes deux de couleur claire; une série à inclusions végétales beaucoup plus importante que la série sans inclusions végétales, une épaisseur le plus souvent supérieure à 10 m, des formes principalement fermées, presque toujours convexes, peut-être des amorces de col, un décor peint sur 4 tessons seulement (sur un total de 1000). Un élément particulier est la présence de nombreux moyens de préhension (cordons, tenons percés ou non et anses). A part ceux-ci, et en particulier les cordons, qui rapprochent ce matériel de celui de Syrie du Nord - Cilicie, tous les autres caractères sont beaucoup trop simples pour permettre une comparaison entre la céramique de Tell Assouad et celle des autres régions céramiques. Par contre cette simplicité, en accord avec les datations, la rapproche des premières céramiques apparues au Proche-Orient.

Tell Damishliyya (P.M.M.G. Akkermans sous presse) (Fig. 1). Dans un sondage, où le sol vierge n'a pas été atteint, deux niveaux acéramiques (1-2) précèdent 5 niveaux céramiques pré-Halaf. Une extension du sondage serait nécessaire pour confirmer l'absence de céramique dans les niveaux inférieurs, le nombre croissant de tessons des niveaux 3 à 7 renforçant l'hypothèse d'une phase acéramique sur ce site. Une occupation Halaf a laissé des vestiges peu importants dans certaines parties du tell. Le matériel céramique est très semblable à celui de Tell Assouad (cf. *supra*): la série à inclusions végétales est beaucoup plus importante encore (99%) à Tell Damishliyya; on n'y trouve pas de cordons et les rares tessons décorés sont imprimés ou incisés et non peints comme à Tell Assouad, mais ce sont les seules variations entre les céramiques des deux sites.

Tell Sabi Abyad (P.M.M.G. Akkermans 1987a; 1987b; sous presse) (Fig. 1). Sur ce site, une séquence continue a révélé jusqu'à présent une importante installation Halaf et des niveaux pré-Halaf. La céramique de ces derniers comporte une série claire à inclusions végétales, parfois polie, avec quelques décors incisés, série la plus nombreuse, ainsi qu'une série plus fine sans inclusions végétales, de couleur gris-noir, souvent polie, et une série de céramiques peintes qui peuvent être également incisées. Il faut signaler aussi dans ces niveaux des fragments de '*husking trays*' ou quelques tessons peints ou peints et incisés 'd'influence' Samarra. Par ailleurs, en surface ont été trouvés des tessons comparables à ceux de Tell Assouad et Tell Damishliyya. Comme le sol vierge n'a pas été atteint et que le dépôt de la période pré-Halaf pourrait être de 5 m d'épaisseur dans certaines parties du tell, on pourrait avoir sur ce site une séquence couvrant la fin du VIIè millénaire et tout le VIè millénaire. Les comparaisons avec les autres régions céramiques sont difficiles à faire en attendant une publication plus précise; cependant les quelques indications dont nous disposons nous orientent plutôt vers la Mésopotamie ('*husking trays*' et céramique Samarra), au moins pour le matériel pré-Halaf le plus récent.

Tulul Breilat (Copeland 1979) (Fig. 1). La céramique trouvée à la surface contient en général des inclusions végétales; elle est claire et parfois polie; les formes sont fermées ou droites, convexes, avec quelques tenons verticaux percés. Elle rappelle donc celle de Tell Assouad et Tell Damishliyya (cf. *supra*) malgré des bords plus élaborés.

Tell Mafraq Slouq (Copeland 1979) (Fig. 1). Ce site de prospection a fourni une céramique dite semblable à celle de Tulul Breilat, c'est-à-dire en général à inclusions végétales, de couleur claire et parfois polie. Mais les formes sont, d'après les figures (Copeland 1979, fig. 5) un peu différentes: droites ou ouvertes et avec des cols. Le matériel est trop peu abondant pour tirer des conclusions importantes mais il semble moins proche de celui de Tell Assouad et Tell Damishliyya que celui de Tulul Breilat.

Tell Mounbateh (P.M.M.G. Akkermans, à paraître; comm. pers.) (Fig. 1). Le tell présente plusieurs éminences et le matériel trouvé à la surface de chacune d'elles n'est pas le même, indépendamment du matériel Halaf ou postérieur. Il y a sur une des éminences une céramique très comparable à celle de Tell Assouad et Tell Damishliyya (cf. *supra*). Sur une autre éminence, la céramique ramassée est différente: moins grossière, elle contient des inclusions végétales plus fines et est moins épaisse; à côté d'une série lissée ou polie, il y a une série engobée rouge; les formes semblent beaucoup plus variées que celles de Tell Assouad et Tell Damishliyya, avec des cols, des carènes et peut-être des cols concaves; enfin plusieurs tessons sont décorés de motifs peints géométriques de couleur rouge, avec même un cas de décor peint et incisé. Cette céramique se rapproche plus de la céramique de Bouqras que celles de Tell Assouad et Tell Damishliyya, notamment la série engobée rouge, dont certains tessons sont tout à fait semblables à ceux de Bouqras; cependant les formes semblent différentes d'un site à l'autre.

Tell Khirbet el-Bassal (Cauvin 1970) (Fig. 1). Le ramassage effectué sur ce tell a fourni à la fois de la céramique brune polie, quelques tessons Halaf et des fragments de vases en pierre polie, qui ont fait penser à J. Cauvin qu'il pourrait y avoir eu une occupation pré-Halaf. Effectivement, à Tell Assouad comme à Tell Damishliyya, il y a une vaisselle de pierre abondante.

Tell Halaf (Schmidt 1943) (Fig. 1). Il faut mentionner ici la céramique '*Altmonochrom*', trouvée tout à fait à la base du tell et dont il n'est pas dit si elle perdure dans les niveaux Halaf. Elle ne comporte pas d'inclusions végétales; la surface est grise, rarement noire, brun-jaunâtre ou rouge, et les couleurs varieraient en fonction de l'enduit, ce qui laisse penser qu'il y a un engobe; la surface est polie; cette céramique est considérée comme très épaisse et la technique de cuisson serait beaucoup moins bonne que celle de la céramique peinte. Les formes sont fermées ou ouvertes et seulement quelques unes droites; elles sont convexes, rectilignes ou concaves avec des carènes et des cols; les bords peuvent être très élaborés; il y a des piédestaux; les moyens de préhension sont des tenons non percés. Si le type de pâte donne l'impression d'une série peut-être comparable au *DFBW*, les formes montrent une élaboration qui empêche de la rapprocher de cette céramique. Une série grossière non polie, dont on ne connaît pas la répartition sur le site, montre des formes encore plus élaborées, avec par exemple des becs-verseurs. Nous attendrons de disposer de nouvelles données sur ce type de céramique et sur sa position stratigraphique et chronologique pour l'interpréter.

Tell es-Sinn (Roodenberg 1979-1980) (Fig. 1). La céramique de Tell es-Sinn se compose en tout et pout tout de 15 tessons, dont la moitié provient de la surface et l'autre moitié du sondage. Il n'est donc pas question de tirer beaucoup de conclusions de ce matériel; néanmoins quelques tessons sont tout à fait semblables à des éléments de la céramique de Bouqras (cf. *infra*), en particulier deux tessons à inclusions végétales, dont un est poli et l'autre engobé rouge poli. Aucune forme particulière ne donne d'indication intéressante.

Tell Bouqras (P.A. Akkermans et al. 1983; Le Mière 1986) (Fig. 1). Le site, qui couvre une séquence de près d'un millénaire (6400-5500 B.C.), n'a pa révélé de céramique dans les trois niveaux inférieurs. Dans les niveaux datés de la fin du VIIè millénaire et du tout début du VIè millénaire le matériel céramique était extrêmement peu abondant.

Très généralement, la céramique de Bouqras se caractérise par une couleur claire, une très grande majorité de tessons à inclusions végétales, la présence d'une série sans inclusions végétales représentant à peu près 10% de l'ensemble, des traitements de surface variés (lissage, polissage, engobe), des formes diverses aussi, fermées, droites ou ouvertes, avec un assez grand nombre de carènes et de panses concaves, des cols rares de même que les moyens de préhension, des décors essentiellement peints, accessoirement imprimés, incisés ou en relief. On insistera sur la présence d'éléments particuliers tels qu'un petit groupe de tessons gris-noir polis à incrustations blanches, sur celle de tessons *Archaic Hassuna* caractéristiques et sur celle d'un tesson *DFBW* tout à fait semblable à ceux de Syrie du Nord - Cilicie quant à la pâte, la forme et le décor.

Les tessons à incrustations blanches et les tessons *Archaic Hassuna* correspondent probablement à la fin de l'occupation du site car ils sont connus ailleurs approximativement au milieu du VIè millénaire. L'ensemble de la céramique de Bouqras se rattache à la région de Mésopotamie du Nord puisqu'on peut la comparer à la céramique *proto-Hassuna* et à la céramique *Archaic Hassuna*; seuls le tesson *DFBW* et les rares tessons à incrustations blanches montrent un rapport avec la Syrie du Nord - Cilicie.

Etant donné la séquence chronologique sur laquelle se répartit cette céramique et le fait qu'elle n'existe pas dans les niveaux les plus anciens, on s'attendrait, d'une part à la voir évoluer de niveau en niveau, d'autre part à retrouver dans les niveaux où elle commence à apparaître les caractères des premières céramiques. Or ce n'est pas le cas. Le matériel des niveaux les plus anciens ne se distingue pas de l'ensemble puisqu'il présente déjà la même variété de pâtes, de traitements de surface et de formes, et il ne montre en rien la simplicité des premières céramiques. Les tessons à incrustations blanches et les tessons *Archaic Hassuna* en sont absents, ce qui peut être dû au petit nombre de ces tessons sur le site, l'effectif de la céramique étant très limité dans les niveaux anciens, mais cela peut aussi correspondre au décalage chronologique existant. On a néanmoins affaire dans les niveaux anciens de Bouqras à un matériel qui serait daté de la deuxième moitié du VIIè millénaire, qui a été trouvé dans des niveaux précédés par des niveaux acéramiques et qui ne représente pas pour autant une des premières phases d'élaboration de la céramique.

Baghouz (Braidwood et al. 1944; de Mesnil du Buisson 1948). On ne fera que mentionner, car cela nous entraîne hors des limites chronologiques fixées, la présence de ce site, qui a révélé une seule occupation, caractérisée par de la céramique Samarra, accompagnée de tessons foncés, noirs ou rouges, polis. Par ailleurs, on parle aussi d'une série grossière non peinte (Braidwood et al. 1944:50). Ce site, qui pourrait dater de la fin du VIè millénaire, est pour le moment isolé dans cette région, de la céramique Samarra ayant été trouvée seulement plus au Nord, sur le Khabour et sur le Balikh (Tell Sabi Abyad, cf. *supra*).

Les sites présentant à la fois des niveaux acéramiques et des niveaux céramiques, et particulièrement ceux où il n'y a pas de hiatus entre les divers niveaux, sont a priori ceux où l'on devrait trouver un matériel témoignant des premières étapes de l'élaboration de la technique céramique. Le Moyen-Euphrate est relativement riche en sites acéramiques puis céramiques, mais ceux-ci sont assez divers: les données fournies par Gritille sont extrêmement succinctes, cependant les quelques tessons présents peuvent éventuellement correspondre aux céramiques les plus anciennes, et ils posent un jalon dans cette région entre les installations acéramiques (Cafer Höyük, Nevala Çori, Hayaz) et les installations céramiques 'confirmées' (Kumartepe). Le matériel de Tell Assouad et Tell Damishliyya semble, lui, bien représenter les débuts de la technique céramique et témoigner d'une élaboration sur place de celle-ci, même s'il manque peut-être encore des traces des toutes premières étapes de l'invention. Ajoutons qu'il ne s'agit pas seulement de ces deux sites mais que du matériel comparable a été trouvé à la surface de Tell Sabi Abyad, Tulul el-Breilat et Tell Mounbateh, et également de nombreux autres sites, une vingtaine environ, dans la vallée du Balikh (P.M.M.G. Akkermans, à paraître). A Abu Hureyra, le fait que les installations des niveaux céramiques soient assez différentes de celles des niveaux acéramiques s'ajoute au fait que la céramique elle-même ne semble pas, dans la mesure où la quantité trouvée permet d'en juger, montrer les caractères de simplicité des céramiques les plus anciennes pour écarter l'hypothèse d'une élaboration sur place de la technique céramique. Quant à Bouqras, nous avons vu que, malgré une séquence de niveaux acéramiques et céramiques sans rupture et des datations de la deuxième moitié du VIIè millénaire pour les premiers niveaux céramiques, le matériel de ceux-ci ne correspond pas à la première phase de la technique céramique. Le matériel de Tell es-Sinn, très réduit, ne donne pour le moment aucune indication d'une céramique ancienne, qui indiquerait, malgré un contexte stratigraphique et chronologique comparable, une situation différente de celle de Bouqras.

Ainsi, alors que ce n'est pas le cas de l'ensemble du Moyen-Euphrate, la vallée du Balikh fait partie des régions où l'on peut suivre les débuts de la céramique, et celle-ci y apparaît non de façon ponctuelle mais à l'échelle de la vallée sur de nombreux sites. Les quelques tessons de Gritille indiquent peut-être, soit une autre région d'invention de la céramique, soit une extension de la région du Balikh plus au Nord, extension suggérée par d'autres types de matériel comme l'industrie lithique. Par contre la partie Sud du Moyen-Euphrate ne semble pas, jusqu'à présent, participer à cette élaboration de la technique céramique.

La deuxième phase de l'évolution de la céramique est celle de la généralisation de la technique et celle de l'apparition de particularités régionales. Pour ce qui est de la généralisation de la céramique, le fait qu'un grand nombre de sites du Moyen-Euphrate n'ont fait l'objet que de prospections ne permet pas de connaître l'existence de niveaux acéramiques sur ces sites, ni d'avoir des datations, et donc limite considérablement l'estimation de ce phénomène. Le matériel trouvé, que ce soit en surface ou en fouille, permet, lui, quand il est assez abondant, d'aborder le problème de l'apparition de particularités régionales. Deux questions se posent: existe-t-il une céramique du Moyen-Euphrate aux caractères spécifiques, et, si ce n'est pas le cas, à quelle région céramique appartient le matériel des divers sites ?

Si on la compare à celle des régions alentour, la céramique du Moyen-Euphrate ne semble pas montrer d'élément (ou association d'éléments) particulier, qui lui conférerait une identité propre. Elle ne forme pas non plus un ensemble homogène dans ses parentés avec le matériel des autres régions. Il semble plutôt que trois zones se distinguent les unes des autres à l'intérieur du Moyen-Euphrate. La première zone se composerait de la vallée de l'Euphrate elle-même jusqu'au confluent du Balikh, avec les affluents de la rive droite (Nizip çay et Sajour). Sa céramique, ou le peu que l'on en sait, peut être rapprochée de la Syrie du Nord - Cilicie: defaçon claire à Kumartepe, de façon moins sûre à Tell Turlu et Abu Hureyra et enfin seulement à titre d'hypothèse pour les sites de la vallée du Sajour.

La deuxième zone est la vallée de l'Euphrate en aval du confluent du Balikh, qui est caractérisée d'abord par la rareté des sites, peut-être due entre autres à la quasi-disparition de la terrasse de l'Euphrate sur laquelle se trouvent les quelques sites repérés (Boerma 1983:365). Cette zone se caractérise également par les affinités nettement Nord-mésopotamiennes de sa céramique, qu'il s'agisse du *proto-Hassuna* ou du *Archaic Hassuna* de Bouqras ou, plus tard dans le VIè millénaire, du Samarra de Baghouz.

Enfin la vallée du Balikh constitue la troisième zone. La phase de généralisation de la céramique et d'apparition des particularismes régionaux y est mal documentée, puisque nous n'avons jusqu'à présent aucune fouille pour la première moitié du VIè millénaire. Cependant nous disposons dans cette zone des tenants et aboutissants de cette phase d'évolution de la céramique avec, d'une part les céramiques du VIIè millénaire de Tell Assouad et Tell Damishliyya, d'autre part les céramiques 'd'influence' Samarra de Tell Sabi Abyad et la céramique Halaf que l'on trouve sur de très nombreux sites de la vallée. Les premières céramiques, malgré leur simplicité, présentent de nombreux moyens de préhension, qui sont une des caractéristiques de Syrie du Nord - Cilicie à la phase suivante, comme l'est le décor imprimé ou incisé des quelques tessons de Damishliyya, cependant pas un site de la vallée du Balikh n'a révélé jusqu'à présent de *DFBW* ou de série vraiment comparable à celui-ci indépendamment de la couleur. Par ailleurs le matériel Samarra ainsi que les '*husking trays*' de la deuxième moitié du VIè millénaire mettent en lumière des relations avec la Mésopotamie du Nord. Ces informations très limitées supposeraient éventuellement une évolution dans les rapports entre la vallée du Balikh et les régions qui l'entourent, évolution dont nous ne connaissons ni le moment, ni le processus. Les rares indications fournies par le matériel de surface sont un peu ambiguës: à Mounbateh, si la pâte et certaines formes sont assez comparables à la céramique de Bouqras et pourraient ainsi se rattacher à l'ensemble Nord-mésopotamien, d'autres formes sont assez différentes et empêchent un rapprochement très étroit; les tessons à engobe rouge, qui sont tout à fait semblables à Mounbateh et à Bouqras, sont probablement caractéristiques à Bouqras de la fin de la période d'occupation du site, vers 5500 environ, et ne sont pas un trait spécifique de la Mésopotamie du Nord. C'est pourquoi plusieurs hypothèses restent possibles quant à la céramique du début du VIè millénaire sur le Balikh: par exemple un ensemble propre à cette vallée aussi bien qu'une évolution qui montrerait selon les périodes des rapprochements plus marqués avec telle ou telle région voisine; la possibilité que la vallée du Balikh ait été, en matière de céramique, une région 'déterminante' est à retenir également dans la mesure où jusqu'à présent, les premières céramiques y paraissent plus anciennes et plus proches de l'invention de la technique céramique que dans les régions alentour. Les quelques éléments que l'on connait pour la fin du VIè millénaire renforceraient plutôt l'hypothèse d'une appartenance de cette zone à la région nord-mésopotamienne.

Le peu de données disponibles nous empêche, pour le moment, d'aller au delà des hypothèses et des éventualités. Néanmoins, comme l'a souligné M. Voigt (1986b) le Moyen-Euphrate est loin d'être déserté dans la première moitié du VIè millénaire. Cependant, alors que dans certains domaines, comme l'industrie lithique, elle constitue un ensemble assez homogène (M.C. Cauvin, Séminaire de Préhistoire, Lyon 1981), dans le domaine de la céramique ce n'est pas le cas et les informations actuelles pourraient suggérer une coupure entre Nord-Ouest d'une part, et Sud et Est d'autre part, tout cela restant à vérifier, notamment dans la vallée du Balikh qui semble détenir un certain nombre de clés pour l'histoire des débuts de la céramique dans cette région.

NOTE

1. Toutes les dates indiquées ici sont des dates B.C. non calibrées.

BIBLIOGRAPHIE

AKKERMANS P.A. ET AL.
1983 'Bouqras revisited: Preliminary Report on a Project in Eastern Syria', *Proceedings of the Prehistoric Society* 49: 335-372.

AKKERMANS P.M.M.G.
1987a 'Tell Sabi Abyad: Preliminary Report on the 1986 Excavations', *Akkadica* 52: 10-28.
1987b 'A Late Neolithic and Early Halaf Village at Sabi Abyad, Northern Syria', *Paléorient* 13: 23-40.
s. presse *Excavations at Tell Sabi Abyad.* Prehistoric Investigations in the Balikh Valley, Northern Syria, Oxford: BAR International Series.
s. presse 'The Soundings at Tell Damishliyya', in: M.N. Van Loon (Ed.), *Hammam et-Turkman I.* Istanbul.
à paraître 'The Neolithic of the Balikh Valley, Northern Syria: A First Assessment', Colloque 'Préhistoire du Levant', Lyon, May 31 - June 4, 1988.

BADER N.O.
1975 'Early Farmproducing Settlement of Tell Sotto, From Excavations of 1971, 1973-1974', *Sovetskaya Arkheologiya*: 99-111.

BOERMA J.A.K.
1983 'Environmental Conditions', in: P.A. Akkermans et al., 'Bouqras Revisited: Preliminary Report on a Project in Eastern Syria', *Proceedings of the Prehistoric Society* 49: 362-365.

BORDAZ J.
1973 'Current Research in the Neolithic of South Central Turkey: Suberde, Erbaba and Their Chronological Implications', *American Journal of Archaeology* 77: 282- 288.

BRAIDWOOD R.J. ET BRAIDWOOD L.S.
1960 *Excavations in the Plain of Antioch* I, OIP LXI. Chicago: University of Chicago Press.

BRAIDWOOD R.J. ET AL.
1944 'New Chalcolithic Material of Samarran Type and Its Implications', *Journal of Near Eastern Studies* 3: 47- 72.

CAUVIN J.
1970 'Mission 1969 en Djezireh (Syrie)', *Bulletin de la Société Préhistorique Française* 67: 286-287.
1972 'Sondage à Tell Assouad (Djézireh, Syrie)', *Annales Archéologiques Arabes Syriennes* 22: 85-88.

COPELAND L.
1979 'Observations on the Prehistory of the Balikh Valley, Syria, during the 7th to 4th Millennia B.C.', *Paléorient* 5: 251-275.
1985 'The Survey of RCP 438 in 1979', in: P. Sanlaville (Ed.), *Holocene Settlement in North Syria*. Oxford: BAR International Series 238. Pp. 67-98.

DE CONTENSON H.
1977 'Le Néolithique de Ras Shamra V d'après les campagnes 1972-1976 dans le sondage SH', *Syria* 54: 1-23.
1985 'Le matériel archéologique des tells', in: P.Sanlaville (Ed.), *Holocene Settlement in North Syria*. Oxford: BAR International Series 238. Pp. 99-161.

FUKAI S. ET MATSUTANI T.
1981 *Telul eth-Thalathat, The Excavation of Tell II: The Fifth Season (1976)*, The Tokyo University Iraq-Iran Archaeological Expedition Report 17, The Institute of Oriental Culture. Tokyo: The University of Tokyo.

GARSTANG J.
1953 *Prehistoric Mersin*. Oxford: Clarendon Press.

KIRKBRIDE D.
1972 'Umm Dabaghiyah 1971: A Preliminary Report', *Iraq* 34: 3-15.
1973 'Umm Dabaghiyah 1972: A Second Preliminary Report', *Iraq* 35: 1-7.

LE MIÈRE M.
1979 'La céramique préhistorique de Tell Assouad, Djézireh,Syrie', *Cahiers de l'Euphrate* 2: 4-76.
1986 *Les premières céramiques du Moyen-Euphrate*, Thèse de Doctorat, Lyon: Université Lumière-Lyon 2.

LLOYD S. ET SAFAR F.
1945 'Tell Hassuna: Excavations by the Iraq Government Directorate General of Antiquities in 1943 and 1944', *Journal of Near Eastern Studies* 4: 255-289.

MALLOWAN M.E.L.
1946 'Excavations in the Balikh Valley, 1938', *Iraq* 8: 111-162.

MELLAART J.
1961 'Early Cultures of the South Anatolian Plain', *Anatolian Studies* 11: 159-184.
1962 'Excavations at Çatal Hüyük, First Preliminary Report, 1961', *Anatolian Studies* 12: 41-65.
1964 'Excavations at Çatal Hüyük, 1963. Third Preliminary Report', *Anatolian Studies* 14: 39-119.
1966 'Excavations at Çatal Hüyük, 1965', *Anatolian Studies* 16: 165-191.
1981 'The Prehistoric Pottery from the Neolithic to the Beginning of EB IV', in: J. Matthers (Ed.), *The River Qoueiq, Northern Syria, and its Catchment*, BAR International series 98. Oxford. Pp. 131-319.

MELLINK M.
1964 'Archaeology in Asia Minor', *American Journal of Archaeology* 68: 148-166.

MERPERT N. ET AL.
1977 'The Investigations of Soviet Expedition in Iraq, 1974', *Sumer* 33: 65-104.
1978 'Soviet Investigations in the Sinjar Plain, 1975', *Sumer* 34: 27-70.
1981 'Investigations of the Soviet Expedition in Northern Iraq, 1976', *Sumer* 37: 22-54.

MESNIL DU BUISSON, LE COMTE DE
1948 *Baghouz, l'ancienne Corsote*. Leiden: E.J. Brill.

MOORE A.M.T.
1975 'The Excavation of Tell Abu Hureyra in Syria: Preliminary Report', *Proceedings of the Prehistoric Society* 41: 50-77.

1979 *The Neolithic of the Levant, Vol. 1-2*. Ann Arbor: University Microfilm International.
1985 'The Archaeological Survey of 1977', in: P. Sanlaville (Ed.), *Holocene Settlement in North Syria*, Oxford: BAR International Series 238. Pp. 45-66.

MORTENSEN P.
1963 'Early Village Farming Occupation, Tepe Guran, Luristan', *Acta Archaeologica* 34: 110-121.

ROODENBERG J.J.
1979-80 'Sondage des niveaux néolithiques de Tell es Sinn, Syrie', *Anatolica* 7: 21-33.

ROODENBERG J.J. ET AL.
1984 'Surveys and Soundings at Kumartepe: An Interim Report', *Anatolica* 11: 1-16.

SCHMIDT H.
1943 *Tell Halaf I: die prähistorischen Funde*, Berlin: Walter de Gruyter and Co.

VOIGT M.M.
1985 'Village on the Euphrates. Excavations at Neolithic Gritille in Turkey', *Expedition* 27: 10-24.
1986 'Excavations at Neolithic Gritille', in: *Table Ronde Le Néolithique acéramique de la Turquie du Sud-Est*, Istanbul.
1986 Compte-rendu de T.C. Young Jr. et al. (Eds.), 1983, The Hilly Flanks and Beyond, Chicago. *Paléorient* 12: 103-113.

A lesser known figurine from the near eastern Neolithic

ERIK LOHOF

The 1976-1978 excavations at Tell Bouqras in eastern Syria, carried out by a joint expedition of the Universities of Amsterdam and Groningen, brought to light numerous figurines and parts thereof (Akkermans et al. 1983).

Our concern here is with a particular figurine that might be called a 'type' in the sense that examples from different sites, occupied during a substantial period of time, fit a specific description.

The figurines under consideration are characterized by a cylindrical or pillar-like body on a splayed base that is either flat or more or less concave underneath. Stump-like arms protrude from the upper part of the body. At this point the figurine is at its widest. An elongated head, either pointed or shaped like a rugby ball, sits low between the shoulders or bends forward from a humped back. Facial features are scarce, but sometimes there is a pinched nose or an elongated chin that may indicate a beard. The possible beard and the complete lack of breasts are the only indications that these may be male anthropomorphic figurines.

In general they are poorly modelled, but posture and details vary, as with the better-known 'mother-goddess' type of figurine. All examples have been made of baked or unbaked clay. They have been variously described as 'clay idols' (Schmidt 1943:101), 'gaming pieces' (Tobler 1950:170), 'pillar figurines' (Dales 1960:169) and, recently, as 'torso type figurines' (Broman Morales 1983:381).

At Tell Bouqras, a few examples came from the northwest room of the burnt house no. 13 (square 14/09), attributed to phase III at the site. It can not be ruled out that the burnt houses nos. 12 and 13 were used as dump areas for households nearby and that refuse material thus became mixed with material originally present in the houses. However, the circumstances in which they were found lead us to assume that these figurines were in situ and that they belong to the original contents of the room. Among these contents were a stylised boot-like female figurine, the torso of a probably male figurine, the torsos of at least eleven animal figurines, together with an assortment of little clay horns, presumably of cattle, and a miniature cup with a bull's head as a handle.

Not all the animal figurines represent cattle. At least one boar, a sheep and a turtle can be recognised. Originally the torsos of at least seven of these 'pillar figurines', as I will call them from now on, were thought to represent birds or turtles. However, the existence of 'real' birds and turtles among the animal figurines of Bouqras showed that whoever made these figurines had no problem in making realistic representations of the creatures mentioned. The flat and concave bases of the pillar figurines show that they were definitely made to stand on end. In view of the presence of a hood, a beard or various indications of clothing on examples from other sites, it seems more likely that the pillar figurines are an anthropomorphic representation, although the possibility remains that they represent stylised animals.

The complete examples from Bouqras are:

– Figurine with pillar-like torso. The height is 4.3 cm whereas the diameter is 2 cm. On one side of the upper part a nose is indicated by pinching. Two tiny stumps protrude at either side. The figurine has a flat base (Fig. 1).

– Figurine with pillar-like torso, height 2.2 cm, pointed head, pointed arm stumps, splayed base, slightly concave underneath (Fig. 2).

– Figurine with pillar-like torso, height 4 cm, rather shapeless head bent forward, humped back, slight indication of arms held backwards. Splayed base, concave underneath. The concave base has an indentation underneath, ca 3 mm deep (Fig. 3).

– Figurine with pillar-like torso, height 4.8 cm, top (head?) bent forward. Tiny arm stumps point backwards. Splayed base, flat underneath (Fig. 4).

– Figurine with pillar-like torso, height 7.4 cm, pointed head bent forward. It came from one of the squares that were excavated down to virgin soil (16/13). Splayed base, slightly concave underneath. Damaged at the base and right arm broken off. The left arm is a small stump held backwards. There is no indication of a face (Fig. 5).

A survey of the literature shows that similar figurines occur at various sites in the Near East. The oldest one seems to come from Tepe Asiab in the Kermanshah valley (Iran), dated in the 8th millennium B.C. This apparently damaged and insufficiently published figurine seems to fit our general description (Schmandt-Besserat 1974, Fig. 1). Examples from the 7th to early 5th millennia B.C are more common. The closest parallels come from the following sites: Jarmo, Gritille, Maghzalia, Çatal Hüyük, Tell Halaf, Tell es-Sawwan, Yarim Tepe III, Tepe Gawra and Tell Mefesh.

Foremost is *Jarmo* in the foothills of the northern Zagros, with a wide variety of this figurine type:

– Figurine with pillar-like torso, height 4.5 cm, pointed head bent forward thus resembling a hunched back. Arm stumps are broken off. The face is indicated by a pinched nose. The body stands on a splayed base which is slightly concave underneath. The figurine is made of lightly burnt clay (Dales 1960, Fig. 4; Broman Morales 1983, Fig. 161, no.3a-b).

– Figurine with pillar-like torso, height 5.8 cm, width at shoulders 3.3 cm. Larger head than usual, with incised mouth and pointed chin, probably indicating a beard. The elongated head is somewhat flattened at the top. The arm stumps are broken off but seem to have pointed forward originally. Splayed base. Made of lightly burnt clay (Broman Morales 1983, Fig. 161, no. 1a-b).

– Figurine with pillar-like torso, height 4.8 cm. Although this one lacks both the humped back and the head bent forward, its head is nevertheless set low between the shoulders. The head is moderately pointed. Eyes and mouth are indicated by incisions. The arm stumps, originally pointing sideways, are broken off. It stands on a splayed base. It is made of lightly burnt clay. An interesting detail is the hem decoration around the base, ca. 6 mm deep, defined by a horizontal line. The hem itself is filled with vertical lines (Broman Morales 1983, Fig. 161, no. 4).

– Figurine with pillar-like torso, height 4.3 cm. This example is similar to the one above in that it shows a hem around the base with vertical incisions (Broman Morales 1983, Fig.161, no. 2).

– Figurine with pillar-like torso, height 3 cm. The upper part does not fit the type description in that the head is not pointed or bent forward and the remaining right arm has a wing-like shape and is held upwards. If the missing left arm had similarly been raised, the total width would have been 3.3 cm. This, however, was not necessarily the case (see below). The figurine has a splayed base. Made of lightly baked clay (Broman Morales 1983, Fig. 161, no. 6).

– Figurine with pillar-like torso. Its base was damaged. Existing height 3.4 cm. The left arm has the same wing-like shape as mentioned above. The right arm, however, is bent in front of the chest. The head is indicated by a folded piece of clay (Broman Morales 1983, Fig. 161, no. 8).

– Figurine with pillar-like torso with broken head and broken arm stumps. Existing height 5.8 cm. Originally probably 7-8 cm tall. Made of lightly baked clay (Broman Morales 1983, Fig. 162, no. 1).

In all, 18 complete figurines similar to the ones above and 61 fragments were found at Jarmo. According to Broman Morales (1983:382) two pillar figurines may have had girdles of clay around their torsos.

A pillar figurine has also been found at *Gritille*, a site in the foothills southeast of the Taurus Mountains. The torso is slightly oval in diameter. The lower part is missing. Existing height 2 cm. Horn-like arm stumps point forward. The elongated head is bent forward which gives this figurine a slightly humped back. It seems that this figurine was originally in a sitting position (Voigt 1985, Fig. 10c).

A few figurines from *Maghzalia*, in the Sinjar area, that might belong to the pillar type, should be mentioned as well (Bader et al. 1981, Pl. XXXIII, top centre). Unfortunately, the publication does not give detailed enough information to say anything more about these pieces. Maghzalia has not yet been firmly dated, but there is little doubt that the site belongs to the 7th millennium.

From *Çatal Hüyük*, in central Anatolia, a few pillar figurines from levels IV-VIII were published, dated in the first half of the 6th millennium (Mellaart 1962, Pl. VIIa; Mellaart 1963, Pl. XVIIIa, centre). These figurines have pointed heads and different postures: some are sitting in a kind of lotus position (height ca. 3 cm) and others are standing erect (height ca. 3.6 cm). The height of the figurine in the 1963 publication is not indicated but it seems to be slightly taller than the other ones. It has a pointed head that is bent forward. The arm stumps point sideways. The figurines were found together with animal figurines of mainly sheep, goats and cattle in debris between the houses (Fig. 6; Mellaart 1963:78). They are all made of unbaked clay.

At *Tell Halaf* in northern Syria similar figurines, called 'clay idols' by Schmidt (1943:101), were found in the early (*Altmonochrome*) and transitional layers:

– Figurine with pillar-like torso with head and arms broken off. Existing height 5.4 cm. Despite the missing head a forward bent neck is perceptible. Splayed base, concave underneath. Made of baked clay (Schmidt 1943, Fig. 146 and Pl. CVI, no. 2).

– Figurine with pillar-like torso (?), height 4.9 cm. This torso is rather plump, but it might belong to our type. Made of baked clay (Schmidt 1943, Pl. CVI, no. 3).

– Figurine with pillar-like torso with damaged head which had apparently been flattened at the back. The latter might indicate a hunched back. Height 4.5 cm. Both sides of the torso are decorated with fingerprint impressions. Splayed base. Made of baked clay (Schmidt 1943, Pl. CVI, no. 4).

It is possible that Joan Oates is talking about similar figurines from *Tell es-Sawwan*,

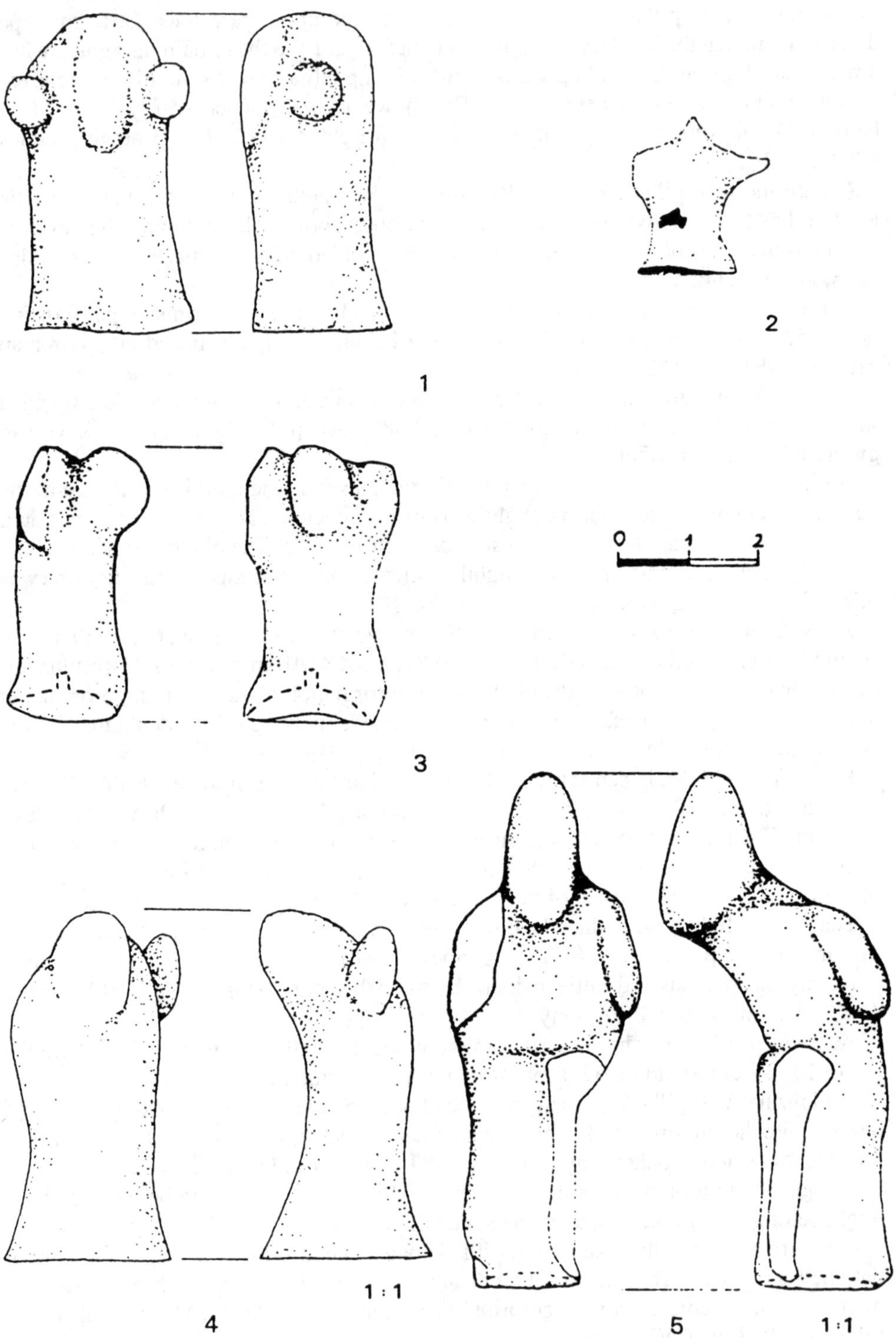

Figure 1-5. Neolithic 'pillar figurines' from Tell Bouqras.

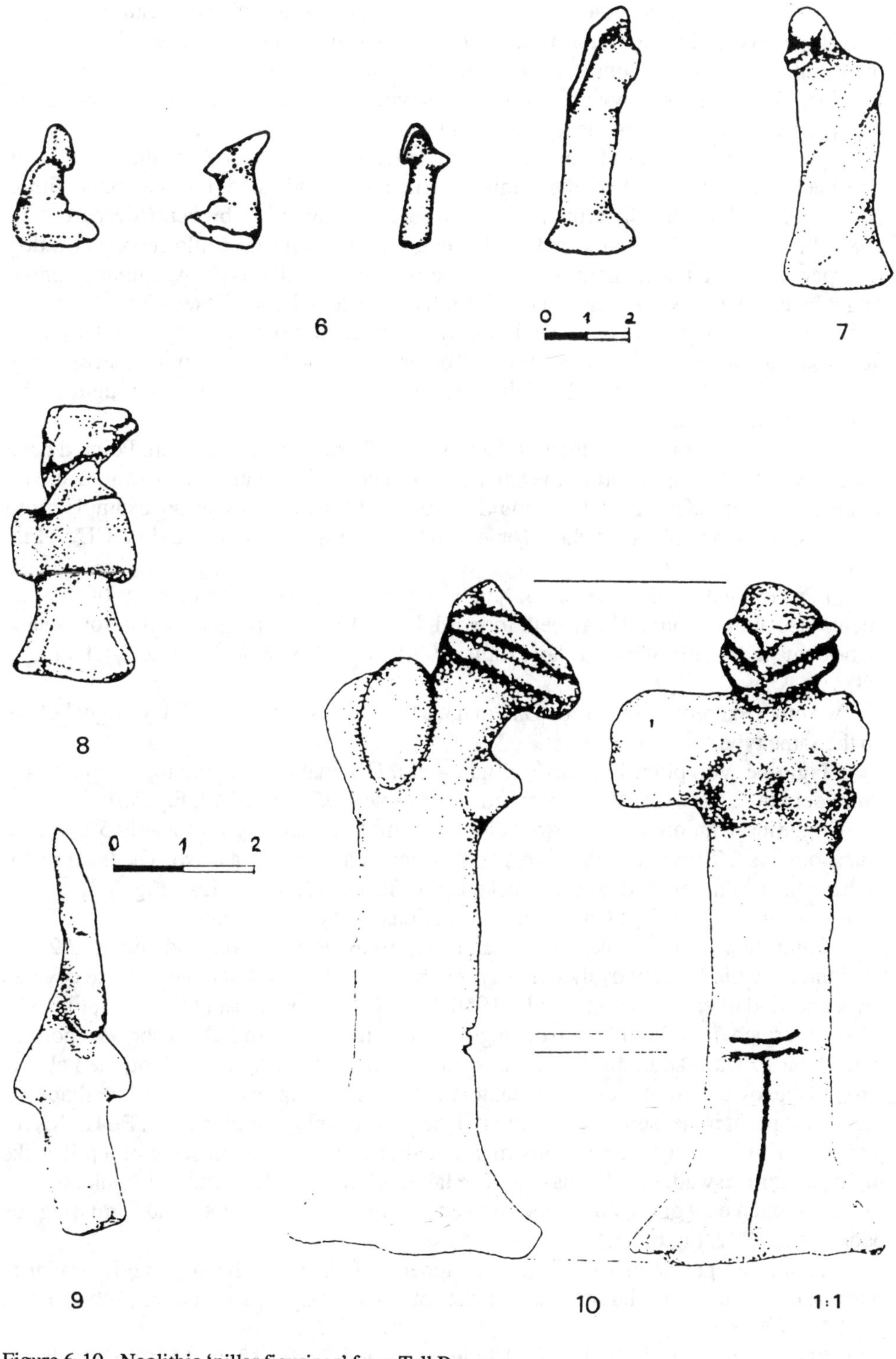

Figure 6-10. Neolithic 'pillar figurines' from Tell Bouqras.

when, for comparison, she refers to some figurines from Jarmo and to the 'gaming pieces' from Tepe Gawra. The figurines from Tell es-Sawwan are described as 'cone-like figures, their pointed tops often bent over' (Oates 1966:148). However, none of the illustrated figurines from Tell es-Sawwan resemble our type. As we shall see, the 'cones' more likely represent stylised female figurines with slantingly pointed heads.

The next example comes from the Halafian level 3 at *Yarim Tepe III* in the Sinjar. This figurine has a pillar-like torso, the height of which is not indicated. It has a pointed head and a broken chin (?). The arm stumps also seem to have been broken (Merpert et al. 1984, Fig. 25, no. 1). The excavators believe that this figurine could represent a dog (Merpert et al. 1984:44), but more likely it belongs to our pillar type. A similar example might be the one shown in Fig. 23, no. 5. Both are made of lightly baked clay.

Several figurines that fit our type description well came from the early Ubaid layers at Tepe Gawra (stratum XVI). According to Tobler, who called them 'gaming pieces', they 'show the head thrust forward' (Tobler 1950:170). They were found throughout the aforementioned stratum:

– Figurine with pillar-like torso, height 3.5 cm. Torso diameter 1.8 cm. Pointed head bent forward. The upper part of the front seems to be slightly damaged. The arms are no more than a broadening of the shoulders, thrust backwards as in an example from Bouqras. It is made of baked clay (Tobler 1950, Pl. LXXXIV, no. b.1 and Pl. CLVI, Fig. 52).

– Figurine with pillar-like torso, height 5.1 cm (Fig. 7). A larger copy of the one mentioned above. Pointed head bent forward. Here also the upper part of the front seems to be damaged. Made of baked clay (Tobler 1950, Pl. LXXXIV, no. b.2 and Pl. CLVI, Fig. 58).

Two other examples from Tepe Gawra have their heads broken off. They might belong to the same type:

– Figurine with pillar-like torso, height 4 cm. Original height probably 5 cm. At the shoulders ca. 2 cm wide. Made of baked clay (Tobler 1950, Pl. CLVI, Fig. 56).

– Figurine with pillar-like torso, height 4.4 cm. Original height probably 5-6 cm. At the shoulders 2.7 cm wide. Two dents at the throat may indicate a clasp. The head seems to have been bent forward. Made of baked clay (Tobler 1950, Pl. CLVI, Fig. 57).

A few others from Tepe Gawra show variations on the same theme:

– Figurine with pillar-like torso, height ca 5 cm (Fig. 8). Torso diameter 2.2 cm. Around the cylindrical body there is a 1.4 cm broad girdle. The head seems to be covered by some kind of cape or hood (Tobler 1950, Pl. LXXXIV, no. b.6 and Pl. CLVI, Fig. 53).

– Figurine with pillar-like torso, height 5.8 cm, from stratum XIII. It shows a pointed head bent forward and a broad girdle, similar to that of the figurine mentioned above, around the lower part of the cylindrical torso. The arm stumps have a wing-like shape. At its widest point it measures 3.6 cm. It is made of baked clay (Tobler 1950, Pl. LXXXIII, no. b.1). The use of girdles, already mentioned for Jarmo, is reminiscent of a pillar-like figurine from Aswad, near Damascus. The latter is 3.5 cm high and has wing-like stumps for arms and a clay girdle above the splayed base. Unfortunately the head is missing (de Contenson 1972, Fig.6, left).

– Finally, a figurine of which only the upper part remains. This figurine is similar to the one described above but its head consists of coils of clay thrust forward (Tobler 1950, Pl. LXXXIII, no. b.2).

A fine example comes from early Ubaid layers at *Tell Mefesh*, in the Balikh valley of northern Syria. This figurine (Fig. 9) has a pillar-like torso and is 5.9 cm high. The head and chin are pointed; the pointed chin probably indicates a beard. The arms are short and stump-like. The figurine is made of unbaked clay (Mallowan 1946, Pl. XXVI, no. 2). It is attributed to Ubaid influence (Dales 1960:186) but, more likely, it fits into a broader and older tradition.

It is in later Ubaid times that the history of the pillar figurine as a separate type becomes obscure. In the east, during the late 6th and the 5th millennia B.C., changes in the symbolic-religious system probably manifested themselves in different ways of modelling figurines. For example, the classical voluptuous female figurines are replaced by slimmer representations that stand upright and have slit-eyes and reptile-like heads. Pillar torsos and splayed bases become popular, and also pointed heads or hats become more common.

Pointed heads on male and female figurines appear in the Levant already in the first half of the 6th millennium. They are known from Ramad, Munhata, Byblos, Shaar-Ha-Golan and Kfar Gilad (Cauvin 1972:84-90). Sometimes these figurines also have coffee-bean eyes. Their heads are on top of long vertical necks. Their long hoods point slantingly upwards. Further to the east pointed heads on female figurines appear in a Hassuna context at Matarrah and Yarim Tepe I. Their shape resembles the cone-like objects with 'their pointed tops often bent over', referred to before in relation to Tell es-Sawwan. These cones may be stylised representations of the predominantly female figurines with pointed heads.

It is difficult to say whether there is a direct relationship between the pointed heads on the figurines from the 5th millennium and the earlier figurines of the pillar type. All we can say is that female (and male) figurines with pointed heads appear both in the east and in the west and that the pillar type remains in use in those areas where they were found before.

At least in Mesopotamia in the 5th millennium male figurines are found as well, whose existence can be traced back to the Neolithic since male figurines, though less common than female, appeared at numerous earlier sites (Çatal Hüyük, Çafer, Bouqras, Jarmo, Tell es-Sawwan, Byblos, Munhata). These earlier male figurines, made of clay or stone, are modelled in a naturalistic style, quite different from the pillar type figurine discussed here. It is for this reason that they can be regarded as the counterpart of the naturalistic female figurines.

Both male and female figurines found in Ubaid contexts show elongated heads and sometimes they have bitumen hats. They are known from 'Oueili, Uqair, Ubaid and Eridu. Similar heads can be found on earlier Samarra figurines from Nineveh or Choga Mami (Oates 1966:150-1).

During the same period similar figurines appear in eastern Anatolia as well, combining ophidian facial features with pillar torsos and splayed bases. Examples come from sites in the Plain of Altinova: Tülintepe (height 14 cm; Arsebük 1983, Fig. 4, no. 1-2) and Körtepe (headless, existing height 6 cm, but it resembles the Tülintepe figurine in all respects and may originally have measured 9-10 cm; Hauptmann 1972, Fig. 6). From Tepecik only a pointed head is known. It is 5 cm tall and, if compared to the Tülintepe example, the complete figurine must have measured 12-14 cm (Esin 1979, Fig. 5). These Anatolian examples are apparently all female representations.

Although in Ubaid times male and female figurines share characteristics with the earlier pillar figurines, such as pointed heads, pillar torsos and splayed bases, there does not have to be a direct relationship between the two types. Pointed heads have their 5th millennium predecessors and the more common use of the pillar shape and splayed base might have been caused by a change in function or in symbolic value which required the figurines to stand upright. This probably explains the absence of pillar torsos and splayed bases from the southern Ubaid figurines found in graves: they were never meant to stand upright.

For a supposed direct relationship between the southern Ubaid figurines and the pillar type figurines (Dales 1960:7) there is only one possible example: a 10 cm tall female figurine from Uruk/Warka (Fig. 10). It combines a pillar torso on a splayed base with a reptile-like head thrust forward, and stump arms, as do many contemporary figurines. Remarkable, however, is the humped back (Nöldeke 1937, Pl. 47h; Dales 1960, Fig. 44).

Only in the Ubaid layers at Tello some figurines were found, all of unbaked clay, that seem to belong to the 'true' pillar type. Unfortunately, it is difficult to distinquish these from a new type of female figurine with bird-like head, wing-like arms and large eyes that became popular in southern Mesopotamia and the Diyala region at about this time (for examples see Dales 1960, Figs. 56, 72, 73). Among the Tello figurines the following might belong to the pillar type:

– Figurine, height 5.6 cm. Head in the shape of a rugby ball and similar round arm stumps held backwards. A narrow girdle is painted around its middle (Barrelet 1968, Pl. I., no. 6).

– Figurine with pillar-like torso, height about 4 cm. Pointed head, slightly bent forward. Arm stumps broken off (Barrelet 1968, Pl. I, no. 4).

– Figurine with rather plump torso. Height ca. 4 cm. The head is broken off, but apparently it was set low between the shoulders and was thrust forward (Barrelet 1968, Pl. I, no. 7).

A similar figurine was found in the Ubaid layers at Ur. This figurine has a pillar-like torso, the base of which is broken off. Existing height 3.5 cm. Small triangular head, slightly bent forward (Woolley 1955, Pl. 23, no. 412500 B).

CONCLUSIONS

There may be some doubts as to whether some of the figurines mentioned above actually belong to the pillar type. Conversely, some figurines which have not been mentioned may, after close inspection, be seen to belong to our type. However, there can be little doubt that these figurines all represent males who are either elderly or have some physical deformity as indicated by the humped back.

Sometimes a beard is indicated and two examples from Jarmo suggest a kind of long dress or robe. Also the girdles from Tepe Gawra suggest clothing.

The average height of the pillar figurines is 4.5 cm. They never exceed 8 cm in height. This is about the same size as most of the 'mother goddesses'. As a rule they stand upright, but some are squatting.

Chronologically, the appearance of the pillar figurines coincides well with that of the 'mother goddesses': from the 8th millennium until at least early Ubaid times. Their geographical distribution, however, seems to be much more limited: they appear in the foothill areas of eastern Anatolia and the Zagros and in settlements in the lowlands of northern and eastern Syria and Iraq during the 7th to 5th millennia. A site where pillar figurines could be expected on account of the location and the extent of the excavations, is Çayönü in the east-Anatolian foothills. Unfortunately, only a few figurines have been published and none of the less usual ones.

The spatial distribution of pillar figurines, especially in the 7th millennium, resembles that of another figurine type: the so-called double-wing bases, also known as T-shaped figurines. They consist of a small round hump centred on an elongated flat base. These bases measure 1.6-5 cm in length and the height of the stalk or stump is 2.5-5 cm. It is difficult to assign them to one sex or the other. Some have facial characteristics like a pinched nose, incised mouth and eyes (10% of the 314 examples from Jarmo; Broman Morales 1983, Fig. 163, nos. 7 and 10; Fig. 166, nos. 5 and 6). Some have girdles or necklaces around the base of their stalks. On account of these features they are generally regarded as anthropomorphic. These figurines are known from Çatal Hüyük, Çayönü, Jarmo, Asiab (where they have been made of two separate pieces of clay), Ali Kosh (where the central stalk is sometimes topped by a clay pellet), Chaga Sefid (where the bases were lightly burnished) and from Sarab (Mellaart 1975:53, 89).

Together with male and female figurines, common in the whole Near East, and T-shaped figurines, the pillar figurines are apparently part of the symbolic-religious system charcteristic of the crescent between north Khuzestan and central Anatolia. The T-shaped figurines are restricted to the 7th millennium. The pillar figurines, however, have a much longer history and continue into the Chalcolithic.

ACKNOWLEDGEMENTS

I am most grateful to Ms. B. Torolsan for making the drawings and to Ms. S. Rollin for correcting my English.

REFERENCES

AKKERMANS P.A. ET AL.

1983 'Bouqras Revisited: Preliminary Report on a Project in Eastern Syria, *Proceedings of the Prehistoric Society* 49: 335-372.

ARSEBÜK G.

1983 'Tulintepe: Some Aspects of a Prehistoric Village', in: R.M. Boehmer and H. Hauptmann (Eds.), *Beiträge zur Altertumskunde Kleinasiens. Festschrift für Kurt Bittel*, Mainz a/R. Pp. 51-57.

BADER N.O. ET AL.

1981 'Soviet Expedition's Survey in the Sinjar Valley', *Sumer* 37: 55-95.

BARRELET M.T.

1968 *Figurines et reliefs en terre cuite de la Mésopotamie antique*, I, Paris.

BROMAN MORALES V.

1983 'Jarmo Figurines and Other Clay Objects', in: L.S. Braidwood et al., *Prehistoric Archeology along the Zagros flanks*, Chicago. Pp. 369-423.

CAUVIN J.

1972 *Religions néolithiques de Syro-Palestine*, Paris.

DE CONTENSON H.

1977-78 'Tell Aswad. Fouilles de 1972', *Annales Archéologiques Arabes Syriennes* 27-28: 207-15.

DALES G.F.
1960 *Mesopotamian and Related Female Figurines: Their Chronology, Diffusion and Cultural Functions*, Ph.D., UMI, Ann Arbor.
ESIN U.
1979 'Tepecik Excavations, 1973', *Keban project 1973 activities*, METU, Ankara. Pp. 97-125.
HAUPTMANN H.
1976 'Die Ausgrabungen auf dem Norsun-Tepe, 1972', *Kebanproject 1972 activities*, METU, Ankara. Pp. 71-90.
MALLOWAN M.E.L.
1946 'Excavations in the Balih Valley, 1938', *Iraq* 8: 111-59.
MELLAART J.
1962 'Excavations at Çatal Hüyük, First Preliminary Report, 1961', *Anatolian Studies* 12: 41-65.
1963 'Excavations at Çatal Hüyük 1962, 2nd Preliminary Report', *Anatolian Studies* 13: 43-103.
1975 *The Neolithic of the Near East*, London.
MUNCHAEV R.M. ET AL.
1984 'Archeological Studies in the Sinjar Valley, 1980', *Sumer* 43: 32-53.
NÖLDEKE A. ET AL.
1937 *VIII. vorläufiger Bericht über die von der deutschen Forschungsgemeinschaft in Uruk-Warka unternommenen Ausgrabungen*, Berlin.
OATES J.
1966 'The Baked Clay Figurines from Tell es-Sawwan', *Iraq* 28: 146-53.
SCHMANDT-BESSERAT D.
1974 'The Use of Clay before Pottery in the Zagros', *Expedition* 16/2: 11-7.
SCHMIDT H.
1943 *Tell Halaf I. Die prähistorischen Funde*, Berlin.
TOBLER A.J.
1950 *Excavations at Tepe Gawra*, vol. 2, Philadelphia.
VOIGT M.M.
1985 'Village on the Euphrates', *Expedition* 27/1: 10-24.
WOOLLEY L.
1955 *Ur-excavations IV: the early periods*, Philadelphia.

Halaf mortuary practices: A survey

PETER M.M.G. AKKERMANS

INTRODUCTION

In recent years, much attention has been given to the sociological impact of burials found in excavations. Following the pioneering work of Saxe (1970) and Binford (1971), virtually all studies of ranking and stratification of ancient societies rely to some degree or other upon analysis of mortuary remains. Simply speaking, it has been suggested that patterns in death bear a direct relationship to patterns of life in society. Any social distinctions in life are assumed to be directly reflected in the burial record, thus enabling the student to reconstruct a society's social organisation by analysing the mortuary evidence. This approach has been severely criticised by, among others, Hodder (1980) and Pader (1980; 1982) as being much too simplistic and as overlooking many of the symbolic actions in human mortuary behaviour. Burial practices do not *a priori* reflect social realities but, in contrast, may give an inverted or disguised picture of the world of the living, due to prevailing societal attitudes towards death.

This paper intends to give an outline of Halaf mortuary practices. On the basis of the burial evidence from several sites, I attempt to draw some general conclusions about Halaf social structures and practices. The period involved ranges approximately from 5500 to 4500 B.C. At this time, Halaf society was widely dispersed along the northern fringes of the Fertile Crescent, including modern northern Iraq, southeastern Turkey and northern Syria. On the basis of the pottery, LeBlanc and Watson (1970) considered Halaf being the first widespread cultural horizon in the Near East. If this view is correct, a highly uniform burial record can be expected, more or less interchangeable from site to site.

THE BURIAL RECORD

Iraq

The first Halaf burials were found in the early 1930s at *Tell Arpachiyah* (Mallowan and Rose 1935:42-43). Mallowan uncovered nine Halaf graves, six of which appeared in the outlying areas whereas three others were found on the mound itself. The burial description is highly incomplete and allows some broad inferences only. All burials consisted of simple and isolated pit inhumations. The dead seem to have been laid in a contracted position either on their left or on their right side. Orientation of the body was either NW-SE or, in one instance, E-W. Two graves contained remains of infants, one of which (G-57) had probably been laid in a mat-covered pit. One grave (G-56) was that of a child whereas the others probably contained adults. In six graves burial gifts were found which consisted of ceramic vessels, placed at the head and occasionally supplemented by bone, obsidian, flint or other stone implements. In one case (G-55), a number of red-painted shell beads were found at the neck, probably forming part of a necklace. Three graves, two of which belonged to infants and one to an adult, yielded no grave goods at all.

In 1976, Ismail Hijara undertook soundings at Arpachiyah, thereby uncovering two fractional adult burials (G1-G2) and one complete interment (G3) in the area south of Mallowan's TT8 tholos (Hijara 1978, 1980; Hijara et al. 1980). Burial G1 consisted of a pit into which a large globular and tall-collared jar containing a skull had been placed. Next to the jar two painted pottery bowls and one of stone were found. Burial G2 represented a collective grave of four skulls, three of which had been interred in pottery bowls whereas the fourth had been placed in a squat jar. In addition, six ceramic vessels and one stone bowl were found in the burial pit. Burial G3 consisted of a complete skeleton, found in a contracted position, lying on its right side. Two pottery bowls and two stone bowls had been placed near the feet. Mallowan (1935:34) suggested that the tholoi found at Arpachiyah represented shrines built within a sacred precinct. Hijara (1978) takes up Mallowan's point, thereby interpreting the skull burials as ritual interments of people of high social rank within a religious centre.

At *Tepe Gawra*, located about 15 km northeast of Arpachiyah, some Halaf graves have been found in the soundings of Area A, at the foot of the mound (Tobler 1950). No building remains have been uncovered here. Area A was probably used as a dump of domestic debris by the inhabitants of Tepe Gawra. In this dump, three graves had been dug, one of which (G36-158) belonged to a child whereas the two others (G36-159 and G36-160) contained adults. The dead were laid in a flexed position either on their left side (child) or on their right side (adults). The child inhumation was oriented N-S and accompanied by a painted bowl. One of the adults (G36-159) was oriented SE-NW and wrapped in a reed mat. Grave goods consisted of two painted jars, one placed in front of the body whereas the other had been placed at the feet (cf. Tobler 1950, Pl. LXIV.b). No description is available for the second adult burial (G36-160).

In addition to these individual graves, some mass burials, containing 24 individuals in total, have been found in a pit in the eastern corner of the sounding. This pit was five metres in depth and maximally 3.10 metres in width. It had probably served as a well or cistern for some time, but rapidly been filled with soil washed down from the slopes of the mound. At a time when this pit was no longer used as a water-supply, it served as a burial pit. Four different levels of interment are distinguished, termed Burials A to D (A being the topmost; Tobler 1950:49).

Burial A consisted of two adults. Tobler (1950:49) suggested that whereas one of these individuals had been formally interred in a contracted position on its left side, the other one may simply have been thrown into the pit. However, judging from a photograph of this burial (Tobler 1950, Pl. C.a), both individuals were found in contracted position at almost the same elevation. Moreover, a bowl was found directly above the so-called cast individual, suggesting some kind of burial ritual. It seems more likely that both individuals were interred at the same time in a similar way. The appearance of impressions of wooden poles between the elbows and knees and underneath one individual is interesting, perhaps indicating some kind of burial construction or coffin.

Burial B was found about one metre below the topmost inhumations and consisted of twelve adults, probably all thrown into the pit. Bones were scattered over the area without any orientation or formal arrangement. A squat, lug-handled jar, two stone pendants and a basalt pestle were associated with this mass interment. One well-preserved skull from Burial B has been analysed and belonged to a female person of about 25 years old (Tobler 1950:217).

Burial C appeared only about 20 cm below burial B and consisted of nine adults. Like in Burial B, these individuals had probably all been thrown into the pit, since no orientation or arrangement of the bones was perceptible. Moreover, some individuals were lying on top of others. No objects were directly associated with this burial but a painted bowl and a painted jar were found at a slightly deeper level (10-15 cm underneath the skeletons). Tobler (1950:49) points out that these vessels must have been cast into the pit as well, since sherds were found scattered all over the area.

Burial D was found about 40 cm below Burial C and consisted of a single young adult. The contorted position of this individual suggested that no formal interment had taken place. No burial gifts were present.

The majority of the mortuary evidence for the Halaf period stems from the Russian excavations at the sites of Yarim Tepe I, II and III, situated in the Sinjar plain west of Mosul. In total, about forty graves have been found at this complex.

Yarim Tepe I belongs entirely to the Hassuna period. No later phases of occupation have been attested. On the basis of some Halaf burials (at least six), sunk into the lower Hassunan levels, Merpert and Munchaev (1969:131) suggest that the site served as a burial ground for the nearby Halaf settlement of Yarim Tepe II.

The dead were buried in a flexed position on their left or on their right side. Orientation of the bodies was E-W, N-S or NW-SE. Grave goods consisted of pottery and stone vessels, beads of bone or shell, and stone axes and maces. Some of the graves showed a vertical shaft giving access to a lateral burial chamber, containing one individual. Another grave consisted of a long but narrow pit, measuring about 200 cm in length, 45 cm in width and 120 cm in depth. Only few human bones were found in this pit but it remains unknown whether these represent a deliberate fractional burial or are, in contrast, due to unfavourable conditions of preservation. A large skull of a bovoid was found above the pit whereas in the pit about 200 astragalus bones were found as well as fragments of three clay vessels and three stone objects, one of which has been identified as a mace. It has been suggested that this grave contained the remains of a hunter (Merpert and Munchaev 1971:17).

Most burials have been found at the Halaf settlement of *Yarim Tepe II*. Here at least thirty graves were uncovered (the exact number is not clear from the various reports), most of which belonged to infants and children up to eight years old. So far, only five or six adult burials have been uncovered. Three types of mortuary treatment can be distinguished at Yarim Tepe II, viz. simple pit inhumation, cremation and skull interment (fractional burial).

Simple primary inhumations in pits are most common at Yarim Tepe II, comprising about two-thirds of the burial sample. These graves all contained infants or children, laid in a contracted position either on their left or on their right side. Orientation of the body was E-W, N-S or NW-SE. One burial (N59) contained two infants whereas another (N61) consisted of two adults and a child (Merpert and Munchaev 1987:25). Infant burials seem to contain no burial gifts but in most other graves one to three pottery vessels were found, occasionally supplemented by stone vessels or bone and stone implements. Thus a burial of a seven-year-old child (N32), uncovered in level 4, yielded a miniature stone bowl, an unfinished alabaster vessel, a painted pottery cup, a flint implement and two bone awls (Munchaev and Merpert 1971:31). Another burial (N58), belonging to the earliest stages of Halaf occupation at Yarim Tepe II (levels 9-8), contained a highly stylised figurine of clay and a small cup filled with hundreds of beads of shell, carnelian and an unidentified kind of stone (Merpert and Munchaev 1987:25-26). One pit burial of level 4 (N36) seems to resemble the shaft graves found at Yarim Tepe I. This burial, described as 'a small catacomb under the floor' (Munchaev and Merpert 1971:31), contained a three-year-old child, lying in flexed position on its right side, facing west. Grave goods consisted of six painted vessels, a necklace made of alabaster beads, and a pendant showing animal heads. Two other graves (N42 and N48) need a short comment. Burial N42 consisted of an irregular, rounded pit about 140/160 cm in diameter. The skeleton was lying on its back with the arms spread and the legs bent and crossed. No grave goods were found. Considering the position of the skeleton, it has been suggested that the corpse had been thrown into the pit without a formal treatment (Merpert et al. 1977:91). The other burial (N48) seems to represent a secondary interment. In a small and shallow pit, measuring about 110 cm in length, 60 cm in width and only 20 cm in depth, the disordered remains of a seven- or eight-year-old child were found. The skull had been laid on top of other bones. No funerary gifts were present (Merpert et al. 1978:40).

Seven cremation burials have been found at Yarim Tepe II, all in the lower levels 7-8. Cremation seems to have taken place either in an open fire-place or in a specially built oven, sunk into lower levels. The latter has been attested in one instance and consisted of a rectangular oven with rounded edges, measuring about 110 cm in length and 35/50 cm in width. The walls of the oven narrowed towards the top (Merpert et al. 1976:51-52). The size of this oven probably reflects the height of the person burnt (apparently a child). During the cremation, three stone vessels and two pottery vessels seem to have been deliberately broken and thrown into the oven. Some other objects were supposed to have been thrown into the fire, too, but it is unclear from the publication of the facts of this burial (Merpert et al. 1976:52) whether all finds listed were actually exposed to the fire or were placed in the oven afterwards. The vast number of beads of various materials (shell, gypsum, obsidian, clay and an unidentified kind of stone) and some pendants found in the oven suggest that the deceased was wearing a kind of necklace at the time of cremation. After the cremation had been completed, the remaining bone fragments were collected and put into a small jar, together with twenty beads. The jar was put in the northwestern corner of the oven whereas in the opposite corner some miniature vessels were placed. Next, the oven was covered with soil.

The other six cremations (N43, N50-N54) uncovered at Yarim Tepe II do not give evidence of specially built structures for cremation, although it should be noted that in some cases the cremation and the subsequent interment seem to have been performed at two separate places, leaving the archaeologist with tangible evidence on inhumation only and without direct information about the circumstances of the burning of the corpse. Burial N43 represents the cremated remains of a child of about ten years old. The burning seems to have taken place in an open fire-place, marked by an accumulation of ashes and charcoal. During the process of burning, two alabaster vessels and three pottery vessels seem to have been deliberately broken and thrown into the fire. After completion of the cremation, the remaining bones were gathered and stored into a lugged jar, buried close to the fire-place (Merpert et al. 1977:91-92).

Burial N50 consisted of a small pit, being 70 cm long and 30 cm wide, in which the cremated remains of an adult had been laid, mingled with ashes and charcoal. Since the base and sides of the pit did not show traces of fire, it is assumed that the actual burning of the corpse took place somewhere else (Merpert et al. 1978:40). Fragments of two painted vessels were found in the northwestern corner of the pit and are thought to have been broken and thrown into the pit during interment (Merpert et al. 1978:40).

Burial N51 represents the remains of another adult cremation. As with Burial N50, the burial ground of N51 did not show any traces of burning. Moreover, no ashes or charcoal were found, except for one piece of burnt wood, and it is again assumed that the actual cremation was performed elsewhere (Merpert et al. 1978:40). The bones were laid in one corner of the pit without any arrangement except for the leg bones which seem to have been properly positioned. Two intentionally broken vessels were found, showing traces of secondary firing (probably resulting from the process of cremation).

Burial N52 was found near N51 and consisted of a small pit measuring 35x25 cm and containing the burnt remains of an infant. No traces of ashes were found nor any burial gifts. Cremation must have been performed elsewhere.

Burial N53 was situated only one metre west of N52 and consisted of an oval pit, measuring 116 cm in length and 53 cm in width. The sides and bottom of the burial pit showed slight traces of burning in some places. The pit contained the cremated remains of a child.

Finally, burial N54 consisted of a pit having a length of 125 cm and a width of about 50 cm. From the centre and the eastern half of the pit the burnt bones of an adult are reported, lying on a ca. 15 cm thick layer of ashes. Mainly in the northern corner of the grave, fragments of one painted and three undecorated vessels have been found, showing traces of secondary firing. Moreover, two clay spindle-whorls and a red-coloured stone were found. Although the excavators of burial N54 are inclined to see this burial as the interred remains of a corpse cremated elsewhere (Merpert et al. 1978:41), the present evidence strongly suggests that burial N54 represents the place both of cremation and of subsequent interment of the remains. If this view is correct, the size of the burial pit will reflect the height of the person cremated, thus pointing towards a child or adolescent instead of an adult.

In the same area as the cremation burials, three fractional burials (N49, N55, N56) have been found, consisting of skull interments. Burials N49 and N55 each contained one skull of an adult, laid on its left side and facing either East (N49) or West (N55). Burial N56 yielded three skulls, buried together in a rounded and shallow grave having a diameter of about 60/65 cm and a depth of about 14/16 cm. All skulls had been laid on their left side, facing northeast. None of the three skull interments contained funerary gifts. Similar burials have been found at Tell Arpachiyah although at the latter site the skulls had been placed in bowls or jars. Moreover, at Arpachiyah the skulls were accompanied by a number of stone and pottery vessels (see Hijara1978).

The mound of *Yarim Tepe III*, situated at a close distance from Yarim Tepe I and II, has yielded only three Halaf burials so far (Merpert and Munchaev 1984). All graves were found in domestic deposits and represented simple pit inhumations of children. Burial N29 contained the remains of a two-year-old child, laid in contracted position on its right side and facing south. Orientation of the body was E-W. No grave goods were present. Burial N31 was found in the close vicinity of N29 and also contained a child of about two years old. The child was lying in contracted position on its left side, facing north. Orientation of the body was E-W. Again, no funerary gifts were found. Burial N30 belonged to a twelve-year-old child, buried in a contracted position with the head westwards. Orientation of the body was N-S. One undecorated and three painted vessels had been placed at the feet. Burial N30 belongs either to the final stages of the Halaf period or to the beginning of the subsequent Ubaid period at Yarim Tepe III (Merpert and Munchaev 1984:57).

East of the Yarim Tepe complex, some Halaf burials have been reported from *Tell Azzo I*. Several decapitated human skeletons were found in a tholos but the relation between building and interments is vague (Killick and Roaf 1983:206). The headless inhumations from Azzo immediately recall the skull interments from Yarim Tepe II and Arpachiyah. Apparently, the practice of dividing the head from the body is widely spread in the Mosul region.

In the Tigris valley north of Mosul, some Halaf graves have been found at *Tell Kutan* and *Kharabeh Shattani*. At the former, an adult inhumation was found but no further details are available yet (Killick and Black 1985:235). At Kharabeh Shattani, south of Tell Kutan, two pit inhumations of adults were found. One interment was accompanied by a bone point and a stone bowl. Moreover, the individual had been wearing a girdle around the hips. On the girdle there were about twenty beads of bone, coloured stone and dentalium shell. A trapezoid bead of black stone was attached to the front of the girdle (Watkins 1987:227).

Syria and the Levant

Inland Syria and the coastal regions along the Mediterranean are poorly understood in terms of Halaf mortuary practices. Halaf burials have been reported from Chagar Bazar, Shams ed-Din and Ras Shamra. It should, however, be kept in mind that Ras Shamra is situated on the extreme western periphery of the Halaf culture. So far, only some of the ceramics found on the site showed Halaf characteristics, leaving the question unanswered whether Halaf influences also entered the more ritual-bound and tradition-based spheres of society such as expressed in the treatment of the dead.

At *Chagar Bazar*, levels 12 to 6 have yielded eight Halaf burials (Mallowan 1936:18). One more burial has been reported from level 12 but apparently no body was found (Mallowan 1936:18). Because further information is lacking, nothing can be said on this feature. The other graves mainly consisted of child inhumations. Adult remains have been unearthed in three cases. All burials represented simple pit interments. The dead were lying in a contracted position on their sides. In most cases, the body was oriented E-W but in two instances the orientation was N-S. All burials except one were accompanied by grave goods, viz. a painted or unpainted pottery vessel (some of them were miniatures).

At *Shams ed-Din* a badly preserved child inhumation was found. This burial consisted of a small and rounded pit, about 80/90 cm in diameter, and contained the skull and some bones of an incomplete child skeleton. No grave goods were present (al-Radi and Seeden 1980:106). It is not known whether the incomplete nature of the skeleton is due to preservation or to secondary interment.

At *Ras Shamra*, levels IV C to IV A are ascribed to the Halaf period. These levels yielded an unspecified number of simple pit inhumations, mainly of infants and children. Apart from a skull in level IV B, adults only appear in the lower level IV C. The dead had all been laid in a contracted position on their sides. No grave goods were present except for the ornaments worn by the deceased at the time of interment (de Contenson 1973:93-94)

Southeastern Turkey

In Turkey graves of the Halaf period have been found at Mersin, Gerikihaciyan and, more recently, at Cavi Tarlasi.

At *Mersin*, levels XIX-XVII are ascribed to the Halaf period (Garstang 1953:101 ff). The pottery of Mersin is of a local, Cilician origin but shows evidence of strong Halaf influences (Garstang 1953:101 ff; see also Davidson 1977). In analogy with Ras Shamra, one may wonder whether the graves unearthed at Mersin also show Halaf traits or, in contrast, fit entirely within local traditions. Evidently, this problem arises whenever one reaches the periphery of Halaf society. At present, the highly restricted knowledge on Halaf society in general, and on the marginal regions of this culture in particular, prevents us from drawing any definite conclusions.

Six burials are ascribed to Mersin XIX-XVII. Two burials belong to level XVII, one to level XVIII and three to level XIX.

The two graves sunk in from level XVII both represent simple pit inhumations. One burial belonged to an infant, lying in a flexed position on its right side and facing north. Orientation of the body was NW-SE. The other skeleton was lying in a contracted position on its left side. Only the leg bones had been preserved. Both graves contained no funerary gifts. The burial ascribed to level XVIII does not seem to represent an intentional interment but an accidental feature. The individual was lying on its left side but not in contracted position. Moreover, stones and mud bricks rested against the shoulder blade and separated the skull from the body. The burials ascribed to level XIX consist of two individual inhumations and one mass interment. One of the single inhumations had an individual lying in a contracted position on its left side. Orientation of the body was E-W. No grave goods were present. The other single burial seems to represent a cremation of a child. The bones showed traces of blackening whereas the area around and underneath the skeleton was red-coloured due to fire. Again, no grave goods were found. The last burial ascribed to level XIX represents a mass cremation of an unknown number of adults. With the bones some sherds and two crushed, imcomplete pottery vessels were found. The ceramics were blackened and apparently accompanied the deceased at the time of cremation. The cremation burials of Mersin immediately recall to mind the occurrence of almost similar graves at Yarim Tepe II in northern Iraq. In this respect, it is interesting to note that cremation burials were unknown at Mersin before the introduction of Halaf at the site.

Gerikihaciyan is a Halaf settlement situated in the hilly country near Diyarbakir. The site is located on the periphery of the known distribution of Halaf society, like Mersin, but in contrast to the latter, Gerikihaciyan is not so much Halaf-influenced as of true Halaf derivation (Watson and LeBlanc, forthcoming). Excavations at Gerikihaciyan in 1968 and 1970 yielded four burials, all simple pit inhumations, three of which were complete and one highly fragmentary. The graves could not be associated with a specific feature or level of occupation.

Burial 1 showed an adult, probably male and aged between 25 and 40 years. The individual was lying on its left side in a contracted position. Orientation of the body was E-W. No grave goods were found. Burial 2 represents a female child, 6-7 years old and lying in a contracted position on its left side. The body was oriented SE-NW. No grave goods were present. Burial 3 belonged to a three-year-old child, lying in extended position. The body was covered by a fragment of a large jar. Orientation of the body was SE-NW. No grave goods were present. The last burial consisted only of five rib fragments and a left humerus. The latter pointed towards a heigth of 160 cm for the person interred, thus suggesting that this individual was an adolescent or an adult. Either this burial was disturbed by later activities of occupation, or it represents a secondary interment. Again, no grave goods were found.

Recent excavations at *Cavi Tarlasi* in the Euphrates valley have yielded 18 Halaf burials (von Wickede and Misir 1985). The majority of these inhumations (16) belonged to children, mainly buried in a contracted position on their right side. None of these child interments contained funerary gifts except for two which each yielded a pottery vessel. Two other graves (nos. 3 and 9) represented double interments, each consisting of an adult male and a child. This is interesting since concomitant inhumations of adults and children, without further determination of sex, are commonly viewed as 'mother-and-child' burials, both of whom are assumed to have died in childbirth. One of the double burials (no. 9) contained a pottery plate, a stone axe and some flint and obsidian implements, all placed near the shoulder of the adult. In view of the limited number of burials uncovered within the inhabited area of Cavi Tarlasi, it is assumed that there must have been a cemetery somewhere in the vicinity of the settlement (von Wickede and Misir 1985:105).

HALAF MORTUARY PRACTICES: SOME REMARKS

From the foregoing, it will be clear that any analysis of Halaf mortuary practices is severely hampered by the fragmentary nature of our present data. The existing burial record is deficient in various respects. Firstly, reports on Halaf burials are often inadequate, either because they generalise the burial evidence or because they omit many details of utmost importance for further analysis (like sex and age). Secondly, vast regions are virtually unknown in terms of Halaf mortuary practices (e.g., Syria). At present, most information stems from sites situated in the Mosul region of northern Iraq. Thirdly, Halaf mortuary behaviour may have changed through time and space. The Halaf period is of long duration and the investigated sites cover various stages in various regions of Halaf development. The overall picture may thus be distorted by temporal and spatial effects (O'Shea 1984:14). Fourthly, as a result of the current emphasis upon settlement archaeology, all burial evidence comes from contexts of occupation. The rare occurrence of adult graves within settlements leads to the conclusion that the present burial sample is highly biased in favour of infants and children. Halaf cemeteries have not yet been found, except perhaps at Yarim Tepe I. On the latter site only six Halaf graves have been found so far and it may very well be the case that these burials belong to a now eroded Halaf occupation level instead of a true cemetery (Breniquet 1987:239). Nevertheless, the striking scarcity of adult interments at the investigated mounds strongly points towards the existence of formal burial grounds outside these settlements. These graveyards will have served to inhume most of the dead of a community, whereas interment within a settlement is largely limited to children and perhaps individuals of a specific status. The present, biased burial record shows only a part of Halaf mortuary variety. Hijara (1980b:209-216) and Breniquet (1987:238) have, among others, suggested that normative Halaf treatment of the dead involves individual inhumation within settlements, a view which needs some modification in the light of the present evidence. Moreover, the common notion that the deceased are interred underneath the floors of inhabited buildings (Breniquet 1987:238) is questionable. The stratigraphic relationship between inhumation and associated building is often vague. In most cases, it cannot be ascertained whether the burial was sunk into a floor of a contemporary building or covered by a floor, thus making any association hazardous. It seems more likely that most burials took place in deserted areas of a settlement, perhaps characterised by collapsed buildings still standing to some height.

The present evidence shows that Halaf mortuary practices widely varied, from single and double pit inhumations, mass interments, single and multiple skull sepultures to individual or mass cremations. Moreover, within each kind of burial group considerable variation is noted, e.g. in grave construction, orientation of the body, and number and kind of funerary gifts. Apparently, the mortuary ritual is complex and defined by belief systems and cultural circumstances, which may vary from case to case. Differences in mortuary treatment may be related to status, age, sex, ethnic entity as well as to the causes of death. Ucko (1969:271) already pointed out that those who died from illness, drowning or lightning may have been treated each in a different way from those who died in battle or through murder. As noted before, the existing burial record is biased, emphasizing the interment of children. Others (e.g. Ucko 1969:270 ff; Saxe 1971:43; Poppa 1978:47) have suggested that child interment generally proceeds along different lines than adult burial. A high infant mortality rate, in combination with some kind of social ranking according to age, may have prescribed specific rites of child burial. Actually, the Halaf burial evidence seems to conform to this view. Infants and children of young age are almost invariably inhumed, whereas the more exotic kinds of mortuary treatment (cremation, skull interment) seem to be largely confined to adolescents and adults.

Within settlements, single and primary inhumations of children are most common. The dead are formally inhumed, whereby the position of the body and the placement of pottery vessels and other goods is taken into account. These single, formal interments assert the individual and suggest that the individual's identity was seen as continuing into death (Braithwaite 1984:104). Even the few double interments found at Tepe Gawra, Yarim Tepe II and Cavi Tarlasi stress the individual by separating each person both spatially and physically. Child inhumations were probably performed within family or close kin circles. In view of social relationships and social implications, child death hardly affected the community as a whole but merely those who were immediately related to the deceased. The dispersed interment of children within settlements, perhaps in courtyards or other family properties instead of in a centralised community burial ground, points towards a family-oriented mortuary procedure, too. The wide variation in positioning and orientation of the body and in number and kinds of grave goods may suggest that interment of children did not entirely proceed according to communal and tradition-bound rules but was open to idiosyncratic wishes or needs of the relatives.

Remains of girdles or necklaces made of beads in burials from Yarim Tepe II and Kharabeh Shattani suggest that at least some of the dead were clothed at the time of interment. From Arpachiyah and Tepe Gawra we have some evidence that matting was occasionally used to wrap the dead. The pottery or stone vessels accompanying the deceased may have contained food, either in reality or in a symbolic manner, for the hereafter. It may, however, also be the case that these vessels should be considered as the deceased's personal belongings, used during his/her lifetime (cf. Ucko 1969:265). In this respect, it is interesting to note that, whereas infant burials in most cases contain no objects at all, children of older age are often accompanied by miniature vessels. The so-called 'hunter burial' from Yarim Tepe I may also indicate that grave objects served to express part of a person's social entity. In recent years, it has been repeatedly tried to infer social ranking from grave wealth in child burials (cf. Brown 1981:30; Flannery 1972:403). Wealth is viewed as an indicator of social differentiation and positioning. Since children, by their low age, are not capable of achieving wealth and associated status by personal efforts, it is assumed that rank is ascribed and inherited. If we accept this view and apply it to the existing Halaf inhumation record, no ranking other than the one probably based on age is observed. Child inhumations show considerable variation in number and kind of grave goods but no exceptional treatment is noted. The few adult inhumations do not suggest any social ranking either.

Mass interments are known from Tepe Gawra and, on the periphery of Halaf society, from Mersin. On the latter site, a mass cremation was uncovered whereas at Gawra two mass inhumations were found in the so-called well of Area A. These mass burials deviate from the other kinds of Halaf graves in various respects. Firstly, the characteristic individual treatment of the dead has been abandoned. Secondly, at Tepe Gawra the appropriate ritual discourse seems to have been largely omitted. Instead of having been carefully positioned, the dead seem to have been simply thrown into the well. However, the presence of a few grave goods, apparently also thrown into the pit, suggests that some rites were performed but perhaps in a simplified manner. Thirdly, mass burials seem to have been confined to adults only. As to the nature of the mass burials of Gawra and Mersin, we can only speculate. Each burial contained a number of individuals who must have died more or less at the same time. Tobler (1950:50) suggests that the dead of Tepe Gawra were victims of famine or plague whereas Garstang (1953:111) considers the cremated adults from Mersin to be victims of war, burnt by their conquerors. Garstang's argumentation (1953:111-112), however, seems to be mere fiction. In case of Tepe Gawra, one may wonder why the regular procedure of interment was not performed. In analogy with the mass graves of the victims of pestilence in medieval Europe, it can be postulated that hasty interment without an appropriate treatment can be neccessitated by hygienic reasons and circumstances beyond one's control. Climatic conditions in the Near East lead to a rapid decomposition of the corpse, thus requiring interment within a short period after death had occurred (cf. Poppa 1978:25). The dead of Burials B and C at Tepe Gawra must have been buried within a short time after each other.The lower mass burial C, consisting of nine adults, was covered by a layer of soil only 20 cm thick, after which the remains of another twelve adults (Burial B) were deposited.

So far, cremations have only been reported from Yarim Tepe II and Mersin. Cremations seem to represent a late Halaf feature. At Yarim Tepe II cremations appear in the lower levels 7 and 8, dated by some radiocarbon samples in the mid-fifth millenium B.C.(Merpert et al. 1977:95; these dates, however, seem to be too late). According to the pottery, Mersin belongs to the latest stages of the Halaf culture and probably dates from around the middle of the fifth millennium (cf. Davidson 1977; Watkins and Campbell 1987). Most cremation graves represent secondary burials, i.e. the burning of the corpse and the subsequent interment were performed at two separate places. Cremation seems to be largely confined to adolescents and adults. The dead seem to have been extended upon a crematory pyre within a shallow depression or, as illustrated by a burial from Yarim Tepe II, in a specially-built oven-construction. The layout of some of the cremation pits, reflecting the size of the deceased, suggests that the dead were cremated within a short period after death had occurred. The occasional appearance of large numbers of beads among the burnt remains points towards necklaces or girdles and suggests that at least some of the dead were clothed at the time of cremation. During the burning of the corpse, some pottery and stone vessels seem to have been deliberately broken and thrown into the fire. Occasionally, some other objects were thrown into the fire, as well. After the burning of the corpse, the remaining bones were gathered in most cases and either laid in a small, shallow pit dug elsewhere or stored in an urn buried at the place of cremation. In some other instances, the remainder of a cremation was simply covered by soil on the spot, without further treatment. From one burial of Yarim Tepe II, we have some indications that during interment of the cremation remains some miniature vessels were placed into the grave as well, thus suggesting that, after the rites of cremation, some rituals associated with inhumation burials were also performed. The deliberate smashing of pottery or stone vessels does not seem to be restricted to the act of cremation only. In the case of a burial from Yarim Tepe II, it has been suggested that two painted vessels were broken and thrown into the pit during interment (Merpert et al. 1977:92). This practice reminds us of Tepe Gawra where, slightly underneath the mass burial C, two painted vessels were found, broken and scattered all over the pit (Tobler 1950:49). The ritual breakage or 'killing' of artefacts in association with funerals is known from many ancient and modern societies (see e.g. Hodder 1980:164; Parker Pearson 1984:72; Tilley 1984:137 ff) and may serve many purposes. Thus, among the Nuba of Sudan the breaking of a dead person's personal belongings is associated with the removal of the impurity and ill effects of death (Hodder 1980:166), whereas in other instances it has been suggested that, by destroying the wholeness of form, the natural order is broken and, ultimately, the existing traditional and 'natural' inequalities in social relationships are denied (Tilley 1984:137 ff). In relation to the latter view, Braithwaite's suggestion is worth mentioning that among the people of Late Bronze Age Wessex the act of cremation itself may have been viewed as the destruction of an individual's entity and as a ritual breakage of the continuity between life and death (Braithwaite 1984:105). In case of the Halaf cremations, one can only speculate about the nature of these graves. Cremations seem to have been performed along lines of ritual behaviour different from simple, primary inhumations. Infants and young children are excluded from cremation. Perhaps cremations are confined to persons of specific rank or status although the present evidence gives few indications in this direction. At Yarim Tepe II some cremation graves contained a considerable quantity of grave goods but other burials yielded no objects at all. Moreover, the act of cremation itself does not necessitate the use of resources restricted in distribution to some groups of a community only. Thus it seems that reasons of a more spiritual nature underlie these cremation burials. Although it can be argued that the inhumation of infants and children of young age will have taken place largely within family or close kin circles, the cremation of older members of a community may have been of a public nature. The latter is illustrated by the mass cremation of adults at Mersin. The ritual burning of these individuals, whatever caused their death, must have gone beyond the level of one family or descent group, thus affecting a large part of the Mersin community in one way or another.

Skull interments are known from Arpachiyah and Yarim Tepe II whereas some decapitated skeletons have been found at Tell Azzo I, all situated in the Mosul region. The skull sepultures probably represent secondary disposals; most likely the head was divided from the body after complete or partial decay of the corpse. Thus, two stages of interment are indicated: firstly, the corpse was temporarily disposed of to allow removal of the soft tissue, and secondly, the remainder was finally interred. At the latter stage, it was decided either to inhume the skeletal remains as a whole, or to divide the skull from the body and bury each separately. A complete secondary interment was found at Yarim Tepe II, and consisted of the disarticulated remains of a seven- or eight-year-old child. The skull had been laid on top of other bones (Merpert et al. 1978:40). Sepultures of skulls only are restricted to adults. At Yarim Tepe II, three such interments were found, each characterised by shallow pits on the floor of which one or more skulls had been laid. Two of these burial pits contained one skull each whereas another yielded three skulls. No funerary gifts were present in any of these graves. At Arpachiyah, the skulls had been treated in a different way. Not only had the skulls been laid in pottery vessels, but they were accompanied by a number of pottery and stone vessels as well. Two cranium graves appeared at Arpachiyah, containing one and four skulls respectively. Hijara (1978:128) suggests that the skull burials of Arpachiyah are associated with people holding an exceptional social position. Concerning the burials of Yarim Tepe II, a similar suggestion has been made by Merpert et al. (1978:40). These interpretations seem merely to have been based upon the rare and seemingly exceptional nature of the skull sepultures. However, not only these graves but adult burials as a whole are rare within settlements and may thus all represent exceptional features, deviating from normative burial treatment. On the other hand, in absence of comparative material (i.e. graveyards yielding a random sample of both population and mortuary practices) any such interpretations seem to be founded on quicksand. At present, no decisive conclusions can be drawn on possible regularities or deviations in Halaf adult burial treatment.

REFERENCES

AL-RADI S. AND SEEDEN H.

1980 'The AUB Rescue Excavations at Shams ed-Din Tannira', *Berytus* 28: 88-126.

BINFORD L.R.

1971 'Mortuary Practices: Their Study and Potential', in: J.A. Brown (Ed.), *Approaches to the Social Dimensions of Mortuary Practices* (Memoirs of the Society for American Archaeology 25). Pp. 6-29.

BRAITHWAITE M.
1984 'Ritual and Prestige in the Prehistory of Wessex c. 2200-1400 BC: A New Dimension to the Archaeological Evidence', in: Daniel Miller and Christopher Tilley (Eds.), *Ideology, Power and Prehistory*, Cambridge:Cambridge University Press. Pp. 93-110.

BRENIQUET C.
1987 'Nouvelle hypothèse sur la disparition de la culture de Halaf', in: Jean-Louis Huot (Ed.), *Préhistoire de la Mésopotamie*, Paris: Éditions du CNRS. Pp. 231-241.

BROWN J.A.
1981 'The Search for Rank', in: Robert Chapman et al. (Eds.), *The Archaeology of Death*, Cambridge: Cambridge University Press. Pp. 25-37.

DAVIDSON T.E.
1977 *Regional Variation within the Halaf Ceramic Tradition*, Edinburgh: University of Edinburgh.

DE CONTENSON H.
1973 'La culture Halafienne du niveau IV de Ras Shamra d'après les campagnes de 1968 à 1972 dans le sondage préhistorique', *Annales Archéologiques Arabes Syriennes* 23: 85-103.

GARSTANG J.
1953 *Prehistoric Mersin*, Oxford: Clarendon Press.

HIJARA I.
1978 'Three New Graves at Arpachiyah', *World Archaeology* 10: 125-128.
1980 *The Halaf Period in Northern Mesopotamia*, London:University of London.

HIJARA I. ET AL.
1980 'Arpachiyah 1976', *Iraq* 42: 131-153.

HODDER I.
1980 'Social Structure and Cemeteries: A Critical Appraisal', in: Philip Rahtz et al. (Eds.), *Anglo-Saxon Cemeteries 1979*, Oxford: BAR British Series 82. Pp. 161-169.

KILLICK R. AND BLACK J. (EDS.)
1985 'Excavations in Iraq, 1983-84', *Iraq* 47: 215-239.

KILLICK R. AND ROAF M. (EDS.)
1983 'Excavations in Iraq, 1981-82', *Iraq* 45: 199-224.

LEBLANC S.A. AND WATSON P.J.
1973 'A Comparative Statistical Analysis of Painted Pottery from Seven Halafian Sites', *Paléorient* 1: 117-133.

MALLOWAN M.E.L.
1936 'The Excavations at Tell Chagar Bazar and an Archaeological Survey of the Habur Region 1934-5', *Iraq* 3: 1-87

MALLOWAN M.E.L. AND ROSE J.C.
1935 'Excavations at Tell Arpachiyah, 1933', *Iraq* 2: 1-178.

MERPERT N. AND MUNCHAEV R.
1969 'The Investigation of the Soviet Archaeological Expedition in Iraq in the Spring 1969', *Sumer* 25: 125-131.
1971 'Excavations at Yarim Tepe 1970, Second Preliminary Report', *Sumer* 27: 9-22.
1984 'Soviet Expedition's Research at Yarim Tepe III Settlement in Northwestern Iraq, 1978-1979', *Sumer* 43: 54-68.
1987 'The Earliest Levels at Yarim Tepe I and Yarim Tepe II in Northern Iraq', *Iraq* 49: 1-36.

MERPERT N. ET AL.
1976 'The Investigations of Soviet Expedition in Iraq, 1973', *Sumer* 32: 25-61.
1977 'The Investigations of Soviet Expedition in Iraq, 1974', *Sumer* 33: 65-104.
1978 'Soviet Investigations in the Sinjar Plain, 1975', *Sumer* 34: 27-70.

O'SHEA J.M.
1984 *Mortuary Variability*. An Archaeological Investigation, Orlando, Fl.: Academic Press.

PADER E.-J.
1980 'Material Symbolism and Social Relations in Mortuary Studies', in: Philip Rahtz et al. (Eds.), *Anglo-Saxon Cemeteries 1979*, Oxford: BAR British Series 82. Pp.143-159.

1982 *Symbolism, Social Relations and the Interpretation of Mortuary Remains*, Oxford: BAR International Series 130.

PARKER PEARSON M.

1984 'Economic and Ideological Change: Cyclical Growth in the Pre-State Societies of Jutland', in: Daniel Miller and Christopher Tilley (Eds.), *Ideology, Power and Prehistory*, Cambridge: Cambridge University Press. Pp. 69-92.

SAXE A.A.

1970 *Social Dimensions of Mortuary Practices*, Michigan:University of Michigan (Ph.D. thesis).

1971 'Social Dimensions of Mortuary Practices in a Mesolithic Population from Wadi Halfa, Sudan', in: James A. Brown (Ed.), *Approaches to the Social Dimensions of Mortuary Practices*, Memoirs of the Society for American Archaeology 25. Pp. 39-57.

TILLEY C.

1984 'Ideology and the Legitimation of Power in the Middle Neolithic of Southern Sweden', in: Daniel Miller and Christopher Tilley (Eds.), *Ideology, Power and Prehistory*, Cambridge: Cambridge University Press. Pp.111-146.

TOBLER A.J.

1950 *Excavations at Tepe Gawra*, Vol. 2., Philadelphia.

UCKO P.J.

1969 'Ethnography and Archaeological Interpretation of Funerary Remains', *World Archaeology* 1: 262-280.

VON WICKEDE A. AND MISIR A.

1985 'Çavi Tarlasi 1984 kazi kampanyasi', *VII. Kazi Sonuçlari Toplantisi*: 103-109.

WATKINS T.

1987 'Kharabeh Shattani: an Halaf Culture Exposure in Northern Iraq', in: Jean-Louis Huot (Ed.), *Préhistoire de la Mésopotamie*, Paris: Éditions du CNRS. Pp. 221-230.

WATKINS T. AND CAMPBELL S.

1987 'The Chronology of the Halaf Culture', in: O. Aurenche et al. (Eds.), *Chronologies in the Near East*, Oxford: BAR International Series 379. Pp. 427-464.

WATSON P.J. AND LEBLANC S.A.

in press *Gerikihaciyan: a Halafian Site in Southeastern Turkey*, UCLA Institute of Archaeology.

The Red Wash Ware ceramic assemblage in Syria: A review[1]

ROGER LEENDERS

INTRODUCTION

The purpose of this article is to direct attention to a group of chalcolithic ceramics, the so-called Red Wash Ware, which has appeared throughout Syria from the Mediterranean coast in the west to the Khabur triangle in the east. Red Wash Ware is always found in connection with Late Halaf or transitional Halaf/Ubaid remains, which probably explains the fact that only little attention has been paid to it; the esthetically more attractive Halaf Pottery seems to have pushed the Red Wash Ware into oblivion.

To counteract this, my paper will present a survey of the sparse data which can be gleaned from the literature on the transitional Halaf/Ubaid period. It does not contain, however, any original fieldwork nor any previously unpublished material.[2] On the following pages a discussion of the general characteristics of the Red Wash Ware will be followed by a survey of its temporal and spatial distribution.

GENERAL CHARACTERISTICS OF THE RED WASH WARE

Wares, forms, decoration. In order to describe the Red Wash Ware properly we would have needed a lavishly published body of sherds, accurately described and illustrated. This, however, is lacking and this study relies exclusively on the meager data recorded in the literature, interspersed in discussions of the Halaf and Ubaid ceramic assemblages. The results are listed here: Red Wash Ware is a handmade ware. The paste is in most cases described as relatively fine. In Hama it is present in a 'medium ware' and in a 'coarse ware' version, but not in the 'fine ware'. As regards temper Red Wash Ware is different from Halaf pottery as it contains only mineral material. Straw is used as secondary material in the Biqa'a and the Palestinian sites only. Generally speaking, Red Wash Ware is a well baked, moderately hard ware with a thoroughly oxidized core. The colour shades of the paste vary from orange to buff. The Red Wash Ware differs therefore considerably from the cream paste of the contemporary Halaf ceramics. The Palestinian sites tend to a greater variety of shades, mostly greyish but including extremes like white or black. Red Wash Ware is no homogeneous ware: actually the only characteristic all variants have in common is the red wash from which it derives its name. Within the Halaf distribution area east of the Euphrates the wash covers the entire surface and is alway burnished. To the west of the Euphrates this is not necessarily so. The Queiq Valley Red Wash Ware is characterized by a red wash applied with a brush and then burnished more often than not (Mellaart 1981:147). Often only the rim of the vessel is covered by the wash. The variety in color is again greater in the southern sites; here brown or even black shades occur. The form repertoire is restricted to bowls and jars. Among the bowls simple hemispherical types or bowls with straight sides predominate. Specimens with inturned rims are less current. Bases are generally flat. Especially within the Halaf Ware distribution area carinated bowls are popular. In the Biqa'a, the coastal area and in northern Palestine flat based dishes occur. Among the jars specimens occur with globular bodies and a short straight or outward-turned neck and hole-mouth jars.

The most distinctive form type of the Red Wash Ware ceramic assemblage, however, is the so-called bow rim jar. This 'Leitfossil' of the Amuq D phase wares in Syria is to my knowledge unique within the ceramic repertoire of the entire ancient Near East. Bow rim jars occur not only in the Red Wash Ware but also in contemporary Halaf wares and even in the earliest 'Northern' or 'Syrian' Ubaid ware.

Red Wash Ware jars were equipped with a variety of handles. They come in three types: simple knobs, small pierced lugs or flat straps with broadened attachments. Spouts were observed on the jars from Tell Khazzami only. Jar bases are generally flat but round bottoms occur also. Apart from the wash, Red Wash Ware can be further embellished by painted or incised motifs. Painting is common in the Khabur area where it occurs, for instance, on many Tell Aqab sherds. In most cases painting is restricted to the rim and consists of simple horizontal bands and sagging lines. Incised decoration is not reported east of the Euphrates. West of the Euphrates, however, the roles are switched: here painting plays a comparatively small role (horizontal lines, shaded triangles, net patterns), while incised decoration takes the lead. Stab marks, fish grates, nail incisions, wavy lines, etc. are reported from every site, especially from the Neolithic 4 sites of the southern reaches of the Red Wash Ware distribution area. Here, too, many instances occur of applied cables that anticipate the coming Ghassul culture.

GEOGRAPHICAL RANGE

Red Wash Ware is found in a large part of Syria ranging from the Khabur valley in the east to the Mediterranean coast in the west. Derivatives of this ware have also been reported from sites in Palestine.

On the following pages, in a broad sweep from East to West, each region of the Red Wash Ware distribution area is discussed separately.

The Tigris Basin and southeastern Turkey. As Red Wash Ware was recorded neither in Tepe Gawra nor in Tell Arpachiyah we may safely assume that the Tigris basin was devoid of this ware (Tobler 1950; Davidson 1977:53; Mallowan 1935; Hijara 1981). The same goes for southeastern Turkey (for instance Sakçe Gözü, Girikihaciyan, Tepecik, Tülin Tepe, Cavi Tarlasi, Kurban Hüyük).

The Sinjar plain. In the 1930s Seton Lloyd collected a prehistoric sherd sample on the Sinjar plain (Lloyd 1938). One of the groups into which the collection was divided, group II, consists of Late Halaf sherds, while group III is called 'Northern Ubaid'. Sherd No. 21, a bow rim jar (Red Wash Ware?), was attributed to group IV (unpainted prehistoric pottery). It indicates the presence of at least a transitional Halaf /Ubaid period in the Sinjar plain. Besides this, Oates was reported to have found Red Wash Ware on a survey in the region, but he had misinterpreted it as Uruk ware (Davidson 1977:156).

Large scale archaeological research started in 1969 when Merpert and Munchaev began their work at Yarim Tepe II. Though several preliminary reports have appeared in the meantime (cf. Merpert 1981), the pottery remains largely unpublished. This has caused conflicting opinions on the early part of the chronological range of the site (Hijara 1980:151; Davidson 1977:195), but both authors agree in assigning the upper levels of Yarim Tepe II to the transitional Halaf/Ubaid period. Typical forms and wares we would expect in this context, like Red Wash Ware or bow rim jars, have not yet been published.

The Khabur Valley. Several sites in the Khabur Valley have a minor Red Wash Ware component: relevant for our subject are the results of sondage T1 at Tell Aqab, not far under the crest of the mound. Here the lowest stratum represents the Later Halaf Period, followed by two transitional Halaf/Ubaid strata and one Ubaid stratum (Davidson 1977:105-157; Davidson and Watkins 1981).

The Late Halaf pottery of T1 can be most easily compared to the Arpachiyah ware of TT6-7 and Chagar Bazar 11-6. It is in this stratum that the first Red Wash Ware sherds were found. This ware is characterized by a thick red wash which covers the entire surface of the vessel, inside and outside. The wash is always burnished and sometimes decorated with simple motifs in black paint, applied over the burnished wash.

The Red Wash Ware of Tell Aqab continues into the Halaf/Ubaid Transitional period, but is absent in the Ubaid period. The paste is different from that of the Halaf pottery and contains both mineral and vegetal temper.

Davidson distinguished three Red Wash Ware form-types: simple, unrestricted hemispherical bowls with flat or rounded bases, simple restricted bowls with flat bases, and beveled-rim jars (Davidson 1977:155-156).

The bowls are a fine example of the gradual nature of the Halaf/Ubaid transition at Tell Aqab; both forms and decoration were already present in the Late Halaf strata and continued into the Ubaid strata, so that it is only on stratigraphical grounds that a clear attribution to the Transitional period could be made.

A very significant form which appears for the first time in the Halaf/Ubaid transitional strata is the 'bow rim jar'. Only two fragments were found at Tell Aqab and both of them are attributed to the Late Halaf pottery, but we shall see that further to the West this intriguing jartype knew an extraordinary career and was usually associated with Red Wash Ware (Davidson and Watkins 1981, fig. 3:6). In fact, we may regard the bow rim jar as the 'Leitfossil' of the Halaf/Ubaid transitional period and especially of the Red Wash Ware. Apart from Tepe Gawra, where one fragment of a bow rim jar was found in Level XVII, Tell Aqab is the easternmost site which yielded this type.

In sharp contrast to the rich repertoire of elaborate painted decoration on the Late Halaf pottery, often polychrome, the Red Wash Ware at Aqab features only two motifs: simple horizontal lines and sagging lines, suspended from horizontal bands.

The most interesting part of the Tell Halaf assemblage from the view-point of this paper is the Ca pottery, belonging to Schmidt's '*Verfallsperiode*'. This is the local Red Wash Ware of Tell Halaf, which closely matches Davidson's Red Ware at Tell Aqab (von Oppenheim 1943:85-87). Here, too, as in Tell Aqab, the red burnished wash is decorated with simple motifs (horizontal bands and sagging lines) in a dull black paint (Von Oppenheim 1943, Pl. 98:5).

The Balikh Valley. Further to the West, in the Balikh Valley, the Halaf/Ubaid transitional material has been recorded in several instances, for the first time at Tell Mefesh, but never in significant quantities.(Mallowan 1946; Copeland 1979; van Loon 1985). Copeland's survey, though very short, augmented the number of transitional Halaf/Ubaid sites with at least four more settlements besides Tell Mefesh: Chahine, Zaidan, Sabi Abyad[3] and Mounbatah all prove that there must have been settlements in the Balikh Valley in the transitional Halaf/Ubaid period. Probably Tell Hammam et-Turkman was also inhabited in this period (Van Loon 1985:22 and fig.3). The stratigraphical trench on the eastern slope of the tell cuts through a series of Ubaid strata which continue under the level of the surrounding plain. The earliest phase (termed IVA in the Hammam terminology, see Akkermans 1988) is comparable to Amuq E. Though some of the sherds of this phase show the characteristics of transitional Halaf/Ubaid pottery, Red Wash Ware plays no role in the ceramic assemblage. Three (!) tiny rim sherds belonging to shallow bowls (grey, evenly coloured ware, mineral temper) with a red polished wash on both the in- and outside are the only Red Wash Ware sherds that have been found. These were, moreover, not found in the earliest strata but somewhat higher up in the Ubaid strata. Among the painted pottery sherds of Hammam IV, several bow rim jars were found.

The Euphrates Valley. The Euphrates Valley sites Yunus and Shams ed-Din had been abandoned before or during the Late Halaf period and did not, therefore, produce any Red Wash Ware sherds (Woolley 1934; al-Radi 1980). In the case of Turlu the stratigraphy extended into the Late Halaf and transitional Halaf/Ubaid period (levels 4-5) and several bow rim jars were found in level 4, but here, too, no Red Wash Ware was found (Perrot 1964a; Davidson 1977:201-214).

Western Syria. Western Syria and Lebanon do not belong to the Halaf distribution area proper: though imported Halaf sherds were found from the Qoueiq valley in the North to the Biqa'a in the South, these form only a very minor part of the total ceramic assemblage. More important is the locally manufactured painted pottery, manifestedly influenced by Halaf pottery. Other assets of the Halaf material culture like the *tholoi*, as found in the Mosul or Khabur areas and in the Euphrates Valley, are absent west of the Euphrates. The Euphrates may therefore be seen as the western frontier of the Halaf core area.

In western Syria, as in southeastern Turkey and Cilicia, we meet an independent culture, influenced by the Halaf ceramic assemblage after this pottery had reached the Euphrates during the Late Halaf period. This influence was no one-way process: one of the western Syrian pottery features which was not subdued by the Halaf influence but, on the contrary, even spread to the Halaf core area, seems to have been the Red Wash Ware. Western Syria is therefore an area of prime importance for the study of this ware. In contrast to the rather meager easteuphratic evidence, Red Was Ware is a major component of the ceramic assemblage of all sites to the west of the Euphrates: the conspicuous surge in number is shown, for instance, by the fact that nearly half of the sherd sample of Tell Kurdu is composed of this ware. In Ras Shamra IVC, the first stratum in which Red Wash Ware appears, this ware accounts for nearly half of the 'luxury' wares of the assemblage. In the following strata this percentage climbs to nearly 80%. In the Qoueiq Valley, too, nearly half of the Phase D sherds belongs to the Red Wash Ware. The Amuq Basin remains, however, the northern boundary of the Red Wash Ware distribution area: neither in Cilicia nor, for instance, in Sakçe Gözü Red Wash Ware has been found.

The Qoueiq Valley. At Tell Maled a non-authorized profile, 4x4 m large, was cleaned and produced a stratified pottery sample (Gustavson-Gaube 1981). The lower 3 m of this profile were composed of a series of earthen floors, some of them chalk-plastered, strewn with sherds. At the lower edge of the profile architecture (not excavated) was indicated. The uppermost meter was heavily disturbed. The lower part of this profile produced local (West Syrian) Halaf pottery, compared by the excavator to Amuq C-D and Ras Shamra IV C-A wares. Alongside this local Halaf ware a Dark Faced Ware (in a burnished and an unburnished variety) continued into the upper range of the profile. Here, the painted ware was replaced by Red Wash Ware, which had been present in small quantities in the lower levels but dominated the sample in the upper meter. This pottery can be compared to Amuq D and to Ras Shamra IVC-IIIC.

Between 1976 and 1978 a team directed by John Matthers conducted a survey in the Qoueiq Valley (Matthers 1981). The analysis of the prehistoric pottery was entrusted to Mellaart (Mellaart 1981), who divided the chalcolithic material into three phases, C-E, roughly according to the Amuq subdivision.

As a reaction to the exaggerated diffusionism of former days, Mellaart stresses the local character of the Qoueiq C pottery, but it is clear that the - locally manufactured[4] - Halaf-like pottery found on 25 Qoueiq sites was influenced by eastern pottery in its manufacturing technique, forms and decoration. Mellaart's view that western Syria must be seen as 'part and parcel of thc Early Halaf province' (Mellaart 1981:143) is therefore not accepted here. What he calls Early Halaf is in our opinion identical with Mallowan's/ Davidson's Late Halaf. It is our view that Halaf influence did not reach the Euphrates earlier than during the Late Half period. Even older Qoueiq wares like the local Dark Face Burnished Ware were influenced in their forms by the painted ware. The same can be said of the Red Wash Ware which now appears as a new strain in the ceramic assemblage of the Qoueiq Valley. The pottery of the next Qoueiq phase, phase D, collected at only 12 sites, is characterized by a much smaller proportion of painted wares. The Red Wash Ware reaches its apogee. In this phase, too, the forms of the Red Wash Ware (flared collar bowls etc.) are clearly influenced by the Phase C painted ware. Generally speaking, the Phase D Red Wash Ware has more in common with the preceding Phase C than with the next 'Ubaid-like' Phase E. Several bow rim jars occur among the illustrated sherds of this phase (Mellaart 1981, fig. 124:506, 507). As in Aqab, a small number of Qoueiq D Red Wash sherds is adorned with simple painted motifs along the rim.

The Amuq Plain. Although the Late Halaf period and transitional period are also present at Tell Judaideh, the scanty material is eclipsed by the much larger sample at Tell Kurdu (Braidwood and Braidwood 1960:157-167): the imported and locally manufactured Halaf sherds from Trench I amount to 33-45% of the total Phase C sample. The standard forms of the Phase C Halaf (or Halaf-influenced) ware are typical for the Late Halaf period and can not be dated earlier than to the end of the Middle Halaf period. The decoration motifs of this painted ware are all Halaf-inspired, although the more intricate patterns are lacking. They, too, can most easily be attributed to the Late Halaf period; typical for this phase are rim ticks and painted ladders. The next phase, Kurdu D, covers the Halaf/Ubaid transitional phase. In ceramic terms it is characterized by the appearance of several new minor pottery groups although the older Halaf (-influenced) wares are by no means lacking. The earliest Ubaidlike sherds also appear in this phase. Most important is the appearance of two red washed local pottery groups: the 'Wiped-Burnish Ware'

(34-39% of the total Kurdu D sample of 528 sherds) and the 'Red Wash Ware' (6-11%). These differ mostly in the technique of their surface finishing, the 'Red Wash Ware' not being burnished. The paste of the two wares does not show any significant differences. In this article they are therefore combined and both will be called Red Wash Ware. Together they account for nearly 50% of the total Kurdu sample. The Kurdu Red Wash Ware is therefore a major component of the local ceramic assemblage.

The forms of the Kurdu Red Wash Ware follow those of the older Dark Faced Burnished Ware, though the sherds are generally thinner. Straight-sided bowls, large bowls with thickened rims, flared collar bowls and small carinated bowls occur. The majority of the bowls, however, are incurving-rim forms and taller hemispherical forms (hole-mouth vessels). Flared collar jars occur also (Braidwood and Braidwood 1960, fig. 126). The forms of the Red Wash Ware proper, according to the Tell Kurdu terminology, indicate only jars. Among them bow rim jars are a conspicuous and numerous innovation (Braidwood and Braidwood 1960, fig. 127:4, 5).

The Orontes Valley. Of the Hama stratigraphical sequence only Phase L is relevant for our purpose. Phase L was reached in two excavation areas: G11x and I 11 (Ingholt 1940; Fugmann 1958). These produced Halaf influenced pottery which the excavators compared to Arpachiyah TT6 and Chagar Bazar 6-8, i.e. Late Halaf period. As the same layers produced Amuq E pottery as well, we are evidently confronted with a mixed stratigraphy containing a mélange of Amuq C-E material. No wonder then that Phase L distinguishes itself also by the appearance of a minor Red Wash Ware component: 'quelques tessons avec enduit rougeâtre' (Fugmann 1958:14). In addition to the published data mentioned above Ingolf Thuessen kindly informed me of his work on Hama K, L and M: in a personal communication he states that the Phase L material from G11x, so far published only to a certain extent, must be dated earlier than the Phase L material from I 11. He divided the G11x trench into 36 strata the topmost 15 of which he attributed to phase L and strata 16 to 36 to Phase M. According to Thuessen there is no Amuq C layer within the Phase L strata: he regards the few Halaf(-inspired) sherds that were found as not significant. Moreover, the most easily attributable Halaf(-inspired) sherds were found in an Iron Age context! Of course, this does not preclude the possibility of an Amuq C settlement in an unexcavated part of the mound. The G11x material also included a quantity of Red Wash Ware sherds in the Amuq D tradition. A bow rim jar was also found, though belonging to the carelessly painted Ubaidlike pottery, the major component of the G11x phase material.

Elsewhere in the Orontes Valley chance finds in the neighbourhood of Idlib and Huwayiz provided several bow rim jars in the Amuq D tradition (Mellaart 1966:17, fig. 6).

The coastal plain. Along with the Amuq sites, Ras Shamra is the single most important chalcolithic site of western Syria. The stratigraphical sequence on the acropolis has been extensively published in the course of several decades (Schaeffer 1935; 1958; de Contenson 1962; 1969;1973; 1982; Courtois 1962; Kuschke 1962) and provides most useful information on the neolithic and chalcolithic periods of the site, although the use of conflicting designations for each of the sondages has complicated the assessment of the excavators' work. De Contenson's most recent terminology has been adopted here (de Contenson 1982).

In period IV Halaf influence is clearly felt: about 15% of the IV C pottery sample is a locally manufactured Halaf-inspired pottery, and about 5% of the same sample is regarded as Halaf import. The ware, types and decoration of the IV C pottery are mostly of late Middle Halaf or Late Halaf affinity and may be compared to Amuq C.

The next Phases IVB and IVA show a markedly reduced though still considerable proportion of Halaf-influenced painted pottery. In IVA the excavators noticed a degraded technique, while several Halaf types disappear from the repertoire. Now Halaf influenced pottery takes up only 7% of the total number of sherds and Halaf imports are mentioned only once. The decoration motifs of phase IV B and IV A show a growing number of typical Late Halaf or even Transitional specimens and the IV A painting style has affinities with the Kurdu D 'transitional painted' pottery. Concomitant with the reduction of Halaf influence in IV B and IV A is an explosive growth of Red Wash Ware. This ware was already present in IV C, but in IV B the Red Wash Ware component dominates the total sample with 70% of all sherds at the beginning of IVB and 80% at the end. During IV A it keeps this dominant position and even in the next Phase III C Red Wash sherds occur in considerable quantities, though the bulk of the material now shows more affinity with the carelessly painted Ubaid-like pottery of Amuq E (globular forms, no handles, no slip, sand tempered, yellow or greenish paste). The yellowish paste of the Ras Shamra Red Wash Ware is tempered with often rather large mineral inclusions. The ware is fired at a moderately high temperature. The red wash is in most cases evenly spread out over the outer surface of the vessels but sometimes broad irregular strokes occur. The wash is mostly restricted to the outside, but several vessels were found with a wash at both in- and outside.

A new phenomenon, not without importance for the pottery cultures further to the South, is the incised decoration on some of the Red Wash Ware vessels, taking the form of broad bands of pointillé patterns or 'shark teeth'. Painted decoration also occurs, but only as simple bands in black or brown.

The form repertoire of the Ras Shamra Red Wash Ware is rather limited: globular or carinated bowls, bow rim jars or jars with a thickened (renflé) rim. Large flat vertical handles with broadened ends or (less often) vertically attached but horizontally pierced little knobs. The flat bottoms of these vessels often show the imprint of the mats they were made on.

The Biqa'a Valley. The Biqa'a Valley sites were heavily affected by the Amuq C and D ceramic traditions. Part of the Amuq D influence eventually reached Palestine. The stratigraphy of Ard Tlaili (Kirkbride 1969), for instance, shows a 4 m deep Amuq C stratum, reaching from virgin soil to the surface of the shallow tell. Ard Tlaili is therewith the southernmost site where Halaf influence could be detected. On the surface Amuq D pottery was found. There are no traces of a subsequent Amuq E phase. According to Davidson the Amuq C pottery of Ard Tlaili is composed of both imported and locally manufactured wares (Davidson 1977:283). The vessel forms are compared by the excavator to Mersin XIX-XVII, Ras Shamra IV B and C, Amuq C, Byblos Néolithique Moyen, but also to chalcolithic sites in Palestine like Jericho PNB, Munhata 2A and Wadi Rabah.

The decoration motifs on the painted ware are rather simple and consist of series of parallel horizontal and vertical lines, somtimes with rows of dots between them, zig-zags, undulating lines, network patterns.

The painted ware was accompanied by a series of other wares, some of which show marked parallels to North Syrian transitional Halaf/Ubaid ceramic styles, notably with Ras Shamra IV B-A. In the upper strata of the tell and on the surface Red Wash Ware appeares. This part of the stratigraphy may be compared with the Amuq D horizon. Most of the Red Wash sherds originate from bowls, prominent among them Davidson's type 16 (flat based bowls with outcurving rims), known from Ras Shamra IV A. Several bow rim jars occur. Another category is formed by flat based dishes. Part of the pottery is decorated with appliqué cables or relief decoration. Pattern burnished and incised sherds, forming another link with Ras Shamra IV, were also noticed. However, especially as regards the unpainted pottery, many parallels can be drawn with Byblos Néolithique Moyen or contemporary sites in Palestine. This is not meant to say that the Biqa'a has no ceramic tradition of its own, but the importance of Ard Tlaili lies in the fact that it provides clear links with both the northern pottery cultures in Syria and the southern ones in Palestine. It is for this reason that the radio carbon data from Ard Tlaili assume considerable importance: three samples from the Amuq C strata were processed and oscillated around 4900 B.C. This fits well with dates we have from Halaf sites further to the North. The Amuq D stratum at the top of the Ard Tlaili sequence did not produce a radiocarbon date, but just below it, in the topmost Amuq C stratum, a radio carbon date around (K-1431) 4710 B.C. (Kirkbride 1969:55) was recorded. The Amuq D stratum at Ard Tlaili must therefore be dated around the middle of the 5th millenium B.C.,which fits very well with the other data.

Byblos. The Néolithique Moyen pottery from Byblos (Davidson 1977:28./81; Dunand 1961; Dunand 1973; Mellaart 1975:232-243; Moore 1979; Perrot 1969), although at least partly contemporary with Halaf-influenced Ras Shamra IV C and Amuq C, clearly belongs to a different ceramic province than the other Western Syrian sites discussed up till now: both in this phase and in the subsequent Néolithique Récent no convincing traces of any Halaf influence could be detected. Amuq D traits like Red Washed Wares and bow rim jars, on the other hand, abound especially in the Néolithique Récent strata.

The Damascus Basin. The Red Wash Ware ceramic tradition of Ras Shamra and the Amuq reached the Damacus Basin through the Orontes valley where the total pottery assemblage of Tell Khazzami and Tell Ramad is composed of this ware (Tell Khazzami: de Contenson 1968; Tell Ramad (surface finds): de Contenson/van Liere 1964; de Contenson 1966b). Though the paste and surface treatment of this ware copy those of the northern sites, the pottery vessels have a form repertoire of their own, without any sign of Halaf influence like flared collar bowls. As in Byblos deep bowls, jars and hole-mouth jars are the most frequent forms. Decoration restricts itself to incised lines or rows of crescents and appliqué cables (cordons indentés).

Palestine. Descending further to the South by the Orontes Valley and the Biqa'a the Red Wash Ware gets its Palestinian guise ('Wadi Rabah Culture', 'Neolithic 4') and advances to the region's standard ware, slowly evolving into the Ghassulian. While the Halaf ceramic culture influenced sites as far south as Ard Tlaili, the West Syrian Amuq D culture reached even further southwards and had, after a considerable time lag a profound and longlasting influence on many chalcolithic Palestinian sites designated by Moore as 'Late Neolithic 2' (Moore 1973) or 'Neolithic 4' (Moore 1979; 1982): e.g. Wadi Rabah

(Kaplan 1958), Ain al-Jarba IV (Kaplan 1969), Teluliot Batashi (Kaplan 1958:153), Munhata 2A (Perrot 1964 b; 1966), Khirbet Sheikh 'Ali (Praussnitz 1957; 1960), Jericho PNB (Kenyon 1957; Moore 1982).

However, as all available dates[5] show that these sites and others were settled well after the latest possible Halaf phase, no direct Halaf influence can be assumed. Kaplan's view to the contrary is therefore rejected here (Kaplan 1960). As the above-mentioned sites belong to a ceramic tradition of their own, in spite of clear West Syrian influences (red wash, bow rims, mat imprints, appliqué cables), they fall outside the scope of this review and are therefore not discussed in detail.

CHRONOLOGICAL RANGE

Chronologically, the Red Wash Ware ceramic assemblage can not be viewed as a monolithic block, either clearly distinguishable from what comes before and after. The data (mostly stylistic comparisons but also a moderate number of radio carbon dates) are summarized in the chronological table at the end of this article. The details are not discussed here.

As far as reconstructable from the extant data, the appearance of the Red Wash Ware seems to have taken place in the following chronological setting: after an incubation period of several centuries the Halaf culture entered a period of expansion near the end of Middle Halaf period, around 5000 B.C. In the Late Halaf period it appears to have reached the Euphrates Valley. Though the Euphrates remains the western border of the Halaf distribution area in its pure form, Halaf influence does not stop here: besides a small quantity of imported Halaf pottery many imitations in local pottery assemblages are a sign of a considerable Westeuphratic Halaf influence.

This phase (Amuq C) ends around 4500 B.C. when the Halaf/Ubaid transitional phase (Amuq D; 4500-4300 B.C.) shows a reduced Halaf influence concomitant with a growing influx of Red Wash Ware, filling the vacuum. This ware, one of the characteristics of the Halaf/Ubaid transitional phase, becomes an important part of western Syrian pottery cultures during several centuries. At the same time a locally produced but hitherto isolated strain of Red Wash Ware forms part of the ceramic assemblage of the Khabur area.

The Red Wash Ware disappears rather suddenly at the beginning of the next phase, Amuq E, when the transitional Halaf/Ubaid pottery is superseded by the so-called 'Northern Ubaid' or 'Syrian Ubaid'. This process starts somewhat earlier in the Khabur region than to the west of the Euphrates where the transformation seems to have been completed near 4000 B.C. Only in the southern Biqa'a and in Palestine Red Wash Ware continues to be produced and becomes an important characteristic of the Neolithic 4 period, gradually evolving into the Ghassul culture around 3750 B.C.

CONCLUSIONS

When seen from a distance, it seems that the Red Wash Ware distribution area can be divided into three distinct regions: the Khabur Valley in the East, Syria west of the Euphrates and the area south of Ard Tlaili.

All three of these complexes show the influence of local ceramic substrata, most conspicuously in the Khabur area where Halaf influence is clearly felt both in formtypes and decoration. To the West of the Euphrates the Halaf influence is much less felt. On the other hand, as no other site shows such a dominant Red Wash Ware component as Ras Shamra it is my conviction that Western Syria must be seen as the region of origin for this ware. Even more so as no other site, up till now, has produced Red Wash Ware sherds in a context as early as Ras Shamra's. The time span during which Red Wash Ware occurred in the Ras Shamra sequence is longer than at any other known site, as it is reported - as a minor component - even in IIIC, at a stage when the Halaf influence had long since been replaced by Amuq E wares.

The Red Wash Ware assemblage proves convincingly that we may regard the Amuq D phase as a genuine phase in its own right, covering the slow and gradual transition from the Halaf influenced Amuq C to the 'Northern Ubaid' of Amuq E. It is therefore my conviction that the Red Wash Ware has a greater importance for the explanation of chalcolithic culture change in Syria than the sparse data in the literature seem to indicate. Anomalies like the isolated Red Wash Ware of the upper Khabur can only be explained by investing more fieldwork in the hitherto neglected Red Wash Ware of Syria.

NOTES

1. This article is an abridged version of my M.A. thesis, submitted to Maurits van Loon at the University of Amsterdam in 1985. It leans heavily on Davidson's *Regional Variation within the Halaf Ceramic Tradition* (Davidson 1977).
2. With the exception of some notes on the Hama L material, kindly communicated to me by Ingolf Thuessen, Copenhagen.
3. This site is currently being excavated by P.M.M.G. Akkermans (cf. Akkermans 1987). The results of the first campaign (Early Halaf) do not seem to corroborate Copeland's assessment of the site as a transitional Halaf/Ubaid settlement.
4. As proved by a misfired specimen from Tell Berne.
5. For instance Ain el-Jarba IV (end of Neolithic 4): 3740 ± 140 B.C.

REFERENCES

AKKERMANS P.M.M.G.
1988 'An Updated Chronology for the Northern Ubaid and Late Chalcolithic Periods in Syria: New Evidence from Tell Hammam et-Turkman', *Iraq* 50.
AL-RADI S. ET AL.
1980 'A Stone Age Village on the Euphrates: The AUB Rescue Excavations at Shams ed-Din Tannira', *Berytus* 28: 87-126.
BADER N. ET AL.
1981 'Soviet Expedition's Surveys in the Sinjar Valley', *Sumer* 37: 55-95.
BRAIDWOOD R.J. AND BRAIDWOOD L.S.
1960 *Excavations on the Plain of Antioch*, Chicago.
CAUVIN J. AND CAUVIN M.-C.
1972 'Sondage à Tell Assouad', *Annales Archéologique Arabes Syriennes* 22: 85-96.
COPELAND L.
1979 'Observations on the Prehistory of the Balikh Valley, Syria, during the 7th to 4th mill. B.C.', *Paléorient* 5: 251-275.

COURTOIS J.-C.

1962 'Contribution à l'étude des niveaux II et III de Ras Shamra (sondages effectués à l'ouest du temple de Baal 1953)', *Ugaritica* 4: 345-397.

DAVIDSON T.E.

1977 *Regional Variation within the Halaf Ceramic Tradition*. Edinburgh: The University of Edinburgh.

DAVIDSON T.E. AND WATKINS T.

1981 'Two Seasons of Excavation at Tell Aqab in the Jezirah, N.E. Syria', *Iraq* 43: 1-18.

DE CONTENSON H.

1962 'Poursuite des recherches dans le sondage à l'ouest du temple de Baal (1955-60)', *Ugaritica* 4: 477-519.

1963 'New Correlations between Ras Shamra and Al-Amuq', *Bulletin of the American Schools of Oriental Research* 172: 35-40.

1964 'A Further Note on the Chronology of Basal Ras Shamra', *Bulletin of the American Schools of Oriental Research* 175: 47.

1966a 'Notes on the Chronology of the Near Eastern Neolithic', *Bulletin of the American Schools of Oriental Research* 184: 2-6.

1966b 'Seconde campagne à Tell Ramad, 1965, rapport preliminaire', *Annales Archélogiques Arabes Syriennes* 16: 167-174.

1968 'Rapport préliminaire sur les fouilles à Tell Khazzami en 1967', *Annales Archéologiques Arabes Syriennes* 18: 55-62.

1969 'Les couches du niveau III au sud de l'acropole de Ras Shamra', *Ugaritica* 4: 43-89.

1970a 'Sondage ouvert en 1962 sur l'acropole de Ras Shamra, Sondage SH. Rapport préliminaire sur les résultats obtenus de 1962 à 1968', *Syria* 47: 1-23.

1970b 'Rapport préliminaire sur le sondage ouvert en 1962 sur l'acropole de Ras Shamra. Campagnes 1962-68', *Annales Archéologiques Arabes Syriennes* 20: 13-24.

1973a 'Le niveau halafien de Ras Shamra. Rapport préliminaire sur les campagnes 1968-1972 dans le sondage préhistorique', *Syria* 50: 13-33.

1973b 'La culture halafienne du niveau IV de Ras Shamra d'après les campagnes de 1968 à 1972 dans le sondage prehistorique', *Annales Archéologique Arabes Syriennes* 23: 85-103.

1982 'Les phases préhistoriques de Ras Shamra et de l'Amuq', *Paléorient* 8: 95-98.

DE CONTENSON H. AND VAN LIERE J.W.

1964 'Sondages à Tell Ramad en 1963, rapport préliminaire'. *Annales Archéologiques Arabes Syriennes* 14: 109-124.

DUNAND M.

1961 'Rapport préliminaire sur les fouilles de Byblos en 1957/58/59', *Bulletin du musée de Beyrouth* 16: 69-85.

1973 *Fouilles de Byblos*, Tome V, Études et documents d'archéologie, tome VI, Maisonneuve, Paris.

FUGMANN E.

1958 *Hama, II, l'architecture des périodes préhellénistiques*. Copenhagen.

GUSTAVSON-GAUBE C.

1981/1982 'Tall Maled', *Archiv für Orientforschung* 28: 200-247.

INGHOLT H.

1940 *Rapport préliminaire sur sept campagnes de fouilles à Hama en Syrie*. Copenhagen.

KAPLAN J.

1958 'The Excavations at Wadi Rabah', *Israel Exploration Journal*: 149-160.

1960 'The Relation of the Chalcolithic Pottery of Palestine to Halafian Ware', *Bulletin of the American Schools of Oriental Research* 159: 32-36.

1969 'Ein el Jarba Chalcolithic Remains in the Plain of Esdraelon', *Bulletin of the American Schools of Oriental Research* 194: 2-38.

KENYON K.

1957 *Digging up Jericho*. London.

KIRKBRIDE D.
1969 'Early Byblos and the Beqa'a', *MUSJ* 45: 45-59.
KUSCHKE A.
1962 'Bericht über eine Sondage im Palastgarten von Ugarit- Ras Shamra', *Ugaritica* 4: 251-299.
LLOYD S.
1938 'Some Ancient Sites in the Sinjar District', *Iraq* 5: 123-142.
MALLOWAN M.E.L.
1946 'Excavations in the Balikh Valley, 1938', *Iraq* 8: 112- 156.
MALLOWAN M.E.L. AND ROSE J.C.
1935 'The Excavations at Tall Arpachiya, 1933', *Iraq* 2: 1- 178.
MATTHERS J. (ED.)
1981 *The River Qoueiq, Northern Syria, and its Catchment.* Oxford: BAR International Series 98.
MELLAART J.
1966 *The Chalcolithic and Early Bronze Age in the Levant and Anatolia.*
1975 *The Neolithic of the Near East.* London.
1981 'The Prehistoric Pottery from the Neolithic to the Beginning of E.B. IV', in: J. Matthers (ED.), *The River Qoueiq, Northern Syria, and its Catchment.* Oxford: BAR International Series 98. Pp. 131-319.
MERPERT N.L. ET AL.
1981 'Investigations of the Soviet Expedition in Northern Iraq', *Sumer* 37: 22-54.
MOORE A.M.T.
1973 'The Late Neolithic in Palestine', *Levant* 5: 36-86.
1979 *The Neolithic of the Levant, Vol. 1-2* Ann Arbor: University Microfilm International.
1982 'A Four-Stage Sequence for the Levantine Neolithic, ca. 8500-3750 B.C.', *Bulletin of the American Schools of Oriental Research* 246: 1-34.
PERROT J.
1964a 'Note on Turlu', *American Journal of Archaeology* 68:156.
1964b 'Les deux premières campagnes de fouilles à Munhata (1962-1963): premiers résultats', *Syria* 41: 323-345.
1966 'La troisième campagne de fouilles a Munhata (1964)', *Syria* 43: 49-63.
1969 'Le 'Néolithique' du Liban et les récentes decouvertes dans la haute et moyenne vallée du Jourdain', *MUSJ* 45: 135-145.
PRAUSSNITZ M.W.
1957 'Khirbet Sheikh 'Ali', *Revue Biblique* 64: 252-253.
1960 'Tell 'Eli (Kh. esh-Sheikh 'Ali)', *Israel Exploration Journal* 10: 119-120.
SCHAEFFER C.F.A.
1934 'Les fouilles de Ras-Shamra, cinquième campagne (printemps 1933), rapport sommaire', *Syria* 15: 105-136.
1935 'Les fouilles de Ras Shamra - Ugarit, sixième campagne (printemps 1934), rapport sommaire', *Syria* 16: 141-176.
1958/59 'Nouvelles découvertes de Ras Shamra - Les assises prehistoriques d'Ugarit, Première partie', *Annales Archéologique Arabes Syriennes* 8/9: 158-178.
1961 'Les fondements pré- et protohistoriques de Syrie du Néolithique Ancien au Bronze Ancien', *Syria* 38: 221-242.
1962 'Les fondements préhistoriques d'Ugarit', *Ugaritica* 4: 151-251.
SEEDEN H.
1978 'Shams ed-Din Tannira', *Berytus* 26: 165-166.
TOBLER A.J.
1950 *Excavations at Tepe Gawra, II.* Philadelphia.
VAN LOON M.N.
1985 'Hammam et-Turkman on the Balikh: First Results of the University of Amsterdam's 1984 Excavation', *Akkadica* 44: 21-40.
VON OPPENHEIM M.
1943 *Tell Halaf, I.* Berlin.

WOOLLEY L.
1934 'The Prehistoric Pottery of Carchemish', *Iraq* 1: 146-162.
1955 *Tell Atchana*. Oxford.
WRIGHT G.E.
1951 'An Important Correlation Between the Palestinian and Syrian Chalcolithic', *Bulletin of the American Schools of Oriental Research* 122: 52-55.

An orthographic peculiarity in the Akkadian letters of Tušratta

W.H. VAN SOLDT

1 INTRODUCTION

Almost exactly a century has elapsed since the discovery of the cuneiform tablets at El-Amarna, the site of the ancient city of Akhetaton, the short-lived capital of Echnaton. In these 100 years many studies have been devoted to all the aspects on which these letters threw new light, language, history, society, politics, *etc*. The letters could be roughly divided in two groups, namely, (a) the correspondence of the so-called Great-Kings (LUGAL GAL) with the Egyptian pharaohs and (b) the letters of the Palestinian vassals to their Egyptian overlord (Moran 1987:13ff.). Among the letters of the first group three letters were discovered which were written in a hitherto unknown language. According to its Akkadian address, EA 24 was written by the Mittannian king Tušratta. Therefore, the language employed in the letter was likely to be the language of the kingdom of Mittanni (Jensen 1890, 1891 and 1899; Sayce 1890 and 1900; Messerschmidt 1899; Knudtzon 1915:180). The other two, EA 31 and 32, represented the pharaoh's correspondence with Tarḫundaraba, the king of Arzawa, and the language of the letters was consequently called after that country (Knudtzon 1902). Later on, the languages in which these texts were written proved to be Hurrian and Hittite, respectively. In this contribution, which is dedicated to Prof. van Loon, I will concern myself with the Hurrian letter and the Akkadian letters of Tušratta. It is the Hurrian component in this contribution which comes closest to one of Prof. van Loon's main interests, the Urartian language.

2 THE ORTHOGRAPHY OF THE HURRIAN LETTER OF TUŠRATTA (EA 24)

In his study of the Mittanni letter EA 24, Bork, building on the grammatical results obtained by Messerschmidt (1899), reached the important conclusion that the Mittanni letter employs only one syllabic sign for the combination stop + vowel (Bork 1909:9, 15ff.). Thus, he was able to show that the dentals were represented by TA TE TI DU. The gutturals, however, were written with KA KI GI KU GU. Bork was able to explain these apparent doublets as pairs of signs expressing different vowels. Thus, KI was always followed by I and GI by E, KU was always followed by U and GU by Ú. Since Bork had also demonstrated that U stood for /o/ and Ú for /u/, the five signs had to be explained as ka, ki, ke, ko, ku, respectively. Therefore, the Mittanni letter used only one sign for stops where the Akkadian orthography used two or more. Bork inferred from this that the Mittanni language had apparently left unused the possibility of distinguishing between voiced and voiceless, at least in the script, and that it was more concerned with establishing the correct vowel quality (Bork 1909:15, 16). This interpretation of the phonological system became the prevalent theory for the next few decades.

An important new element was added to Bork's reconstruction after the decipherment of the Hittite language by Hrozny in the 1920s. In 1932 Sturtevant published an article in which he followed a suggestion of a student of his, C.L. Mudge (Sturtevant 1932). This student had pointed out that original voiceless stops were usually represented in Hittite by doubled consonants wherever the cuneiform writing made this possible, while single consonants tended to be written to indicate original voiced stops and voiced aspirates. The rule was subsequently incorporated in Sturtevant's grammar, which was published in the following year (Sturtevant 1933).

A significant step forward was made in the thirties after the publication of recently discovered syllabic and alphabetic cuneiform texts from the ancient city of Ugarit, modern Ras Shamra on the Phoenician coast. In 1931, Thureau-Dangin published a tablet that contained the Ras Shamra version of ḫarra-ḫubullu II. Instead of the usual Akkadian, however, the tablet had a column with Hurrian words as translation for the Sumerian. From the way in which these words were spelled Thureau-Dangin inferred that Bork's hypothesis was no longer tenable (Thureau-Dangin 1931:252). For various reasons he concluded that the lack of opposition between the voiced and voiceless consonants was only apparent and that it was a consequence of the Hurrian syllabary, which, just like the Old Akkadian syllabary, did not distinguish between voiced and voiceless. The main argument for Thureau-Dangin's hypothesis, however, would not come from the syllabic texts but from the alphabetic texts found at Ras Shamra. Already in 1932 Hrozny published an article in which he demonstrated the language of one of these tablets to be Hurrian (Hrozny 1932, RS 2.-- = KTU 1.60). Since the Ugaritic alphabet was more consistent in its spellings of voiced and voiceless consonants than the syllabic script, this discovery paved the way for a new approach to the Hurrian phonological system.

In the following years (1933-37) only a few articles on the subject were published. Friedrich (1935) was the first one to demonstrate the importance of the alphabetic spellings. He noted the consistent use of voiced stops in certain cases, especially between vowels and after liquidae. At the beginning of words only the voiceless consonants were used and Friedrich even hinted at the possible validity of the Sturtevant rule for Hurrian (Friedrich 1935:130, 131 note 1). More outspoken was von Brandenstein in his study of the alphabetic material two years later (1937:574). He confirmed Friedrich's observations, including the Sturtevant rule. Due to the uncertainty in the interpretation of some of the alphabetic signs, especially ǵ, ḏ and ẓ, Friedrich and von Brandenstein were unable to unravel all the orthographic peculiarities. However, during the latter half of the thirties a sudden increase in interest for the subject took place and, as a consequence, by 1941 most of the mysteries were solved. Mainly responsible for this development was the discovery in the late twenties of a large number of texts on the site of the ancient city of Nuzi. Not a single text in Hurrian was unearthed at Nuzi, but a large percentage of the population bore Hurrian names and many features in the Akkadian of the Nuzian tablets had to be explained as interference from the local tongue, Hurrian. In 1937 Berkooz published his study of the Nuzi dialect of Akkadian. He concentrated on the orthography and phonology and his most important conclusion was, that the confusion of the signs for stops could have resulted from the fact that the Hurrian stops were different from those in Akkadian (1937:41). It is during the years 1938-1941 that the greatest progress on Hurrian phonology was made. Two names in particular are to be mentioned in this connection, Speiser and Purves. In one of his first articles dealing with Hurrian phonology, Speiser made valuable observations on the Hurrian sibilants (1938). Most of his deductions, however, have more relevance for the Akkadian of Nuzi and will be taken up again in section 3. As to the alphabetic signs from Ugarit, Speiser concluded that there was a pair ḫ-ǵ (the quality of the last character was still debated)[1]. In his discussion of the sign ḏ, he followed Harris, who had been the first to determine the voiced nature of this sibilant but who regarded it as a Hurrian affricate. The voiceless counterpart of ḏ, the sign ṯ, was transcribed as θ by Speiser, thereby indicating its cognation with Arabic ṯa (Speiser 1938:179). A year later he published his thorough study on Hurrian morphology, in particular the verbal system (Speiser 1939; cf. also Goetze 1940). The years 1940-41 meant a real break-through for Hurrian orthography and phonology. First, there was Purves' study on scribal idiosyncracies at Nuzi (1940). Purves found that the Akkadian written by the earliest generation of scribes, represented by Apil-Sin, Amurru-šar-ilāni and Ṭāb-milk(i)-abi, was good Babylonian in paleography, orthography and language (Purves 1940:171ff.). As opposed to scribes of later generations they did not confuse the stops and in their spelling of Hurrian personal names they followed the alphabetic orthography from Ugarit. This led Purves to the important conclusion that the voicing of single Hurrian consonants was only attested in texts written by Semitic scribes and should be attributed to their perception of Hurrian sounds (Purves 1940:172, 180 and especially 184ff.; for the chronology, see Friedman 1987).

The final verdict on the orthography came from Speiser (1940). In just over 20 pages he showed that in Hurrian the Sturtevant rule not only applied to stops, but also to sibilants and the velar ḫ, in his words: 'Where sounds were differentiated according to voice, double writing was used to represent the voiceless sounds. This method was applied to *p t k*; to a spirant (written *-šš-*) which was intermediate between Semitic [ṯ] and a sibilant; in all likelihood to *ḫ*, and probably also to the voiceless counterpart of *z* .' The same conclusions were reached by Purves in an article published a year later (Purves 1941). These studies, as well as the continuing study of the Ugaritic material also helped to determine the likeliest phonetic shape of Ugaritic consonants. Ugaritic was shown to possess the consonantal pairs ḏ[2]-ṯ (Herdner 1942/3, 1944/5 and 1949; see also Rainey 1971:157 and Rustum Shehadeh 1968:18ff.) and ǵ-ḫ (Speiser 1940:334ff.; Purves 1941:395ff.).

All these newly attained views were finally put together by Speiser in the second chapter of his monumental *Introduction to Hurrian* (1941). With this work, the debate about the Hurrian orthography came to a sudden end and later research has only been able to offer new suggestions for the phonetic shape of the Hurrian consonants. The rules as summarized by Speiser have not been questioned.

Twenty-three years after the publication of Speiser's grammar a new description of the Hurrian language was undertaken by Bush (1964). In general, Speiser's views were confirmed, but Bush, taking up Bork's discovery of the limited sign-inventory in the Mittanni letter (see above), emphasized that EA 24 indeed uses only one set of signs to represent the stop phonemes (Bush 1964:19ff., 293 note 27), and the same goes for the sibilant š and the velar ḫ. Bush' successors were mainly concerned with the origin and the pronunciation of the consonants which were subject to the Sturtevant rule. Thus,

Diakonoff suggested that Hurrian had lost one set of stops which were still present in Urartian, the emphatics. Moreover, while Urartian had the opposition voiced - voiceless, the Hurrian consonants showed an opposition single - double (Diakonoff 1971:40ff.; cf. Diakonoff/Starostin 1986, 13ff., 'non-intensive' vs. 'intensive'). That the opposition was one of quantity, rather than of quality had already been advocated by Purves and Speiser (see above), followed by Bush (1964:47ff.).

A different approach to the problem was followed by Thiel. In his description of the Hurrian grammar he defined the oppositions emanating from the Hurrian texts in terms of tense and lax (Thiel 1975:116ff.). Thus, a single consonant was to be regarded as lax, a doubled consonant as tense. Thiel reconstructed a 'minimal system' and a 'maximal system' for the Hurrian consonants. Of these two, the first one more or less complies with traditional views of the Hurrian phoneme inventory, be it that the labial fricatives were regarded by Thiel as positional variants of the labial stops (Thiel 1975:120, 134ff.).

3 THE AKKADIAN LETTERS OF TUŠRATTA

The interest shown for the letters of Tušratta written in Akkadian has mainly been historical. The letters represent the correspondence of the Mittannian king with the Egyptian pharaohs Amenophis III and IV and shed light on the period just before the downfall of the Mittannian state (Kühne 1973:17ff.; Wilhelm 1982:41ff.). The Akkadian in these letters has been more of an aid in deciphering the Hurrian letter of Tušratta (EA 24) than a subject of study in itself (see, for instance, Messerschmidt 1899:3; Speiser 1941:183). The situation is quite different with regard to the Akkadian written by Hurrians at Nuzi, on which subject already in the thirties a whole series of articles was published (Wilhelm 1970:11). Some scholars went as far as regarding this Akkadian as 'a third Akkadian dialect' (Goetze 1938:143). Of course, it was quite clear to all scholars that the daily tongue of the Nuzian population was Hurrian, rather than Akkadian (Berkooz 1937:5; Wilhelm 1970:8, 11).

Speiser gave a first outline of the Akkadian of Nuzi (1936). Berkooz published a study on the orthography and phonology of the Nuzi tablets and noticed a striking resemblance in the treatment of the stops with the Hurrian letter of Tušratta (Berkooz 1937:40, see above, 2.). Goetze observed that the S-signs were used occasionally for š and that the syllabary as a whole could easily have been derived from the Old Akkadian syllabary (Goetze 1938:137), despite the fact that the language was close to Middle Babylonian (Goetze 1938:138). Both Goetze and Oppenheim noticed strong Hurrian influence on morphology and syntax (Goetze 1938:140; Oppenheim 1936/7:60). Speiser pointed out that wherever S-signs were used for Akkadian š, this was always done where the reconstructed Proto-Semitic form requires š, not ṯ (Speiser 1938:187ff.). On this basis, Speiser also claimed an Old Akkadian ancestry for the Nuzi syllabary.

A big step forward was the study on the early scribes of Nuzi by Purves, mentioned in section 2. As opposed to the Akkadian written by the earlier scribes (see above), the Akkadian of the later scribes (from the generation of Teḫip-tilla) showed strong Hurrian influence (Purves 1940:185ff.) and many of the characteristics of the Akkadian letters of Tušratta can already be seen in the texts from Nuzi.

Interest in the Akkadian from Nuzi suffered the same fate as the interest in Hurrian (see section 2). After 1940 few studies on the Akkadian from Hurrian-speaking areas were published. Only the publication of the Alalakh tablets in the fifties induced the publication of a few articles (Wilhelm 1970:12). Interest started coming back in the sixties and seventies. Dietrich and Loretz published interpretative articles on Hurrian terms from Alalakh (1969; 1969/70; 1970) and in 1970 Wilhelm published his collection of studies on the 'Hurro-Akkadian' of Nuzi. Many features of the Nuzi dialect with a Hurrian background were discussed in detail in his book and a special chapter dealt with the Hurrian influence on Akkadian syntax (Wilhelm 1970:61ff.).

Then, in 1976 a study by Adler especially devoted to the Akkadian written in the letters of Tušratta appeared. Adler provided a grammar of the letters, new transliterations and a glossary. In his first chapter he gave a survey of the syllabary and the orthography. In general, Adler followed the views of earlier scholars with regard to the spelling of voiced and voiceless consonants. He also noticed that differences in preference for signs incorporating these consonants existed. Thus, for instance, he observed that the sign PA occurs more often than BA (Adler 1976:4). However, no attempt at a systematic count of the attestations of signs used for stops was undertaken by Adler and the impression one gets is still one of total confusion. It is this state of apparent confusion which I hope to clarify in the following section.

4 THE SYLLABARY OF THE TUŠRATTA LETTERS

As observed by Adler, spellings of Akkadian words varied greatly in the Tušratta letters (see above). Thus, one word could be spelled with different signs even within the same text. Compare for instance: *da-an-ni-iš* and *ta-an-ni-iš* (EA 19:8 and 31); *ú-ub-ba-lu* and *ú-ub-pa-lu* (EA 20:25 and 20); *da-ma* and *ta-ma* (EA 22 I:20 and 25, *etc.*); *na-da-ni* and *na-ta-a-ni* (EA 27:49 and 45); *li-iš-gu-un* and *iš-ku-na-an-ni* (EA 29:181 and 111, 160).

Apparently, some words were more susceptible to 'faulty' spellings than others. Thus, for instance, the verb *ma'ādu* with its derivatives *ma'diš* / *mādiš*, *ma'du* / *mādu* is regularly spelled with the signs TI and TA (Adler 1976:297).

In order to determine what might be the reason for this confusion, I will give a cumulative table of occurrences from all Akkadian Tušratta letters with the exception of the two lists of presents, EA 22 and 25. These two texts have been excepted in view of the many uncertainties concerning the nature of the consonants of many of the listed items. The tables are organized according to Akkadian phoneme.

A The dentals

/d/		DA	TA	(DE)[3]	TE	DI	TI	DU	TU
EA	17	–	2	–	–	–	1	7	–
	19	8	3	–	1	2	6	6	1
	20	9	8	–	–	2	–	3	3
	21	1	1	–	–	–	1	2	–
	23	–	–	1	–	–	–	2	
	26	–	3	–	4	–	2	1	–
	27	1	9	–	1	1	–	8	–
	28	2	–	1	–	–	–	1	1
	29	2	5	–	1	13	6	8	–
	30	–	–	–	–	–	–	1	–
Total		23	31	2	7	18	16	39	5

/t/		DA	TA	(DE)	TE	DI	TI	DU	TU
EA	17	–	16	–	6	–	17	1	3
	18	–	1	–	–	–	–	–	–
	19	2	30	–	20	–	19	3	1
	20	–	16	–	12	–	23	1	–
	21	–	9	–	3	–	3	–	–
	26	–	17	–	8	–	2	1	1
	27	2	46	–	16	–	20	12	–
	28	–	14	–	1	–	12	–	–
	29	–	75	–	37	–	66	6	5
	30	–	2	–	–	–	1	–	–
Total		4	230	–	105	–	183	26	10

/ṭ/		TA	DA	(DE)	TE	DI	TI	DU	TU
EA	17	3	–	–	1	–	–	–	–
	19	–	2	–	–	–	1	–	–
	20	1	–	–	–	–	–	–	–
	21	–	1	–	1	–	–	–	–
	23	–	1	–	–	–	–	–	–
	26	–	1	–	–	–	–	–	–
	27	–	4	–	–	–	–	–	–
	29	–	2	–	4	–	2	1	–
Total		4	11	–	6	–	3	1	–

B The gutturals

/g/		GA	KA	QA	GI	KI	QI	GU	KU	QU
EA	21	–	–	–	–	–	–	1	–	–
	28	–	1	–	–	–	–	–	–	–
	29	–	–	–	1	–	–	–	–	–
Total		–	1	–	1	–	–	1	–	–

/k/		GA	KA	QA	GI	KI	QI	GU	KU	QU
EA	17	–	14	1	–	6	–	–	5	–
	18	–	2	–	–	–	–	–	–	–
	19	–	17	–	–	20	–	–	8	–
	20	–	18	–	–	14	–	2	3	–
	21	–	14	–	–	4	–	–	–	–
	23	–	15	–	2	3	–	–	–	–
	26	–	15	–	–	8	–	2	1	–
	27	–	32	–	–	15	–	6	4	–
	28	–	20	–	–	3	–	2	1	–
	29	–	57	–	1	31	–	2	15	–
	30	–	–	–	–	1	–	–	–	–
Total		–	204	1	3	105	–	14	37	–

/q/		GA	KA	QA	GI	KI	QI	GU	KU	QU
EA	17	–	–	2	–	1	–	–	–	–
	18	–	–	–	–	1	–	–	–	–
	19	–	–	1	1	1	–	–	–	–
	20	–	2	2	1	1	–	–	–	–
	21	–	–	–	–	1	–	–	–	–
	23	–	–	–	–	1	–	–	–	–
	26	–	–	–	–	1	–	–	–	–
	27	–	–	–	–	–	–	–	1	4
	28	–	–	–	–	1	–	–	–	–
	29	–	1	2	–	4	–	3	5	–
Total		–	3	7	2	12	–	3	6	4

C The labials
Knudtzon, in his edition of the Mittanni letter EA 24 (1902b: 134ff.), read the sign PI as pi. Messerschmidt took over this transliteration although he was convinced that the actual reading had to be different (Messerschmidt 1899:2). Already in 1890, Jensen had established the correct reading on the basis of writings like ḫur-ru-u-ḫé and ḫur-PI-u-ḫé, where obviously a value *wu* had to be assumed (176ff.). His results were taken over by Bork (1909:21ff.) and the reading wA (i.e. wa, we, wi, wo, wu) has since been generally accepted (Thureau-Dangin 1931:256ff.). This means that the Mittanni letter only used five signs to indicate labial stops: BA, PA, BE, BI and BU. The same holds true for the Akkadian letters of Tušratta (Adler 1976:18ff.) and the Akkadian texts from Nuzi (Berkooz 1937:13; Gordon 1938:34). Thus, we are left with but one pair of signs, PA and BA.

/b/		BA	PA	/p/		BA	PA
EA	17	4	3				
	18	–	1				
	19	1	1	EA	19	–	3
	20	8	7		20	–	8
	21	3	–		21	–	1
	26	–	1		26	–	7
	27	–	2		27	–	4
	28	4	2		28	–	1
	29	2	22		29	–	18
	30	–	2				
Total		22	41			–	44

A number of interesting conclusions can be drawn from these tables:
1. The distribution of signs to write oppositional pairs of stops does *not* show a balanced picture. On the contrary, within a pair one of the signs is usually preferred.
2. The preference of certain signs over other signs leads to a difference in treatment of the Akkadian stop phonemes. The preferential sign was used for the Akkadian stop that it was supposed to stand for almost without exception. However, it also penetrated into the orthography of related stops and was used side by side with the signs to be expected.

A good case in point is the spelling of /b/ and /p/ (c). The table shows that, of the two signs PA and BA, PA is the sign preferred by the Mittannian scribes (85 occurrences to 22). However, all the 22 attestations of BA apply to cases where Akkadian grammar prescribes /b/. Whenever the scribes wanted to write /p/, they wrote PA consistently. To isolate the preferential signs more clearly, we can draw up a table of grand totals.

Dentals	DA	TA	(DE)	TE	DI	TI	DU	TU	
	31	**272**	2	**118**	18	**202**	**65**	15	
Gutturals	GA	KA	QA	GI	KI	QI	GU	KU	QU
	–	**208**	8	6	**117**	–	18	**43**	4
Labials	BA	PA							
	22	**85**							

This table shows that the signs preferred are TA TE TI DU, KA KI KU and PA. The choice of these signs is not fortuitous and reminds one of the syllabary employed in the Mittanni letter EA 24 (Bork 1909:18; Bush 1964:20):

	-a	-e	-i	-o	-u
Dental	TA	TE	TI	DU	DU
Guttural	KA	GI	KI	KU	GU
Labials	PA	(BE	BI	BU	BU)

It is obvious that the source of the preference of certain signs in the Akkadian letters is to be found in the Mittanni letter. The preferential signs prove to be identical with the signs used in this letter. Thus, we are able to deduce the following general rule for the Akkadian letters: *Whenever a scribe deviated from the spelling tradition current in contemporary Mesopotamia, he almost always did so by substituting the nearest sign of his own Hurrian syllabary.*

A few points still have to be cleared up. In the first place, there is the problem of the vowels e and i. For the gutturals, the Mittanni letter EA 24 used the signs GI = /ke/ and KI = /ki/, respectively. The same is true for the Akkadian letters, GI is used to indicate a guttural + e, while KI consistently stands for a guttural + i. This is especially clear in cases where an extra vowel sign is added, as in KI-*i* and KI-*i-me-e* (passim, cf. Adler 1976:288ff.) and *a-na-ak*-GI-*e-er* (29:74) and *le*-GI-*e* (20:9)[4]. Therefore, the rules which determined the use of GI and KI ran along other lines than those for, for instance, DA and TA and the two signs are better left out of the discussion.

Different is the treatment of the sequence guttural + u. The Mittannian syllabary used the signs U = /o/, Ú = /u/, KU = /ko/ and GU = /ku/ for EA 24. As shown above, the sign KU occurs much more frequently than GU. Does this mean that Akkadian had a vowel /o/? Proof for this could be gathered from attestations of the sign U. However, this sign does not occur in the Akkadian letters. Wherever the vowel /u/ is written plene, the sign Ú is used. Apparently, the Mittannian scribes recognized only the vowel /u/ in contemporary Akkadian and no allophones which came close to their U = /o/. The only solution to the problem is, that the sign KU could also be used for a guttural + u, as against Mittannian practice. Proof for this is the sequence KU-ú, a spelling impossible in EA 24, cf. *il*-KU-*ú* (29:23) and *up*-KU-*ú-tu*$_4$ (29:50). The frequent use of GU (often combined with ú) for /k/ and /q/, however, reminds us that the Mittannian rule was not totally abandoned[5].

A last word on another characteristic of the Mittanni letter EA 24, the opposition of single and double spellings for stops and fricatives (see above, 2). No such feature can be found in the Akkadian letters. The rules prescribed by Akkadian grammar are generally followed, the Sturtevant rule does not apply (Adler 1976:16ff.).

5 THE MITTANNIAN SYLLABARY

The origin of the syllabary used for the Mittanni letter EA 24 cannot be traced with certainty. As observed by Purves, the early scribes at Nuzi were of Babylonian origin and the idiosyncracies typical for the Nuzi texts only gradually came into being and were in full use from the generation of Teḫip-tilla on (Purves 1940). One of the most outstanding characteristics of the Nuzi syllabary is the use of GI for a guttural + e and KI for a guttural + i, a phenomenon well-known from the Tušratta letters (see above, 4, and Berkooz 1937:10ff.; Purves 1940:171). Thus, the development at Nuzi could have been one of local origin, which made its way into neighbouring (Hurrian-speaking) areas and was taken over by the Hittites (Riemschneider 1973:276ff.). However, the phenomenon is earlier attested outside Nuzi and occurs, for instance, in a number of Old Babylonian letters[6]. This may serve as a warning that the origin will have to be sought elsewhere, possibly in Mesopotamia itself[7]. However this may be, the scribes of Tušratta's letters seem to have made a deliberate choice, using only certain syllabograms for their Hurrian syllabary while discarding others, apparently on arbitrary grounds[8].

Is the origin of the Mittanni syllabary rather obscure, not much better is our knowledge of the period after the disappearance of the Mittannian state. Nevertheless, we have two texts from Ugarit which still show the characteristics of the Mittannian syllabary, RS 11.853 (PRU 3, 327ff.) and RS23.31 (unpublished). Especially the latter shows a close adherence to Mittannian practice[9]. Both letters probably date to the latter half of the 13th century, thus bearing witness that the syllabary was still in use more than a century after Mittanni's downfall. Since letter 11.853 originates at Karkemish, it is possible that Hurrian traditions were carried on at this Syrian capital. The origin of the second letter is unknown.

6 THE SPREAD OF THE MITTANNIAN SYLLABARY

Although a survey of all the text groups that show features characteristic of the Mittannian syllabary (Nuzi, Boghazköy, Alalakh, Amarna) is beyond the scope of this contribution, I will make a few remarks on one particular group of texts, those found at Ugarit.

A large number of the Akkadian texts from Ugarit can be dated with some precision due to the mentioning of the king in all legal texts preserved at the Central Palace. As I have tried to show elsewhere (van Soldt 1986:368ff.), there is a marked difference in orthography between texts written under early Ugaritic kings (especially Niqmaddu II and Arḫalba) and those written under later kings like Niqmepa' and especially his son 'Ammiṯtamru II. Some of the characteristic features of this early phase are relevant for our topic (van Soldt 1986:382):

1. The occasional use of TA TI (and sometimes TU) for /d/ and DU for /t/.
2. The frequent use of GI to write /qe/ and, occasionally, /ke/.
3. The frequent use of PA for /ba/.

The later texts use a syllabary much closer to the one current in contemporary Mesopotamia (Assyria as well as Babylonia). As an example, I have listed the numbers of attestations for the various signs used to write the syllable /qe/ in forms of the verb leqû.

/qe/	GI	KI	QI
Niqmaddu II (? -1315[10])	31	–	–
Arḫalba (1315-1313)	4	–	–
Niqmepa‘ (1313-±1263)	15	5	–
‘Ammittamru II (±1263-±1220)	39	12	–
Ibirānu (±1220- ?)	3	–	1
Niqmaddu III (?-?)	–	–	4
‘Ammurapi’ (?-±1175)	–	–	–

The table shows an interesting distribution of the signs GI, KI and QI. During the early years of Ugarit's history only GI was used, in accordance with the Mittanni syllabary. Under Niqmepa‘ Babylonian KI was introduced and was used alongside GI. The same situation existed under ‘Ammiṯtamru II. Under Ibirānu we find the first occurrence of the Assyrian QI and this sign becomes prevalent during the final phase of Ugarit's existence.

In view of this development and the other characteristics mentioned at the beginning of this section, it appears that in the earliest phase the influence of the Mittannian syllabary was strong. After the first generation this influence started to wane and to give way to new writing habits, probably under the influence of new material originating in Babylonia and Assyria. The disappearance of the Mittannian state, which had served as an intermediary between Mesopotamia and the west, is almost certainly to blame for this development. However, the exact ways through which the teachers and their study material reached peripheral areas such as Ugarit remains largely beyond our grasp.

NOTES

1. The same opposition was deduced for Hittite by Sturtevant (1940).
2. Of course, in Ugaritic ḏ had shifted to d in most of the cases (see Gordon 1965:26ff).
3. Although there is no special sign DE, I have listed those cases where the sign DI is used to write /de/.
4. Compare also the two forms of *kabāsu*, EA 23:23,24.
5. For instance, *ib-te-* GU *-ú* (20:50), *ša-ap-* GU *-ú-ti* (26:37) and *dâm-* GU *-ú-tu*$_4$ (29:185). The sign U occurs twice in the list EA 25, *lu-u-re-me-ti* (*nurmêti*, II:4) and *ku-u-ḫa-as-sú* (*guḫassu*, II:29). This could mean that the Hurrian scribes perceived the back-vowel in these words as [o]. Note that both words are of unknown origin.
6. AbB 2, 132:13; 9:19; AbB 11, 4:6'; BM 97562:24 (all *leqûm*) and AbB 9, 119 R:8' (*qerēbum*).
7. See Gelb (1936/37:40), for parallels in Old Akkadian. As stated above (section 2), the Akkadian of Nuzi was linked to Old Akkadian on paleographic as well as orthographic grounds (Goetze 1938:135ff.; Speiser 1938:189ff.; Speiser 1941:13; Bush 1964:28ff.).
8. Why, for instance, choose the sign KA instead of the much shorter QA, a sign often used for /ka/ in peripheral texts (Kühne 1974:7, note 35).

9. Cf. forms like *ḫa-u-ša-aš-šu* (ḫa.ot.a.tt.u, 7), *ḫa-še-e-ta* (ḫat.et.a, 10) and *in-na-at-ta-ni-in* (in-na.tta.nin, 17).
10. For these dates,see van Soldt 1987:73.

REFERENCES (ABBREVIATIONS ACCORDING TO CAD)

ADLER H.-P.
1976 *Das Akkadische des Königs Tušratta von Mitanni*, AOAT 201. Neukirchen-Vluyn.

BERKOOZ M.
1937 *The Nuzi Dialect of Akkadian, Orthography and Phonology*. Language Dissertations (Supplement to Language) no. 23, Philadelphia.

BORK F.
1909 *Die Mitannisprache*, MVAG 14/1-2.

BRÜNNOW
1890 *Die Mitâni Sprache*, ZA 5, 209ff.

BUSH F.W.
1964 *A Grammar of the Hurrian Language*. Ann Arbor.

DIAKONOFF I.M.
1971 *Hurrisch und Urartäisch*. München.

DIAKONOFF I.M. AND STAROSTIN S.A.
1986 *Hurro-Urartian as an Eastern Caucasian Language*, Münchener Studien zur Sprachwissenschaft, Beiheft 12, Neue Folge. München.

DIETRICH M. AND LORETZ O.
1969 'Die soziale Struktur von Alalaḫ und Ugarit (V). Die Weingärten des Gebietes von Alalaḫ im 15.Jahrhundert', *Ugarit-Forschungen* 1: 37-64.
1969/70 'Die soziale Struktur von Alalaḫ und Ugarit (II). Die sozialen Gruppen ḫupše-namê, ḫaniaḫḫe-ekû, eḫele-šūzubu und marjanne nach Texten aus Alalaḫ IV', *Welt des Orients* 5: 57-93.
1970 'Die soziale Struktur von Alalaḫ und Ugarit (IV)', *Zeitschrift für Assyriologie* 60: 88-123.

EINARSSON S.
1932 'Parallels to the Stops in Hittite', *Language* 8: 177- 182.

FRIEDMANN A.H.
1987 'Toward a Relative Chronology at Nuzi', in: *Studies on the Civilization and Culture of Nuzi and the Hurrians*, Volume 2: 109-129.

FRIEDRICH J.
1935 'Zum Subaräischen und Urartäischen', *Analecta Orientalia* 12: 122-135.

GELB I.J
1936/37 'Additional Akkadian Values', *American Journal of Semitic Languages* 53: 34-44, 180-187.

GOETZE A.
1938 'Some Observations on Nuzu Akkadian', *Language* 14: 134- 143.
1940 'The Ḫurrian Verbal System', *Language* 16: 125-140.

GORDON C.H.
1938 'The Dialect of the Nuzu Tablets', *Orientalia* 7: 32-63, 215-232.
1965 *Ugaritic Textbook*, Analecta Orientalia 38. Rome.

HARRIS Z.S.
1935 'A Hurrian Affricate or Sibilant in Ras Shamra', *Journal of the American Oriental Society* 55: 95-100.

HERDNER A.
1942/3 Review of J. Pedersen, *Die Krt Legende*, *Syria* 23: 281- 282.
1944/5 Review of C.H. Gordon, *Ugaritic Grammar*, *Syria* 24: 113- 114.
1949 'Note Lexicographique: Ras-Shamra mr', *Syria* 16: 161.

HROZNY B.
1932 'Une Inscription de Ras-Šamra en Langue Churrite', *Archiv Orientální* 4: 118-129.
JENSEN P.
1890 'Vorstudien zur Entzifferung des Mitanni', *Zeitschrift für Assyriologie* 5: 166-208.
1891 'Vorstudien zur Entzifferung des Mitanni II', *Zeitschrift für Assyriologie* 6: 34-72.
1899 'Zur Erklärung des Mitanni', *Zeitschrift für Assyriologie* 14: 173-181.
KNUDTZON J.A.
1902a *Die zwei Arzawa-Briefe, die ältesten Urkunden in indogermanischer Sprache*.Leipzig.
1902b 'Ergebnisse einer Kollation der El-Amarna–Tafeln, *Beiträge zur Assyriologie,* 101-154.
1915 *Die El-Amarna-Tafeln*, VAB 2. Leipzig.
KÜHNE C.
1973 *Die Chronologie der internationalen Korrespondenz von El-Amarna*, AOAT 17. Neukirchen-Vluyn.
MESSERSCHMIDT L.
1899 *Mitanni-Studien*, MVAG 4/4.
MORAN W.L.
1987 *Les Lettres d'El Amarna, Correspondance Diplomatique du Pharaon*. Paris.
OPPENHEIM A.L.
1936/7 'Zur Landessprache von Arrapḫa-Nuzi', *Archiv für Orientforschung* 11: 56-65.
PETERSEN W.
1933 'Hittite and Tocharian', *Language* 9: 12-34.
PURVES P.M.
1940 'The Early Scribes of Nuzi', *American Journal of Semitic Languages* 57: 162-187.
1941 'Hurrian Consonantal Pattern', *American Journal of Semitic Languages* 58: 378-404.
RAINEY A.F.
1971 'Observations on Ugaritic Grammar', *UF* 3: 151-172.
RIEMSCHNEIDER K.K.
1973 'Zur Untersuchung der Vokale e und i in der hethitischen Orthographie', in: *Festschrift H. Otten*, Pp. 273-281.
RUSTUM SHEHADEH L.A.
1968 *The Sibilants in the West Semitic Languages of the Second Millenium B.C.*. Cambridge.
SAYCE A.H.
1890 'The Language of Mitanni', *Zeitschrift für Assyriologie* 5: 260-274.
1900 'The Language of Mitanni', *PSBA* 22: 171-225.
SPEISER E.A.
1936 'One Hundred New Selected Nuzi Texts', *Annual of the American Schools of Oriental Research* 16: 131-142.
1938 'Notes on Hurrian Phonology', *Journal of the American Oriental Society* 58: 173-201.
1939 'Studies in Hurrian Grammar', *Journal of the American Oriental Society* 59: 289-324.
1940 'Phonetic Method in Hurrian Orthography', *Language* 16: 319-340.
1941 *Introduction to Hurrian*, AASOR 20, New Haven.
STURTEVANT E.H.
1932 'The Development of the Stops in Hittite', *Journal of the American Oriental Society* 52: 1-12.
1933 *A Comparative Grammar of the Hittite Language*. Philadelphia.
1940 'Evidence for Voicing in Indo-Hittite γ', *Language 16, 81ff.*
THIEL H.J.
1975 'Phonematik und grammatische Struktur des Hurrischen', in: *Das hurritologische Archiv* (Corpus der hurri(ti)schen Sprachdenkmäler) des Altorientalischen Seminars der Freien Universität Berlin, Pp. 98-239.
THUREAU-DANGIN F.
1931 'Vocabulaires de Ras-Shamra', *Syria* 12: 225-266.
VAN SOLDT W.H.
1986 *Studies in the Akkadian of Ugarit: Dating and Grammar*. Leiden.

1987 'The Queens of Ugarit', *Jaarbericht Ex Oriente Lux* 29: 68-73.
VON BRANDENSTEIN C.-G.
1937 'Zum Churrischen aus den Ras-Schamra-Texten', *ZDMG* 91: 555-576.
WILHELM G.
1970 *Untersuchungen zum Hurro-Akkadischen von Nuzi*, AOAT 9. Neukirchen-Vluyn.
1982 *Grundzüge der Geschichte und Kultur der Hurriter*. Darmstadt.

Tišpak, his seal, and the dragon mušḫuššu

F.A.M. WIGGERMANN

A. THE *LABBU-MYTH*

Tišpak, his seal, and a monster play a part in a myth imperfectly preserved on a tablet from the library of Assurbanipal, Rm. 282, copied by Delitzsch (*AW* 390f.) and King (*CT* 13:33f.), and translated and annotated by A. Heidel, *The Babylonian Genesis*[2] (1951):141f., who entitled it 'The Slaying of the *Labbu*'. Later Jacobsen (1932:53f.) and Lambert (1986:55f.) contributed important insights towards its understanding, but an up-to-date edition is yet to appear (for older editions see Heidel 1951: 141[1]). The entry 107, Le Mythe du *Labbu*, in the bibliography of J. Bottéro (1985:345) refers to a brief paragraph in the original text of *Annuaire* 1978-79, 86, omitted from the re-edition of his article in *Mythes et Rites de Babylone*, 279ff. The text attracted some attention from scholars outside the field of Assyriology, notably J. Fontenrose, (1959:608, Index), and N. Forsyth (1987:44f).

Thanks to the fortunate fact that beginning and end of the myth are preserved, the main line of the plot is not in dispute. The remaining problems that interest us here concern the identity of the monster, the identity of the god that slew it, and the purpose of the myth. In order to solve these problems, we must present the text:

1. *i-ta-an-ḫu* URU.MEŠ KALAM.MEŠ *d[a-* ...]
in-da-ṭa-a UN.MEŠ *e-*[li er-ṣe-ti]
a-na ik-kil-li-ši-na u[*l* i-qa-al MUŠ.ḪUŠ]
a-na rim-ma-ti-ši-na ul i-ṣab-[bat rĕ-e-ma]
5. *man-nu-um-ma* MUŠ.[ḪUŠ ib-ta-ni]
tam-tu-um-ma MUŠ.[ḪUŠ ib-ta-ni]
^{d}En-líl *ina* AN-*e i-te-ṣir* [u-ṣur-ta-šu]

1. 'The cities have gone to ruins, the nations [have become ...],
the people have decreased in number o[n the earth],
to their noise [Furious-Snake does] not [pay attention],
on their cries he has no [pity]'
5. 'Who has [created Furious]-Snake ?'
'Sea has [created Furious]-Snake,
but Enlil in heaven designed [its form]:

Lines 14ff. make it clear that Enlil's action against humanity is not supported by the other gods. On the analogy with the *Atra-ḫasīs*-myth[1], where it is human noise depriving him of sleep that inspires Enlil to self-willed measures, Jacobsen (1932:53[8], also Lambert 1986:55) restores 3-4 as: 'because of their noise [Enlil] cannot [...]; because of their roaring he cannot [sleep] (*ul i-ṣab-*[*bat šit-tu*])'. There are reasons, however, not to accept this interpretation:

– in 1-2 we would expect humanity to multiply, making more and more noise; instead their cities decay, and their numbers decrease. Clearly some kind of disaster has befallen them, the disaster of Enlil's monster. Later, by killing the monster, Tišpak saves the nation. Obviously the 'noise' and 'cries' of 3-4 are a reaction on this disaster, and thus cannot describe Enlil's motive for causing it.

– Since in 5 the question 'who [transitive verb] [Furious]-Snake ?' cannot be posed unless Furious-Snake has been identified before as the source of misfortune, we must restore Furious-Snake in 3 or 4 (*ṣabātu* in 4 must have an object, so that space for this restoration is available only in 3). This tallies well with the preceding observation.

– If Enlil's motive is described in the present tense (4, *i-ṣab-*[*bat*]), his reaction, the creation of an exterminator, should not be expressed with a perfect (7).

Thus the text does not give us a motive for Enlil's behaviour, but the analogy with the *Atra-ḫasīs*-myth still favours the assumption that it was human noise depriving him of sleep. The absence of a divine ruler must have been the cause of the derailment of the nation[2].

In 5 someone poses a question that is answered with an account of what has happened (and a description of the monster). Lines 1-4 give an account of what has happened and is happening still (3-4), and we may safely assume that the informed speaker that answers the question of 5 is identical with the one reporting on the situation in 1-4. In 14ff. it becomes clear that we are witnessing an assembly of the gods (minus Enlil) in heaven; the one that poses the question in 5 must be one of these gods, the one that answers it, the reporter of 1-4, must be a divine messenger sent by the gods in heaven to find out what is happening on earth[3]. Since the answer on the question in line 5 involves a description of the monster, the monster must have been named or referred to in the question. In the break of 5 (also 6) we must restore a predicate balancing *i-te-ṣir* in 7, leaving space for one or two signs after MUŠ, together with MUŠ denoting the monster. Since *muš-*[*x-(x)*] or *ṣer-*[*x-(x)*] can hardly be restored as an epithet of the monster, MUŠ.[x-(x)] must be its name, it is a dragon type. Possible restorations are MUŠ.[HUŠ] and MUŠ.[ŠÀ.TÙR] (or MUŠ.[*ba-aš-mu*]), but, as we will see below, MUŠ.[HUŠ] is to be preferred.

The problem of the double designation of the monster, as MUŠ.[*x-(x)*] in 5f. and as LAB-*bu* (Lambert 1986:55, note 1 considers readings other than *lab-bu* not excluded) in 17, 20, 24, r. 4, 7, 9, has received little attention (Heidel 1951:141) and can only be solved by assuming that one of the two is an epithet. Since LAB-*bu* (in any reading)[4] designates a mythological beast only in this text, while *mušḫuššu* (and *bašmu*) is attested as the name of a monster subservient to Tišpak, we assume that LAB-*bu* is the epithet, meaning perhaps 'the Raging One' (*CAD labbu* B) rather than 'lion' (*labbu*)[5].

The monster is designed by Enlil (7) and brought forth by Sea (*Tâmtu*, 6) and River (*Nāru*, 23). The two acts of creation, designing and bringing forth, are commonly attested in cuneiform literature (see provisionally Pettinato 1971:29ff.), and so is Sea as a breeding place of monster snakes (Cooper 1978:139; Wiggermann 1986:281f).

The text continues (8-13) with a description of the monster for which we refer to the translation of Heidel (1951:142). This description can hardly be correct: the body of a mythological snake (with or without legs) could conceivably be 500 kilometers long and 10 kilometers [thick], but with a mouth of only 3 meters, and ears (?)[6] 6 meters in circumference, it certainly would look ridiculous; and then, being 10 kilometers [thick], it would stretch out its [neck][7] 3.5 meters, an imposing sight[8].

After they have thus been informed on the nature and cause of human misfortune, 'All the gods in heaven [...]. In heaven the gods lay prostrate before [Aruru] [9], and hast [ily grasped][10] the hem of Nannaru (reading *^{d}30-rum* to avoid the preposition before TÚG.SÍG-*šú*)' (14-16). Then the fearful gods speak (17-19): 'who will go, [kill] the Raging One, sa[fe] the wide land, and exercise kingship ?'. The kingship that is offered here to the victor is certainly not that over the gods[11], but that over the nation he saves from peril. Then someone, probably Nannaru, answers (20-22): 'go, Tišpak, k [ill] the Raging One, safe the wide land, and exercise kingship'. Tišpak does not refuse, but apparently asks for Nannaru's advice (23-24); 'you have sent me, lord, [to kill] the offspring of River, but I do not know the [...] of the Raging One'.

After a break of unknown length, the text continues on the reverse:

1. [^{d}X] *pa-a-šú i-pu-uš-ma a-na* ^{d} [Tišpak *iz-za-kar*]
šu-uš-ḫi-iṭ ur-pa mi-ḫa-a [...]
ku-nu-uk-ku na-piš-ti-ka i-na pa-ni-ka [...]
us-kám-ma lab-ba du-[*uk*]
5. *ú-šá-áš-ḫi-iṭ ur-pa mi-ḫa-*⌈*a*⌉ [...]
ku-nu-uk-ku na-piš-ti-šú i-na pa-ni-šú [...]
is-su-kám-ma lab-bi [*i-du-uk*]
3 MU.MEŠ *3* ITI.MEŠ UD *1*.KÁM *ù* G[I$_6$1.KÁM]
šá lab-bi il-la-ku da-mu-šú
[...] (colophon Assurbanipal)

1. [The god X] opened his mouth and [spoke] to [Tišpak]:
'stir up the clouds, [create (?)] a storm,
[raise (?)] your very own seal in front of you,
(then) hurl it down, and ki[ll] the Raging One'.
5. He (Tišpak) stirred up the clouds and [created (?)] a storm,
[he raised] his very own seal in front of him,
(then) hurled it down, and [killed] the Raging One.
For three years, three months, a day and a n[ight],
the blood of the Raging One flowed, [...].
(After Heidel 1951:143, with modifications)

Although doubts on the identity of the god slaying the Raging One are justified as long as the text remains incomplete, there is a strong case for the assumption that it was Tišpak:
– The analogy with *Enūma Eliš* and the *Anzû*-myth in which several gods fail or retire before the enemy is defeated, does not prove that the *Labbu*-myth followed the same pattern (so Heidel 1951:142). There are other important differences between *Enūma Eliš* and the *Anzû*-myth at one hand, and the *Labbu*-myth at the other. In the former two the gods fight forces (*Tiamat*, *Anzû*) that threaten their existence, in the latter they are threatened indirectly (the direct threat is against humanity); the victory of the gods in *Enūma Eliš* and the *Anzû*-myth helps to shape the cosmos, while the battle against the Raging One takes place after the creation of the human race, and is an historical incident.

– It is difficult to see how a local god like Tišpak could play the part of a failing or refusing god, unless the myth originates in Ešnunna. In that case, however, we expect him to play the part of the victor, since he is the head of the local pantheon, and 'the warrior of the gods' (OAkk.)[12] and thus resembles the monster slayer Ninurta, 'the warrior of Enlil',with whom he later is equated (*CT* 25 8:15).
– There is no other suitable candidate (below c).
– Tišpak is commonly associated with the dragon *mušḫuššu* (below b), and the identification of the protagonists as Tišpak and *mušḫuššu* (so already Jacobsen 1940:183) gives the myth a definite meaning (below c).
– (the seal used by the victor to kill the Raging One may have belonged to Tišpak, but this cannot be established beyond doubt (below d and note 43)).

B. TIŠPAK AND *MUŠḪUŠŠU*[13]

Tišpak replaces Ninazu as city god of Ešnunna from Old Akkadian times onward (Jacobsen 1932:51ff.; 1934:20ff.). He is a god of foreign origin about whose original character nothing is known. Jacobsen at first identified him with Tešub, the Hurrian weathergod (1932:51ff.), but later (1963) abandoned this identification (1970:34) and explained Tišpak as 'the Down-pouring One' (derived from *šapāku*), perhaps originally a rain-god. Even if it is conceded that Tišpak is the nameless victor of the *Labbu*-myth, his use of clouds and storms would not make him a weather god. Ninurta and Marduk too use storms against their enemies. Also van Dijk (1969:539ff.) rejects the identification with Tešub, and holds that Tišpak could replace Ninazu because both were snake gods. Indeed, both gods are snake gods, but their common traits may just as well be the result of their identification as its original incentive.

In *An-Anum* V 274f. Tišpak has two wives, ᵈÚ-kul-lá and ᵈÚ-ka-lá, undoubtedly variants of the same name. In the OB forerunner her name is spelled ᵈKul-áb (*TCL* 15 10:405), and in a SB god list (*CT* 25 8:17) as ᵈKul-lá-[(ab)]. Normally the name of the wife of Ninazu is Ningirida, but in the OB Nippur god list (*SLT* 122 iii 9) he is followed by ᵈKulla, elsewhere a name of the brickgod 'Clay-pit', but here perhaps the name of the wife of Tišpak, replacing her predecessor. Uncertain is ᵈGul-lá var. ᵈKulla in the *Weidner list* (Cavigneaux 1981 98f.:242, var.from *AfK* 2 pl. VIII 4). Outside lists the name is rare; it is OB attested in Ešnunna (ᵈÙ-gul-lá, Jacobsen 1940, 184:87, with further references), in the name of a small settlement, *Tu-ur*-ᵈÙ-gul-lá^{ki} (*RGTC* 3 240), and perhaps in a Mari personal name (ᵈGul-lá-*ḫa-zi-ra-at*, quoted in *RLA* 6:305b). In SB texts it occurs twice: ᵈÚ-kul-ab, 'queen of Ešnunna' in a litany from Nimrud (Wiseman 1969, 177:33), and in the theological text KAR 109 r. 2: ᵈú-kul-la *ba-nit ri-i-ti ḫa-i-ṭa-at kul-la-ti* [...], 'Ukulla, who creates meadows, who surveyes the universe'[14].

Although it is possible to suggest a Sumerian (but not an Akkadian) etymology for the name of Tišpak's wife (Jacobsen 1940, 184:87), the variety of spellings and the certainly non-Sumerian names of her husband and son (ᵈNa-an-ša-ak, *An-Anum* V 277, cf. van Dijk 1982:104) support the assumption of a non-mesopotamian origin.

In Old Akkadian Ešnunna, Tišpak's animal is the dragon *mušḫuššu*. The evidence for the identification of the god associated with the dragon as Tišpak was recently discussed by Lambert (1985:92; Boehmer 1965, figs. 565-572; Amiet 1980:174, note 84); that the name of this dragon was *mušḫuššu*, as expected, appears from the name of Tišpak's boat in an OB year name from Ešnunna: m á - g u r$_8$ - m a ḫ - m u š - ḫ u š - a (Jacobsen 1940:161, 183:83)[15]. That Tišpak did not take the dragon with him from his original home, but inherited it from his predecessor Ninazu, is shown by the fact that in Lagaš (Gudea), where Tišpak is not attested, it is associated with Ninazu's son, Ningizzida[16]. To the evidence collected by Lambert (1985:92ff.) an Ur III seal from Lagaš should be added (AO 4359, see Douglas van Buren 1934, Pl. Xa), dedicated to Ningizzida for the life of Šulgi, and showing the god seated on a (slightly divergent) dragon, and with snakes instead of dragon's heads growing from his shoulders.

From the time of Melišipak onwards the dragon is associated with Marduk and his son Nabû in art (Lambert 1985:89f.)[17]. The earliest attestation in the texts is in a copy of an inscription of Agum-kakrime, where the *mušḫuššu* is related to Marduk as the *ru-[ku-ub] i-lu-ti-š[ú]*, 'his divine mount' (*5R* 33 iii 13ff.). The authenticity of this inscription, however, remains disputed (Brinkman 1976:95f., Sommerfeld 1982:172, note 4). Although there is no exact proof, the god holding a scimitar and standing on a dragon on OB seals can hardly be dissociated from Marduk (Lambert 1985:90f.), especially not on OB seals from Babylon (Klengel-Brandt 1983:90), where some god certainly must be Marduk [18].

The mechanism behind Marduk's usurpation of the *mušḫuššu* may also secure its date. Without definite proof we suggest that Hammurabi's conquest of Ešnunna, the city of Tišpak, was translated into mythology (as reflected in the cult) and freed the animal of Tišpak for use by his adversary Marduk. A similar mechanism lies behind the appearance of the *mušḫuššu* as symbolic animal of Aššur after Sennacherib's conquest of Babylon (Lambert 1985:89).

Perhaps exactly because Marduk usurped the *mušḫuššu*, later sources associate Tišpak with other dragons:

– *bašmu*:

The 'Göttertypentext' (*MIO* 1 78 v 52ff.) describes a representation of ^{d}*Ti-iš-pa-ak*, armed with mace (*ḫurpalû*, cf. note 12), bow and arrow, and standing in a walking pose (*āliku*, cf. *CAD* A/1:347a), with both feet on a *bašmu* (*ba-áš-ma ka-bi-is*). A ^{d}*Ba-aš-mu* is the vizier of Tišpak in *An-Anum* V 278, and it is probably the same vizier that is described in *CT* 14 7 'r.' 2'f., if 3' is read correctly after Thompson's copy as SU]KKAL ^{d}*Tišpak*, instead of Landsberger's emendation (1934:53, note 6) M]UŠ ^{d}Tišpak[19]. The description gives him a form '(drawn out long?) like wi[re] (GIM *gu-ḫa-a[ṣ-ṣi]*)'. Another (?) vizier of Tišpak is ^{mul}BAL.UR.A (*CT* 33 1:14; *VACh* Išt XXV 30; cf. *ŠL* IV/2 no 44).

– ^{d}U š u m - u r - s a g:

The councellor of Tišpak in *An-Anum* V 279. Since *bašmu* is the Akkadian word for Sumerian u š u m (perhaps a loanword from Akkadian), we may assume that in fact only one dragon is meant. The 'warrior' (u r - s a g) u š u m is originally one of the beings defeated by Ninurta (Cooper 1978:33; van Dijk 1983:129; *Gudea* Cyl. A XXVI 2). These beings are designated as u r - s a g (Wiggermann 1986:276).

Above we argued that before he was succeded by Tišpak, Ninazu must have been master of the *mušḫuššu*. Only once is a m u š - ḫ u š mentioned in connection with Ninazu and his cult centre EN.DÍM.GIG^{ki} (= Enigi; Wilcke 1974:197 note 85; Klein 1981:114; Steinkeller 1981:86f.), in a broken passage in a Šulgi-hymn (*D* 307ff., Klein 1981:82). The *ušumgallu* of Ninazu [20] descending from the mountains together with the ferocious dogs of other gods in a bilingual SB incantation of older origin [21] probably denotes the same monster, since also the dragon of Nabû can be called *ušumgallu*[22].

Elsewhere he is associated with snakes: an ED macehead with snakes is dedicated to ^{d}N i n – [a] – z u (Amiet 1970:9f.)[23]; he is king of the snakes in Old Babylonian incantations (*PBS* I/2 131, *YOS* 11 32:4, 34:4, see van Dijk 1969:545f.). Ninazu himself spits venom in a SB incantation (Geller 1980:179)[24].

If Ninazu is indeed the original master of the *mušḫuššu*, the nature of the master might be used to speculate about the nature of the servant and the reason for their association. There is, however, a problem with the nature of Ninazu. Van Dijk (1960:77; 1969:539f.) constructs two gods Ninazu, one Ninurta-type god with parents Enlil and Ninlil and at home in Ešnunna, and one underworld god with parents x (several different fathers) and Ereškigal (also his wife in Geller 1985 36:308) and at home especially in Enigi. There are reasons to disregard van Dijk's split, and to retain only one god Ninazu:

– In Ešnunna Ninazu is identified with Tišpak. The god lists (see van Dijk 1960:71ff. for quotations) treat Ninazu in a clear underworld context, and the Ninazu of the god lists should thus be the Ninazu of Enigi (so van Dijk). The godlists, however, treat the successor of Ninazu ofEšnunna, Tišpak, in the same context.

– The Ninazu with parents Enlil and Ninlil has clear underworld connections in the myth *Enlil and Ninlil*, where Enlil impersonates the l ú - i_7 - k u r - r a, 'the man of the underworld river' (Cooper 1980:183) when begetting Ninazu (see Behrens 1978:191, 195f.). The hymn to the temple of Ninazu in Ešnunna (Sjöberg 1969, 42:425ff.), the key reference for the Ninurta-type Ninazu, does certainly not exclude the possibility that this Ninazu is also in origin an underworld god, as admitted by van Dijk (1960:75, note 45).

– Both Ninazu have the same brother; Ninmada, the snake charmer of An (*An-Anum* I 346) or Enlil (*MSL* 4 5:29), is a brother of Ninazu and a son of Enlil (Falkenstein 1962:368; Römer 1978:182) or perhaps of An (Falkenstein 1966:105f.) in the fragmentary Sumerian myth *TMHNF* 3 5; in the litany treated by Bergmann (1964:26f) he is related to the Ninazu of Enigi (identified by his temple é - g í d - d a), whose father should not be Enlil or An.

– Both Ninazu have the same wife, Ningirida [25] (for the underworld Ninazu of Enigi see the godlists quoted by van Dijk, and for the son of Enlil see e.g. van Dijk 1960, 58:29).

– Both Ninazu are associated with the dragon *mušḫuššu*, the one from Enigi in *Šulgi* D 307ff. (quoted above), and the one from Ešnunna as predecessor of Tišpak.

– In addition to Enlil and x, also Sin (Sjöberg 1969:88, *KAR* 88 Frgm. 4, right r. 10) and perhaps An (Falkenstein 1960:88, 106, note 5; Sjöberg 1967:210, Lambert 1985:93) are attested as the father of Ninazu. Should this be taken to imply the existence of a third and perhaps a fourth Ninazu?

There is only one god Ninazu, an underworld god. As a god of the younger generation and chief god of Ešnunna his theology was patterned after that of Ninurta; he became a warrior and son of Enlil and Ninlil. This change of descent was made possible by uncertainty about the identity of the ruler of the underworld, the x, the father of Ninazu (cf. Lambert 1980:60f.).

Tentativily the association of the *mušḫuššu* with Ninazu may be explained with the assumption that the former was originally an angel of death, killing with his venom on the command of his master. To make him a prized possession of the successors of Ninazu, Tišpak, Marduk, and Aššur, he must have had other qualities too, presumably as a warrior. If we have correctly assumed that the relation of the *mušḫuššu* with his original master was harmonious, his relation with his later masters must have been more or less tense, depending on the measure in which their characters differed from that of Ninazu.

C. MEANING OF THE *LABBU-MYTH*

The identification of Tišpak as protagonist of the *Labbu-myth* partly depends on the existence of counter candidates. In fact, there is only one counter candidate, Ninurta. Among the trophies of Ninurta (Cooper 1978:141ff.) we find the dragons u š u m and m u š - s a g - i m i n (or u r - s a g - i m i n), on his mace a m u š - m a ḫ is depicted (Cooper 1978:138). The former is probably identical with the MUŠ.*ba*-[*aš-mu*] (restored by Landsberger 1934:58, note 3, also *CAD* B 141a) defeated by Nergal, here and elsewhere replacing Ninurta (Cooper 1978:146, note 3), in the provincial Assyrian myth [26] *KAR* 6, the latter two are to be identified with the seven-headed dragons fighting a god in third millennium art[27]. Ninurta, however, the 'warrior of Enlil', and the 'avenger of his father', cannot be the protagonist of a myth in which the gods and their champion oppose Enlil instead of saving him from peril. For the same reason, if Tišpak is the protagonist of this myth, it cannot be assumed that Tišpak replaces the dragon slayer Ninurta of a postulated original *Labbu-myth*[28].

The myth is certainly older than *Enūma Eliš*, in which the dragons and other monsters are made into children and soldiers of Sea (*Tiamat*), archenemy of Marduk, and all are finally defeated by him (Wiggermann 1986:279f.).Since it is hardly likely that a myth in which Tišpak plays the part of divine champion (or indeed any part) was composed after Hammurabi's conquest of Ešnunna and its subsequent decay, we assume that it was composed before that date (1755 B.C.). If we take the curious omission of phrases introducing direct speech on the obverse (one is preserved on the reverse) as an original feature of the text, the failure to adapt it to the symmetrical taste of the first millennium (cf. Vogelzang 1986:61ff.) perhaps indicates a lack of interest in the text on the part of later editors. Anyway, since the myth was preserved in only one copy and is never alluded to in the extensive first millennium literature, it cannot have been very important in this time[29]. Why, like other antiquated myths (*Angim, Lugale, Anzû*)[30], it was handed down at all, remains an open question.

If then the myth is composed in Ešnunna before Hammurabi destroyed it and Marduk succeded Tišpak as master of the *mušḫuššu*, the MUŠ.[*x*-(*x*)] of the myth can hardly have been any other dragon than the Ešnunnian dragon par excellence, the MUŠ.[HUŠ], the 'Furious Snake'. The only other dragon that is associated with Tišpak, the u š u m̆ /*bašmu*, is originally and still during the Old Babylonian period a defeated enemy of Ninurta (see above) only.

Since the myth does not treat cosmological questions, and the divine protagonist plays his part as a city god and warrior and not as a representative of some cosmic force, Ninazu may have been its original protagonist just as well as Tišpak. Some sort of combat myths involving dragons certainly existed when Ninazu was city god of Ešnunna (Frankfort 1955, no. 497; Muscarella 1981, no. 28), and Ninazu of Ešnunna is a fearless warrior with Ninurta features (Sjöberg 1969, 42 425ff.; van Dijk 1960:77f.). Tišpak succeded Ninazu as city god, master of the *mušḫuššu*, warrior[31] and seal owner (below d), so why should he not also succeed Ninazu as a dragon slayer? There are reasons to assume that he did not:

– The divine protagonist descends from heaven to slay the monster and become king. This fits Tišpak better than Ninazu, who was a newcomer only in the remote past.

– The victor becomes king without the consent of Enlil (he is covered by Nannaru). This fits the upstart Tišpak better than the venerable Ninazu, king of Ešnunna from times immemorial.

– On the assumption made above that the relation of the *mušḫuššu* with his original master was harmonious, the enimical relation between god and monster in the myth is better explained with Tišpak as its original protagonist than with Ninazu.

With Tišpak and the *mušḫuššu* as its original protagonists, the purpose of the myth at once becomes transparent. It translates history, the Old Akkadian overtake in Ešnunna, into theology, and justifies Tišpak's accession as king and the servitude of the *mušḫuššu* as a consequence of his 'liberation' of the nation, sanctioned by the decision of a divine council. The myth, though probably not in its present form, belongs to the Old Akkadian period[32].

The later history of the *mušḫuššu* is determined by the model of mythological explanation of fact established in the *Labbu*-myth; when the city of his master is defeated, the *mušḫuššu* comes into the possession of the city god of the victor (Hammurabi - Marduk, Sennacherib - Aššur). Finally, in the restructured mythology of *Enūma Eliš*, he and the monsters once slain by Ninurta are the children and soldiers of Sea (*Tiamat*), defeated by Marduk in a cosmic battle (Wiggermann 1986:263ff.; *VAS* 24 97, a new MB? text enumerating monsters in Esagil).

It cannot be objected that on Old Akkadian seals the dragon of Tišpak is alive, not dead as the myth gives to expect. He does not differ in this respect from the first millennium dragon of Marduk (and Aššur), killed by him in the battle against *Tiamat*. Through the ages iconography simply reproduced (with minor adaptations) what once had been established, the dragon consorts and identifies his master, or stands watch. Generally speaking, monsters once alive stay alive [33], even battle scenes are rare and restricted to certain types. Mythology apparently is allowed a certain freedom in using the fixed types of iconography; it interprets the relation of monster and god as one of servitude, and fixes the monster in the pantheon as a courtier or defeated enemy. From the defeated enemies an extremely powerful scheme of mythological explanation is developed, the combat myth. Good and evil can be distributed among the protagonists of a story, and the triumph of good establishes order in a moral universe.

As is shown by iconography, the combat myth is certainly older than the Old Akkadian *Labbu*-myth. Presumably the combat myth was developed in relation with Ninurta, the warrior of his father, who defeated the enemies of god and man, the chance products of Heaven, Earth and Sea that they encountered while colonizing the world from the d u$_6$ - k ù, the first tell, their original home[34]. Only much later, in *Enūma Eliš*, were good and evil promoted to Good and Evil, and became the victory in a cosmic battle the foundation of the ordered universe[35].

D. THE SEAL OF TIŠPAK

Interesting, but of minor importance for the chain of reasoning produced above, is the seal Tišpak uses as a weapon to kill the Raging One.

The seal of Ninazu, Tišpak's predecessor, appears in a ḫ u l - b a - z i - z i incantation of uncertain age (*KAR* 88 Fr. 4 r. iv 12, dupl. *STT* 217:29, both SB): *na4*KISIB dNin-a-zu *šá* GABA.RI NU TUKU-*u*, 'seal of Ninazu, that cannot be withstood'[36]. He probably possesses this seal on account of his function in the divine world, he is the k i š i b (text ŠID)- g a l : *kanniku*, 'seal-keeper' of the underworld in a collection of SB z i - p à incantations going back to older originals (Ebeling 1953, 388:60f., dupl. *STT* 210 r. 8', ref. to be added to *CAD kanniku*).

Tišpak does not have the title k i š i b - g a l (or - g á l), but in an Old Akkadian incantation (?) from Ešnunna he is called *abarak Tiamtim*, 'steward of Sea (or the sea)' (Westenholz 1977:102). Charpin, discussing these functions (1986:244, 388) concludes that 'on a bien l'impression que ce titre de *kišibgallum* correspond, dans la tradition d'Eridu, à la fonction appelée ailleurs a g r i g 'intendant', - l'intendant étant celui qui contrôle les magasins en scellant leur porte'.

Thus, when in Ešnunna Tišpak was identified with Ninazu, he inherited not only his city and his dragon, but also, at least in Ešnunna in the Old Akkadian period, his office in the divine world, and the seal that went with it[37]. Outside Ešnunna, in the mainstream of Babylonian theology, Ninazu kept his office and his seal (the SB texts cited above).

In the *Labbu*-myth Tišpak's seal is defined by 'your/his very own', as we have translated *napištika/šu*. What the text wants to point out is, it seems, not so much the location of the seal on Tišpak's body (translating 'at your/his throat', Jacobsen 1932:53f.; *CAD* K 544b, N/1 303a), but the fact that it was his own (cf. *CAD napištu* mng. 6b) seal, not just any seal.

It is clearly impossible to consider the seal a votive seal, as some authors want[38]. Not only is it impossible to assign a function to votive seals on a mythological level, but also grammar protests; since the pronoun in *napištika/šu* refers to Tišpak, he would have dedicated the seal to himself for his own life[39]!

When the use of seals to kill mortal kings with needs an explanation (Hallo 1983:13), there must be a very special reason for the use of a seal against a monster of the dimensions of the Raging One[40]. Jacobsen's explanation of the seal as symbolic for thunder (1932:54) loses its strength when Tišpak is not a weather god (see above). Although the possibility that magic made the seal an offensive weapon cannot be excluded[41], it seems more likely that the seal is chosen to explain something, that it provides an etiology of a lost historical or natural fact[42].

Etiology is what one expects in a myth commenting on history, and indeed, beside the bleeding (see note 32) and the seal etiologies, there is a third, this one explaining something known also to us: Enlil's original design remained visible in heaven as the constellation *mušḫuššu*. Since the constellation *mušḫuššu* appears to have been renamed after the Old Babylonian period (*CAD* M/2 270a), we take this etiology as an original feature of the text. That the Akkadians needed this shallow explanation shows that they were no longer conversant with the original mythological notions that named the stars and constellations.

E. CONCLUSIONS

During the Old Akkadian period an Akkadian speaking population infiltrated the Sumerian city of Ešnunna. Among it was a small contingent of foreigners[43] whose god Tišpak became the chief god of the infiltrators. Tišpak was identified with the Sumerian chief god of the city, Ninazu, an underworld god with Ninurta features, and adopted part of his functions and attributes: he became seal-keeper (though of Sea, and not of the underworld) and a Ninurta-type warrior, and took over the death demon, the dragon *mušḫuššu*. His original character, whatever it was, seems to have prevented him from becoming an underworld god, and consequently his relation with the death demon *mušḫuššu* must have differed from its relaton with Ninazu.

Tišpak's new task as chief god of a local pantheon and master of a death demon needed theological justification, and to this end, after the ways of the period, a myth of origins was composed, based on a combat theme. A SB version of this myth is preserved on a tablet from the library of Assurbanipal, and known today as the 'Slaying of the *Labbu*'. The text is not only damaged, but also short-spoken, and our understanding of it must be based on restorations and conjectures.

The myth solves the puzzle as follows: Before Tišpak, the land did not have a divine ruler, and derailed. Enlil, angered, has Sea create a death dealing monster, the dragon *mušḫuššu*, that is able to restore order at a high prize: it is killing off the population. The other gods, lead by the [mother goddess Aruru] and the moongod Nannaru, fear that soon their human servants will be unable to supply them with food (conjectured after the Flood myths), and appoint a champion whom is offered kingship as reward for killing the *mušḫuššu*. The champion, Tišpak, succeeds, and becomes king. Undoubtedly he restored order, and kept his human servants in check, so that the cause of Enlil's anger was removed; although thwarted at the outset, he will have been pleased with the final result.

When Ešnunna declines (probably as a consequence of its defeat by Hammurabi) and practically disappears from history after the Old Babylonian period, Tišpak, his family, and his mythology become obsolete. For the dragon *mušḫuššu*, however, a great future laid in store. When Marduk defeated Tišpak, the *mušḫuššu* became his beast, and in the course of time as national as its master.

I hope the jubilarian has enjoyed these speculations.

NOTES

1. For human noise (*ḫubūrum, rigmum*), see *CAD* H 220, G 157; Finkelstein 1956:328-331; Frankena 1965:9ff.; Pettinato 1968:182ff.; von Soden 1973:353.
2. If for some reason an earlier divine king was deposed or unable to deal with the terror of the Raging One, so much would have been left unexpressed in the text, that without a 'first tablet' it could not have been understood by anyone. Moreover, if Tišpak was identified with Ninazu (see below), it is hard to see who could have preceded him as ruler of Ešnunna.
3. The gods may have noticed that something was wrong when their food supplies failed, or they no longer smelled incense.
4. The possibility of a reading *Reb-bu* (cognate with Hebrew *Rahab*) was revived by Lambert (1986:55, note 1). Roberts (1972:116, note 444) adduces Hebrew *lwytn* and Ugaritic *ltn* (see Emerton 1982:327ff.) and derives *labbû* (sic) from *lawû (parras*-formation): 'the twisting one'. Lexicon (**labbû* is a hapax), orthography (the *-û* is not expressed) and phonology (Babylonian *-bb-* cannot be derived from *-ww-*) plead strongly against this derivation.
5. Unclear is the relation between lion and dragon in an unpublished Ur III incantation quoted by Steinkeller (1984:6, p i r i g m u š - ḫ u š a b - š à - g a l u$_5$ - g a, 'the lion, the *mušḫuššu*-dragon, which lives in the midst of the sea'. The roaring p i r i g of the obverse of *TDr* 1 seems to be identical with the u š u m - g a l (another type of dragon) of reverse 7. Other roaring m u š-dragons are found in *UET* 6 70:4 (*mušḫuššu*), Heimpel 1968:86.4, 86.8 (m u š - s a g - k a l; the questionable sign 'ŠÀ' in 86.8 source A is š e, a syllabic spelling for KA[xLI] in source B), *VAS* 17 38:12 (m u š - š à- t ù r). Dragon types (like the *mušḫuššu*) may have legs, and sometimes leonine heads. Thus, it seems, p i r i g/*labbu*, 'lion' and m u š -dragons are not always distinguished, so that *labbu* in the *Labbu*-myth may mean 'lion' after all. See also notes 16 and 27.
6. The round ears seem strange (the parallel *KAR* 6 ii 24 has '[his] eye[s]'), but I am unable to suggest another restoration for *u*[*z*-...] in line 10.
7. Reading 12f., on the basis of parallelism: *i-na* A.MEŠ *9* KÚŠ *i-šad-da-*[*ad kišāssu* (or *rēšēšu-*] *ú-še-eq-qí zi-im-bat-su i-*[*na tam-ti*] 'in the water it stretches out [its neck (or head)] nine cubits, [in the sea] it raises its tail'.

8. The description of the *ba[šmu]* (see below) in *KAR* 6 ii is much more 'realistic'. The description in the *Labbu*-myth is perhaps based on the constellation, rather than on the actual monster.
9. If we restore [Nannaru] in 15 (so Heidel 1951:142 and others), we would expect him to be referred to with a pronoun in the next line, which is not the case. The restoration *Aruru* is based on *KAR* 6 ii 9ff., where she plays an important part in choosing the champion of the gods.
10. While Enlil, angered, is absent, some other god has to take decisions. If there must be a specific reason for the choice of Nannaru, it might be sought in the fact that he is the firstborn son of Enlil. The monster is active in the sea, and operates against humanity; the gods probably fear that they will lose their human servant and have to work again, but be this as it may, there is no indication that Nannaru is threatened more than the other gods. Thus we are not forced to find a reference to such a threat to Nannaru in 16, as does *CAD* S 325a (and Lambert 1986:55).
11. In *Enūma Eliš* the gods are threatened and saved, and kingship over the gods is offered to the victor, here the land is threatened and saved, and kingship over the land left in chaos is offered to the victor. More generally; the reward has to be balanced by the deed.
12. Tišpak is addressed as UR.SAG *ì-li* on Old Akkadian seals (Boehmer 1965:100), as *qurādum azzum*, 'fierce warrior' in an Old Akkadian incantation (?) from Ešnunna (Westenholz 1977:102). Ninazu is u r - s a g (Sjöberg 1969, 43:445). On OAkk. seals, Tišpak often holds a mace (Boehmer 1965, figs. 565-572), probably identical with the *ḫurpalû* of the Göttertypentext cited below.
13. See Lambert 1985; Wiggermann 1986:293f. Still useful are Dougas van Buren 1934 and 1946, both with many photographs and drawings. For the later periods see Seidl 1968, with previous literature.
14. Etymological explanations of the type discussed by Bottéro 1977:5ff.: see the dictionaries for k u l= *banû*, ù = *rîtu* and u_6 = *ḫâṭu*.
15. Such a boat may be depicted on an OAkk. seal (Meijer 1977-78, 20:23), if he *mušḫuššu* in boat form is not simply an unfinished regular *mušḫuššu*.
16. The dragon *mušḫuššu* got the form it would have to the end in the OAkk. period. Thus, if we have correctly concluded that Ninazu was master of the *mušḫuššu* before Tišpak, we should not expect this original pre-Tišpak dragon to be exactly like the fully formed OAkk. one. Forerunners, looking more leonine than their successors (whose hind pawns are the talons of a bird of prey) and without the crown, are perhaps the monsters on an albaster group (Frankfort 1943, pl. 70), and those on the seals Amiet 1980:1389-1392 (for 1389 see also Vanel 1965:15f.; the monster is perhaps rather the lion-dragon of Adad, without wings) and Boehmer and Finkbeiner 1984:100, no.6.
17. For the later history of the dragon see also Madhloom 1970:107; Rittig 1977:114; Kolbe 1981:123ff.; Reade 1979:40; 1977:39, 42. For the dragon of Nabû see Seidel 1968:192; Kolbe 1981:128f.; Pomponio 1978:208, note 6, 213, 215f.; below note 18. The first to identify the dragon with the *mušḫuššu* was Koldewey (1903:14ff.), for his arguments see now Lambert 1985:87.
18. See also Collon 1986, 47:C 19 (add 625); Frankfort 1939:168f. A god holding a 'wedge', which is called a stylus by Seidl (1968:121ff.) and other authors, and standing on a dragon appears on the Old Babylonian seal Collon 1986, no. 382. But for the scimitar he holds in his left hand he looks exactly like the Nabû of an early NAss. seal (Porada 1948, no. 691) and of a NAss. 'amulet' (BM. 118796, Budge 1930:98; Saggs 1962, fig. 56B, 66) dedicated to Nabû (read from photograph, uncertain). Thus on OB seals the dragon is associated not only with a god looking like later Marduk, but also with a god looking like later Nabû. Although it cannot be logically excluded, it is extremely unlikely that these gods were not already Marduk and Nabû but two other gods, one of them a writer, from whom Marduk and Nabû later adopted the iconography. In SB texts Tišpak is identified with Marduk (*CT* 24 50, 47406:11) and Nabû (*KAR* 94 i 49).
19. Landsberger looked for the *bašmu* expected in this description of snakes in the preceding section (1934:54, note 1), where he read (15) [MUŠ.A.AB.] BA. The equation MUŠ.A.AB.BA =*bašmu* is limited to the lexical texts, however.
20. [u š u m - g a]l ^{d}N i n - a - z u - k e_4 : ⌈*ú*⌉-*šum-gal-lum šá* ^{d}*Nin*-⌈*a*⌉-[*zu*] (var. ^{d}MIN), *KAR* 54 r. 2, dupl. *STT* 178:13f.
21. Duplicate of *STT* 178 is *TMHNF* 4 67 (OB)!
22. 4 R^2 20/3:15f.; *KAR* 104:29; *ABL* 951:12f. (cf. Lambert 1978:93; Parpola 1983:266 *ad* 16). He is also called *mušḫuššu* (Lambert 1978, 86:15).

23. In view of the snakes a restoration ᵈN i n - [g í r] - s u is less likely.
24. In *An-anum* V 240 ᵈMU Š is another spelling of (or identification with) ᵈNin-a-zu of the preceding line.
25. For this deity see Römer 1975:61; Ferrara 1972:2; *TIM* 9 86 iii 8 (ᵈNin-g̃ìri-da), *UET* 6 395 r. 14' (ᵈNin-giri$_{16}$-du), Civil 1973:57, note 2 (*ᵈNin-kir$_4$-ù-tu*), cf. Krecher 1978:46.
26. The tablet stems from the 'library' of Tiglath-pileser I (Weidner 1952-53:207), and has some Assyrian forms (*at-ti-din* in 15, *li-bit* for *limītu* in 24, *e-kúl* in 26ff.).
27. These beings need a full discussion. Briefly: a) m u š - m a ḫ, a seven-headed snake (Douglas van Buren 1946:18f.; Heimpel 1968:480ff.), fighting a god on a seal (Frankfort 1955, no. 497). b) m u š - s a g - i m i n and u r - s a g - i m i n, seven-headed snakes with lion's paws, fighting gods on Boehmer 1965, fig. 292, and on the shell inlay of the collection Borowski (Muscarella 1981, no. 28, with previous literature, cf. Amiet 1980:136a). c) The possibility that the m u š - s a g - k a l is another seven-headed dragon cannot be excluded (Falkowitz 1983:110a). The reading of the latter is assured by the continuation with - l a in *VAS* 17 1 iv 21, and by ᵈSag-gal = *ka-at-te-el-lum* in *CT* 25 22:40 (cf. *ᵈSag-kal* in *CT* 24 47 ii 18 and Sjöberg 1969 142).
28. Asaluḫi defeats a dragon in an Ur III text from a totally different sphere *TDr* 1. The incantation uses elements of a combat-myth for its own purposes.
29. Perhaps Sennacherib, who must have had a special interest in the *mušḫuššu* after he took it from Marduk (see above), thought of Tišpak as its victor when he had him depicted on his *akītu* house, ready to do battle with *Tiamat* together with Aššur and other gods (Luckenbill 1924, 142:11; *KAV* 49 ii 13).
30. That these texts were still important goes forth not only from quotations and the number of manuscripts in late libraries, but also from the Catalogue of Texts and Authors (Lambert 1962, 46 I 3) in which *Angim* and *Lugale* are ascribed to Ea.
31. See note 12. The association of Ninazu with Ninurta and his transformation into a warrior attested in the hymn to his temple (Sjöberg 1969 42:425ff.) must be dated prior to his replacement by Tišpak in the Old Akkadian period.
32. The curiously precise and mythologically useless detail of the duration of bleeding at the end of the myth obviously contains an historical allusion that cannot be interpreted without more explicit historical sources.
33. For examples with a long history see Wiggermann 1986:297.c (*ugallu*/Lion Demon), 303 with note 18 (*kusarikku*/Bull man). The combat myth is also applied to thoroughly peaceful beings like the *suḫurmāšu*/Goat-Fish and the *kulullû*/Fish-man, and even to inanimate things like 'king palm' and 'gypsum', probably only because they were apotropaic features of gates that needed a mythological explanation (Wiggermann 1986:275f.).
34. I hope to make this the subject of a future study.
35. That the earlier combat myths did not have cosmological implications can be deduced from the *Labbu*-myth, the Ninurta mythology (the *Anzû*-myth seems to have something to do with the water supply of the world, but does not have a bearing on creation), and the battles between gods on Old Akkadian seals, that seem to take place in a created universe.
36. The context indicates that the seal of Ninazu is identified with the object, probably also a seal, on which the incantation is to be written.
37. Presumably the change of territory (Ninazu is seal-keeper of the underworld, Tišpak is steward of Sea) is related to the fact that Tišpak never became and underworld god.
38. Jastrow 1908:280, note 8 end; Hallo 1983:13 (leaves the option 'at your/histhroat'), with further references. Against this interpretation see also Watanabe 1985:390.
39. To be an argument in the defense of Tišpak as the dragon slayer of the reverse of *CT* 13 33f., we would also have to prove that the seal could not have belonged to any other dragon slaying god. The divine seal-keepers of the earlier texts have been collected by Charpin (1986:388) and among them only Ninurta, whom we have excluded above as protagonist of the *Labbu*-myth, is a dragon-slayer. Other references to divine seals generally concern votive seals, or seals adorning a divine statue (Vieyra 1961:50f.; Leemans 1952:8f.; Thureau-Dangin 1922:86; Smith 1926:442f.; and in general Hallo 1983:7ff.). Marduk and Assur own seals to seal the tablet of destinies (Watanabe 1985:377ff.; George 1986:133ff.), but generally speaking, seals belonging to divine

functions are rare. It cannot be excluded, however, that more gods owned seals than we know of, and the argument concering Tišpak can therefore not be closed.

40. Tišpak could have used his mace (note 12), or one of the weapons used by other dragon slayers, a boomerang (Muscarella 1981, no. 28), a spear (Boehmer 1965, fig. 292), a thunderbolt (always used against a kind of *mušḫuššu* without hind legs; the scene occurs about ten times on NAss. seals, see for example Heidel 1951, fig. 8) or an arrow (always used against a monster looking like a rearing cobra with horns; attested more than ten times on NAss. seals, see Teissier 1985, no. 173, with literature. The identity of these NAss. dragons cannot be discussed here).
41. See note 39, where, however, the seal is used defensively.
42. Like a mountain (Etna still covers Typhon) or an impressive building. Conceivably *kunukku*, 'seal' was used for an 'etymological' explanation of a noun composed with the name of the city, properly Eš*nunak (šitpuk/Tišpak Išnunak: kunuk napištišu?).*
43. Van Dijk suspects that the name of Tišpak's son, Nanšak, is Elamite (1982:104). Jacobsen detects a small foreign element in personal names of the Old Babylonian period (1932:43).

REFERENCES

AMIET P.

1970 'Une masse d'armes présargonique de la collection M. Foroughi', *Revue d'Assyriologie* 64: 9-16.

1980 La Glyptique mésopotamienne archaique. Paris.

BEHRENS H.

1978 *Enlil und Ninlil. Ein Sumerischer Mythos aus Nippur*. Rome.

BERGMANN E.

1964 'Untersuchungen zu Syllabisch Geschriebenen Sumerischen Texten', *Zeitschrift für Assyriologie* 56: 1-43.

BOEHMER R.M.

1965 *Die Entwicklung der Glyptik während der Akkad Zeit*. Berlin.

BOEHMER R.M. AND FINKBEINER U.

1984 'Uruk - Warka XXXVII, Survey des Stadtgebietes von Uruk', *Sumer* 43: 91-107.

BOTTERO J.

1977 'Les noms de Marduk, l'écriture et la 'logique' en Mésopotamie ancienne', in: M. de Jong Ellis (Ed.), *Essays on the Ancient Near East. In Memory of Jacob Joel Finkelstein*. Hamden, Connecticut. Pp. 5-28.

1985 *Mythes et rites de Babylone*. Geneva and Paris.

BRINKMAN J.A.

1976 *Materials and Studies for Kassite History*, Vol. I. Chicago.

BUDGE E.A.W.

1930 *Amulets and Superstitions*. Oxford.

CAVIGNEAUX A.

1981 *Texts from Babylon. Textes Scolaires du Temple de Nabû ša ḫarê*. Baghdad.

CHARPIN D.

1986 *Le clergé d'Ur au siècle d'Hammurabi (XIX-XVIIIe siècles)*. Geneva and Paris.

CIVIL M.

1973 'From Enki's Headaches to Phonology', *Journal of Near Eastern Studies* 32: 57-61.

COLLON D.

1986 *Catalogue of Western Asiatic Seals in the British Museum. Cylinder Seals III. Isin-Larsa and Old Babylonian Periods*. London.

COOPER J.

1978 *The Return of Ninurta to Nippur*. Rome.

1980 Review of H. Behrens, *Enlil und Ninlil. Ein Sumerischer Mythos aus Nippur*, *Journal of Cuneiform Studies* 32: 175-188.

DELITZSCH F.

1887-1890 *Assyrisches Wörterbuch zur gesamten bisher veröffentlichten Keilschriftliteratur unter Berücksichtigung zahlreicher unveröffentlicher Texte*. Leipzig.

DOUGLAS VAN BUREN E.

1934 'The God Ningizzida', *Iraq* 1: 60-89.

1946 'The Dragon in Ancient Mesopotamia', *Orientalia N.S* 15: 1-46.

EBELING E.

1953 'Sammlungen von Beschwörungsformeln teils in Sumerisch-akkadischer, teils in Sumerischer oder Akkadischer Sprache', *Archiv Orientální 21: 357-423*.

EMERTON J.A.

1982 'Leviathan and *ltn*: the Vocalization of the Ugaritic Word for the Dragon', *Vetus Testamentum* 32: 327-331.

FALKENSTEIN A.

1962 Review of S.N. Kramer, *Sumerische Literarische Texte aus Nippur*, *Orientalistische Literaturzeitung* 57: 366-373.

1966 *Die Inschriften Gudeas von Lagas I. Einleitung*. Rome.

FALKOWITZ R.S

1983 'Notes on 'Lugalbanda and Enmerkar', *Journal of the American Oriental society* 103: 103-114.

FERRARA A.J.

1972 'The Itinerary of Nanna-Suen's Journey to Nippur, *Orientalia N.S*. 41: 1-4.

FINKELSTEIN J.J

1956 'Hebrew and Semitic -ḪBR', *Journal of Biblical Literature* 75: 328-333.

FONTENROSE J.

1959 *Python: a Study of Delphic Myth and its Origin*. Berkeley.

FORSYTH N.

1987 *The Old Enemy*.

FRANKENA R.

1965 Kanttekeningen van een Assyrioloog bij Ezechiël. Leiden.

FRANKFORT H.

1939 *Cylinder Seals. A Documentary Essay on the Art and Religion of the Ancient Near East*. London.

1943 *More Sculpture from the Diyala Region*. Chicago.

1955 *Stratified Cylinder Seals*. Chicago.

GELLER M.J.

1980 'A Middle Assyrian Tablet of Utukkū Lemnūtu Tablet 12', *Iraq* 42: 23-51.

1985 *Forerunners to* UDUG-HUL. *Sumerian Exorcistic Incantations*. Stuttgart.

GEORGE A.R.

1986 'Sennacherib and the Tablet of Destinies', *Iraq* 48: 133-146.

HALLO W.W.

1983 '"As the Seal upon Thine Arm": Glyptic Metaphors in the Biblical world', in: L. Gorelick and E. Williams-Forte (Eds.), *Ancient Seals and the Bible*. Malibu. Pp. 1-17.

HEIDEL A.

1951 *The Babylonian Genesis, the Story of Creation, 2nd edition*. Chicago.

HEIMPEL W.

1968 *Tierbilder in der Sumerischen Literatur*. Rome.

JACOBSEN TH.

1932 *Tell Asmar and Khafaje. The First Season's work in Eshnunna*. Chicago.

1934 *Philological Notes on Eshnunna and its Inscriptions*. Chicago.

1940 *The 'Gimilsin' Temple and the Palace of the Rulers at Tell Asmar*. Chicago.

1970 *Toward the Image of Tammuz and Other essays on Mesopotamian History and Culture*, W.L. Moran (Ed.). Cambridge, Massachusetts.

JASTROW M.

1908 'Urumuš', *Zeitschrift für Assyriologie* 21: 277-282.

KLEIN J.
1981 *Three Šulgi Hymns*. Ramat-gan.
KLENGEL-BRANDT E.
1983 'Siegelabrollungen auf altbabylonischen Tontafeln aus Babylon', *Altorientalische Forschungen* 10: 65-106.
KOLBE D.
1981 *Die Reliefprogramme religiös-mythologischen Characters in neu-assyrischen Palästen*. Frankfurt and Bern.
KOLDEWEY R.
1903 'Das Istartor in Babylon', *Mitteilungen der Deutschen Orientgesellschaft* 19: 7-28.
KRECHER J.
1978 'Das Sumerische Phonem /g̃/', in: G. Komoróczy (Ed.), *Festschrift Lubor Matouš, Vol. II*. Budapest. Pp. 7-73.
LAMBERT W.G.
1962 'A Catalogue of Texts and Authors', *Journal of Cuneiform Studies* 16: 59-77.
1978 'Nabû Hyms on Cylinders', in: G. Komóroczy (Ed.), *Festschrift Lubor Matouš, Vol. II*. Budapest. Pp. 75-111.
1980 'The Theology of Death', in: B. Alster (Ed.), *Death in Mesopotamia*. XXVIe Rencontre Assyriologique Internationale. Copenhagen. Pp. 53-66.
1985 'The History of the m u š-ḫ u š in Ancient Mesopotamia', in: *l'Animal, l'homme, le dieu dans le proche orient ancien*. Actes du Colloque de Cartigny (1981). Leuven. Pp. 87-94.
1986 'Ninurta Mythology in the Babylonian Epic of Creation', in: K. Hecker et al. (Eds.), *Keilschriftliche Literaturen*.
LANDSBERGER B.
1934 *Die Fauna des Alten Mesopotamien nach der 14. Tafel der Serie Ḫar-ra = ḫubullu*. Leipzig.
LEEMANS W.F.
1952 *Ishtar of Lagaba and her Dress*. Leiden.
LUCKENBILL D.D.
1924 *The Annals of Sennacherib*. Chicago.
MADHLOOM T.A.
1970 *The Chronology of Neo-Assyrian Art*. London.
MEIJER D.J.W.
1977-78 'Mesopotamian Cylinder Seals in the Allard Pierson Museum, Amsterdam', *Jaarbericht Ex Oriente Lux* 25: 7-34.
MUSCARELLA O.W.
1981 *Ladders to Heaven*. Toronto.
PARPOLA S.
1983 *Letters from Assyrian Scholars to the Kings Esarhaddon and Assurbanipal*. Neukirchen-Vluyn.
PETTINATO G.
1968 'Die Bestrafung des Menschengeslechts durch die Sintflut', *Orientalia N.S.* 37: 165-200.
1971 *Das Altorientalische Menschenbild und die Sumerischen und Akkadischen Schöpfungsmythen*. Heidelberg.
POMPONIO F.
1978 *Nabû. Il Culto e la Figura di un Dio del Pantheon Babilonese ed Assiro*. Rome.
PORADA E.
1948 *Corpus of Ancient Near Eastern Seals in North American Collections. I. The Collection of the Pierpont Morgan Library*. Washington.
READE J.
1977 'Shikaft-i Gulgul: its Date and Symbolism', *Iranica Antiqua* 12: 33-47.
1979 'Assyrian Architectural Decoration: Technique and Subject Matter', *Baghdader Mitteilungen* 10: 17-49.
RITTIG D.
1977 *Assyrisch-babylonische Kleinplastiek magischer Bedeutung vom 13.- 6. Jh. v. Chr.* München.

ROBERTS J.J.M.
1972 *The Earliest Semitic Pantheon. A Study of Semitic Deities attested in Mesopotamia before Ur III*. Baltimore.
RÖMER W.H.PH.
1975 Review of A.J. Ferrara, *Nanna-Suen's Journey to Nippur, Bibliotheca Orientalis* 32: 57-62.
1978 Review of C. Wilcke, *Kollationen zu den sumerischen literarischen Texten aus Nippur in der Hilprecht-Sammlung Jena, Bibliotheca Orientalis* 35: 181-183.
SAGGS H.W.F.
1962 *The Greatness that was Babylon*. London.
SEIDL U.
1968 'Die Babylonische Kudurru-Reliefs', *Baghdader Mitteilungen* 4: 7-220.
SJÖBERG A.
1967 'Zu einigen Verwandtschaftsbezeichnungen im Sumerischen', in: *Heidelberger Studien zum alten Orient*. Pp. 201-231.
1969 *The Collection of the Sumerian Temple Hymns and Keš Temple Hymn*. New York.
SMITH S.
1926 'The Seal before the God', *Journal of the Royal Asiatic Society* (1926): 442-446.
SOMMERFELD W.
1982 *Der Aufstieg Marduks. Die Stellung Marduks in der Babylonischen Religion des zweiten Jahrtausends v. Chr*. Neukirchen-Vluyn.
STEINKELLER P.
1981 'More on the Ur III Royal Wives', *Acta Sumerologica* 3: 77-92.
1984 'The Sumerian Verb l u g_x (LUL)', *Studi Epigrafici e Linguistici* 1: 5-17.
TEISSIER B.
1985 *Ancient Near Eastern Seals from the Marcopoli Collection*. Berkeley.
THUREAU-DANGIN F.
1922 'Une Donation à la Déesse Uṣur-amâtsa d'Uruk', *Revue d'Assyriologie* 19: 86-87.
VAN DIJK J.J.A.
1960 *Sumerische Götterlieder II. Teil*. Heidelberg.
1969 'Vert comme Tišpak', *Orientalia N.S.* 38: 539-547.
1982 'Fremdsprachige Beschwörungstexte in der südmesopotamischen literarischen 'Überlieferung', in: H.-J. Nissen and J. Renger (Eds.), *Mesopotamien und seine Nachbarn*. Berlin. Pp. 97-110.
1983 *LUGAL UD ME-LÁM-bi-NIR-GÁL. Le récit épique et didactique des travaux de Ninurta, du déluge et de la nouvelle création*. Leiden.
VANEL A.
1965 *l'Iconographie du dieu de l'orage*. Paris.
VIEYRA M.
1961 Review of Fr. Wetzel, E. Schmidt, A. Mallwitz, *Das Babylon der Spätzeit, Revue d'Assyriologie* 55: 50-51.
VOGELZANG M.E.
1986 'Kill Anzu! On a Point of Literary Evolution', in: K. Hecker (Ed.), *Keilschriftliche Literaturen*.
VON SODEN W.
1973 'Der Mensch bescheidet sich nicht. Überlegungen zu Schöpfungserzählungen in Babylonien und Israel', in: M.A. Beek (Ed.), *Symbolae Biblicae et Mesopotamicae Francisco Mario Theodoro de Liagre Böhl dedicatae*. Leiden. Pp. 349-358.
WATANABE K.
1985 'Die Sieglung der "Vasallenverträge Asarhaddons" durch den Gott Aššur', *Baghdader Mitteilungen* 16: 377-392.
WEIDNER E.
1952-53 'Die Bibliothek Tiglatpilesers I', *Archiv für Orientforschung* 16: 197-215.
WESTENHOLZ A.
1977 'Old Akkadian Schooltexts', *Archiv für Orientforschung* 25: 95-110.

WIGGERMANN F.A.M.
1986 *Babylonian Prophylactic Figures*. Amsterdam.
WILCKE C.
1974 'Zum Königtum in der Ur III-Zeit', in: P. Garelli (Ed.), *Le Palais et la Royauté*. Paris. Pp. 177-232.
WISEMAN D.J.
1969 'A lipšur litany from Nimrud', *Iraq* 31: 175-183.

The battle of Qadesh: Some reconsiderations*

M.J. DE BRUYN

1 INTRODUCTION

Since the publication of Breasted's pioneer study on the Battle of Qadesh (Breasted 1903) a continuous flow of publications has appeared ever since and is still continuing. One would expect that by now the last words on this battle had been said and that nothing new could be added. It appears, however, that there are still a number of uncertainties which thwart a detailed reconstruction of the course of this battle. Some of these uncertainties concern the geography of the area where the battle took place and, directly related to this, the positions of the various army units at a given time (Who, Where, and When, as they say). Other uncertainties concern the exact chronologial order of events because the various written sources do not all seem to tell exactly the same story. Since clarity on these points is a prerequisite for a reconstruction of this battle (a reconstruction from a military point of view being the primary interest of the present author), it will be attempted to solve some of these problems below (cf. Kuschke 1983; Fecht 1984a,b; Helck 1971).

2 THE GEOGRAPHY OF THE AREA

2.1 *Evidence from the Egyptian textual sources*

The combined evidence from the Egyptian sources, pictorial as well as textual, provides enough information in order to reconstruct the major part of the positions and movements of both the Egyptian and the Hittite army[1].

Since the Qadesh texts, at least P and B, are in a basically chronological order, it was my idea to make the best of this by attempting to 'synchronize' these texts, i.e. they can either be juxtaposed or be dovetailed into one another. Admittedly, this idea is not a completely new one (cf. Gardiner 1960). However, Gardiner did not use the literal text of P, B, and R, but paraphrazed the texts instead. The complete result of this synchronization would occupy to much space here, therefore only those sections relevant for the discussion below are presented in Figs. 1.1 - 1.5.

*The present article presents parts of a larger study on the Battle of Qadesh (de Bruyn forthcoming).

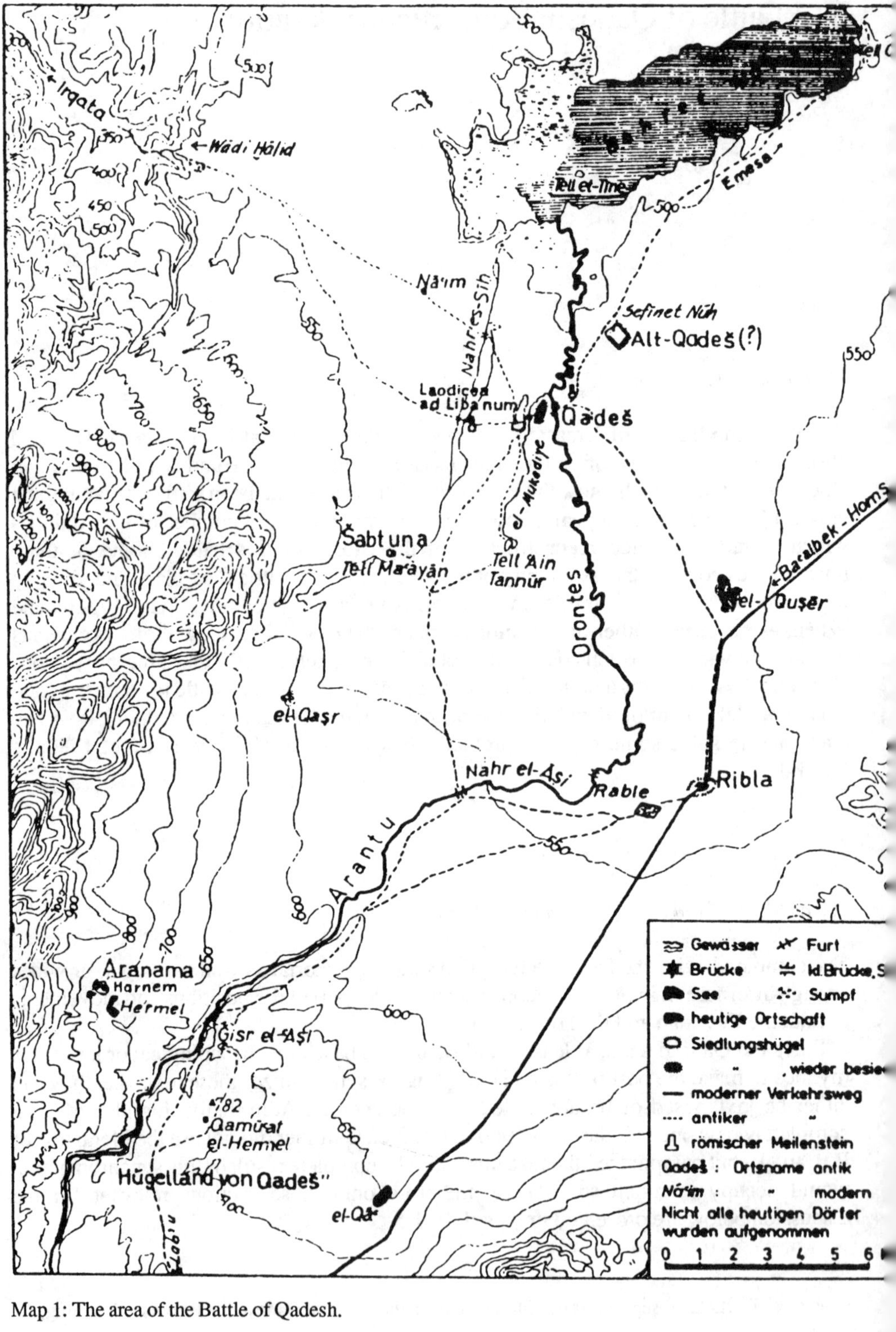

Map 1: The area of the Battle of Qadesh.

In the Qadesh texts there are three important enumerations of places in combination with the position of army units: the first one deals with the route of march taken by Ramesses II and reflects the situation during the morning, the second one is connected with the first and describes the positions of the other army units, the third reflects the situation later on, when the Hittite attack is in full swing.

2.1.1 *The marching route of the Egyptian army*

From the texts of P and B we are able to follow the march of Ramesses II accompanied by his Royal Guard, with the first division Amun immediately following behind, until they reach the place where they will start to pitch the Egyptian camp. The relevant passages can be found in P §§ 36-64, B §§ 1-32, and R §§ 1-4 (cf. Figs. 1.1 1.2)[2]. The following places occur in the texts and can be identified as follows (cf. Map 1):

– the Hill-country of Qadesh (P § 36).

– the Hill-country south of Qadesh (B § 4).

Without any doubt the point of departure for Ramesses II and those army units which had spent the night with him there. Already identified by Breasted as Qamu^c^at el-Hermel and since then generally accepted (Breasted 1903:18-21; Kuschke 1979:28 ff.).

– (the) ford of the Orontes (P § 38).

A fordable place on the Orontes (of which several existed and still exist) more than once mentioned in Egyptian documents (cf. e.g, Urk IV, 1302,7; 1310,18; 1311,1). It probably was of strategical importance as it can be connected with the reconstructed ancient roads (Kuschke 1979:30). The same ford is later referred to as being 'in the neighbourhood south of the town of Shabtuna' (P §§ 58-59), see below.

– the neighbourhood south of the town of Shabtuna (B § 7).

Most likely the area to the north of the ford. The identification of the town of Shabtuna has played a crucial role in any reconstruction of the battle because it was not exactly known until recently. Originally Shabtuna was located at Riblah (Breasted 1903:99 ff.) although this was not quite in agreement with the description of Shabtuna and the ford in the Orontes as stated in the text of P § 59. A more likely location for Shabtuna at Tell Ma^{c} yan was argued by A. Kuschke, who rightly rejected the older view.

Kuschke (1979:31-32); his down-to-earth commentary to the older view is worth repeating: 'Die opinio cummunis äußert sich nun dahin, daß Šabtuna mit Ribla identisch ist, und daß der eben erwähnte, 3 km westlich vom alten Ribla befindliche Orontesdurchgang nichts anderes sei als die "Furt im Suden von Šabtuna"! Ich frage erstens, wie man das mit unseren, immer noch gültigen Himmelsrichtungen in Einklang bringen soll; und zweitens: Ribla ist Ribla und nicht Šabtuna; drittens wäre es im damaligen Wegenetz ein Umweg gewesen, über Ribla zu dem flachen Gelände westlich von Qadeš zu gelangen.'

A look at Map 1 will make clear what the location of Shabtuna means for a reconstruction of the battle, the difference between the two locations is about 12 km as the crow flies and even more when measured over the road(s). But it really becomes significant when one has to place the division Pre^{c} 'in the neighbourhood south of the town of Shabtuna at a distance of 1 *jtr* (=10.5 km) from where His Majesty was;' (P §§ 59-60), or '(still) on the march to the south of the town of Shabtuna' (B § 73), only slightly before it will be attacked, while Ramesses II and the division Amun are already established at the new campsite (see further 3.ff).

From the synchronization of the texts it is more than clear that the enumeration of places and locations shows that the Egyptian army first crosses the ford on the Orontes

Table 1.1 Synchronized texts of Poem, Bulletin, and Reliefs

Poem	Bulletin	Reliefs
(§§ 1-35 take place before the area near Qadesh is reached) § 36: His Majesty proceeded northward. After His Majesty reached the Hill-country of Qadesh,		
	§ 1: Year 5, 3rd month of Summer, day 9, under the Majesty of Re^{c}-Harakhti. The-Strong-Bull-Beloved-Of-Truth. § 2: the King of Upper and Lower Egypt, Usima^{c}re^{c}-Setpenre^{c}, the Son of Re^{c}, Ra^{c}messe-Miamun, given life eternally. § 3: Now His Majesty was in Djahi on his second expedition of victory. § 4: A good awakening in Life, Prosperity, Health, in the tent of His Majesty in the Hill-country south of Qadesh. § 5: After this at the time of morning. § 6: His Majesty appeared in glory even as when Re^{c} arises, and he assumed the accoutrements of his father Monthu. § 7: The Lord proceeded northward, and	
§ 37: then His Majesty went ahead like Monthu, Lord of Thebes, § 38: and he crossed the ford of the Orontes, § 39: with the first army of Amun-gives Victory-to-Usima^{c}re^{c}-Setpenre^{c}.		
	His Majesty arrived at the vicinity south of the town of Shabtuna. (§§ 8-28, relate the 'Shashu-incident')	
§ 40: His Majesty arrived at the town of Qadesh, § 41: and now the wretched enemy of Khatti had come § 42: and had collected together all the foreign countries as far as the end of the sea.	§ 29: Travelling northwards by His Majesty;	

Table 1.1 (continued)

Poem	Bulletin	Reliefs
(§§ 43-53, list of Hittite allies) § 54: Now the wretched enemy of Khatti, together with the many foreign countries that were with him, § 55: stood there concealed and ready to the north-east of the town of Qadesh, § 56: while His Majesty was alone by himself with his followers, § 57: the army of Amun on the march behind him, § 58: the army of Pre^{c} crossing the ford § 59: in the neighbourhood south of the town of Shabtuna § 60: at a distance of 1 jtr from where His Majesty was, § 61: the army of Ptah being to the south of the town of Irnama, § 62: the army of Sutekh on the march on the road; § 63: and His Majesty had made a first battle-force from the foremost of his army, § 64: and they were upon the shore in the land Amurru.		
	arriving at the north-west of Qadesh.	
	§ 30: The camp of the army of His Majesty was pitched there.	§ 1: The first army of Amun-gives-victory-to-Usima^{c}re^{c}-Setpenre^{c}, given Life; in which Pharoah was. In course of pitching the camp. § 2: The living lion, follower of His Majesty, 'Slayer-of-his-enemies'. § 3: We (?) bring it into the presence of Pharaoh. § 4: (cartouches)
	§ 31: His Majesty took his seat on a throne of fine gold § 32: to the north of Qadesh on the west side of the Orontes.	
	§ 33: The coming of a scout who was in His Majesty's following	§ 8: The coming of Pharaoh's scout bringing two scouts of the enemy of Khatti into the Pharoah's presence.

Table 1.1 (continued)

Poem	Bulletin	Reliefs
	§ 34: with two scouts of the enemy of Kharri.	They beat them to make them say where the wretched enemy of Khatti was.
	§ 35: Then His Majesty said to them: 'What are ye?' § 36: And they said: 'We (belong to) the Chief of Khatti; § 37: he it is who sent us to look where His Majesty is. (§§ 38-50, the Hittite scouts enumerate the Hittite allies, cf POEM §§ 43-53) § 51: Look, they stand equipped and ready to fight behind Qadesh the Old.'	
	§ 52: Thereupon His Majesty caused high officers to be called into the Presence	§ 5: (cartouches)
		§ 6: The goodly god, powerful of strength, great of victories, slaying all foreign countries, making a great slaughter among their chiefs, the King of Upper and lower Egypt, Usima^{c}re^c-Setpenre^c, Son of Re^c Ra^{c}messe-Miamun, given life eternally. § 7: The first great horse of His Majesty 'Victory-in-Thebes' belonging to the great stable of Usima^{c}re^c-Setpenre^c, beloved of Amun.
	§ 53: to let them hear all the words which the two scouts of the enemy of Khatti who were in the Presence had spoken. § 54: Then said His Majesty to them: 'See ye the case § 55: wherein are the governors of foreign countries together with the chiefs of the lands of Pharoah. § 56: They stand and say to Pharoah every day: (§§ 57-71, Ramesses II complains about the unreliable information) § 72: Then the Vizier was ordered to hasten the army of His Majesty	

Table 1.1 (continued)

Poem	Bulletin	Reliefs
	§ 73: while they were still on the march to the south of the town of Shabtuna, § 74: so as to bring them as to where His Majesty was. § 75: But while His Majesty sat speaking with the high officers, § 76: the wretched enemy of Khatti had come with his infantry and his chariotry, § 77: and likewise the many foreign countries that were with him,	
§ 65: But the wretched enemy of Khatti stood in the midst of the army that was with him § 66: and he did not come out to fight through fear of His Majesty. § 67: But he had made to come men and horses exceeding many and numerous like the sand, § 68: and they were three men on a chariot § 69: and they were equipped with every weapon of warfare. § 70: Behold, they had been made to stand concealed behind the town of Qadesh,		
§ 71: then they came forth from the south side of Qadesh	§ 78: and they crossed the ford south of Qadesh	

and then moves further northwards into the direction of Shabtuna which is passed by on the way to Qadesh and the later campsite.

– (the) town of Qadesh (P § 40).

The town of Qadesh with its rulers was well known to the Egyptians, already during the XVIIIth Dynasty, from records and representations of earlier campaigns into this area (the most recent for them being a campaign of Ramesses II's father, Sethi I, during which there was also fighting with the Hittites in the neighbourhood of Qadesh (cf. Helck 1972[2], ÄA5, passim; Klengel 1969, Teil II, Abschn. 5; Kitchen 1982:20 ff.) Since it is clear from the texts that Ramesses II did not conquer the town the Egyptians did not even lay siege to it the text in P § 40 means clearly that he passed by the town and marched further northward.

– to the north-east of the town of Qadesh (P § 55).

– behind 'Qadesh the Old' (B § 26, § 51)

While Ramesses II is still marching northward (after) bypassing Qadesh, the texts give

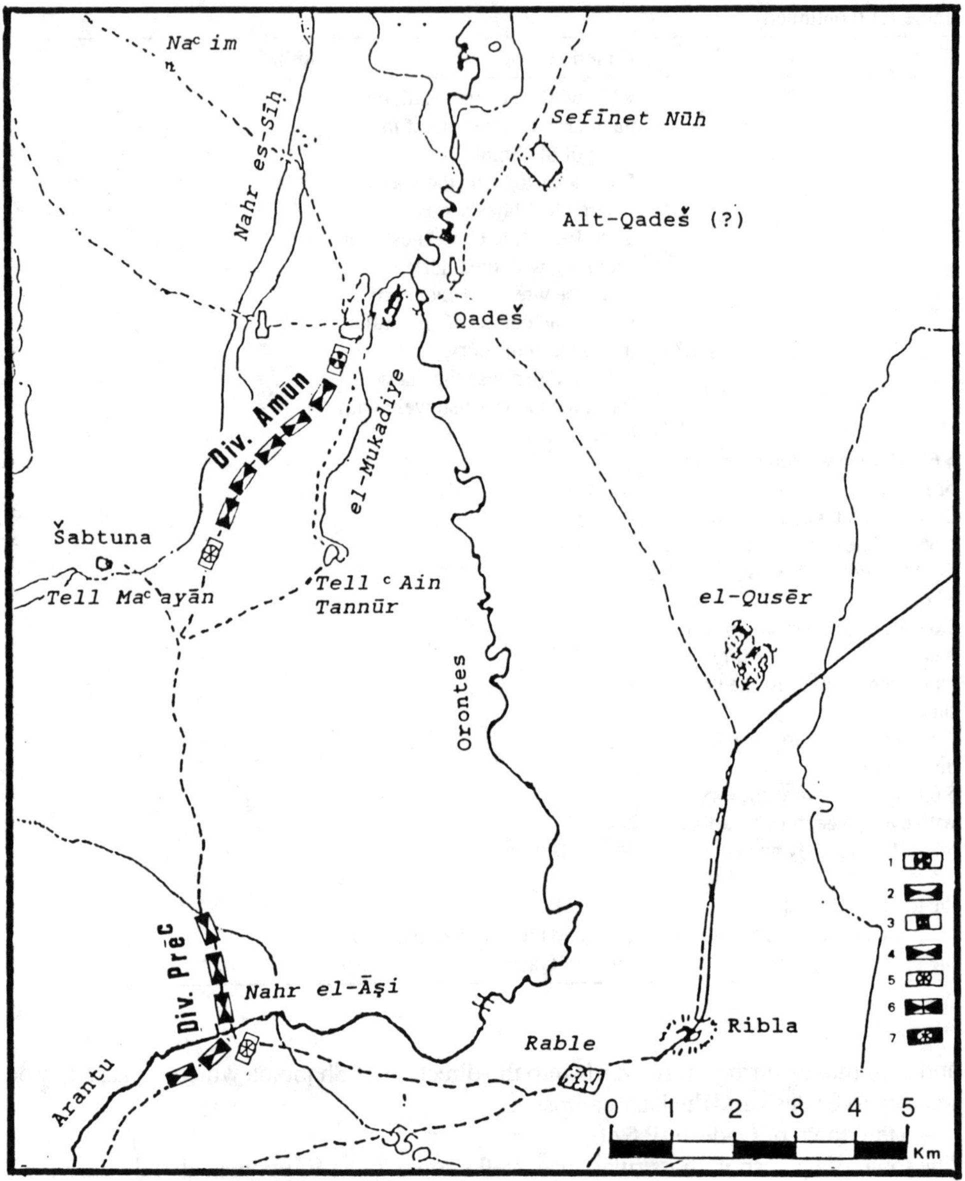

Map 2: Positions of Ramesses II/Div. Amun and Div. Pre' (P § 56- 64).

two indications that the Hittite army is present in the neighbourhood, apparently on the other side of the river Orontes. The majority of the Hittite army was probably stationed (and hidden) in the area to the east and north-east of Qadesh itself; the mound upon which the town was situated must have helped to keep them out of sight.

The location of 'Qadesh the Old' has caused some confusion in the past. The identification with Sefinet Nuh (cf. Map 1) seems the best solution. It seems to have been a campsite measuring 550 x 450 m with earthen walls (Kuschke 1979:33-34). A considerable part of the Hittite army could have been hidden inside and 'behind' it.

This passage is followed by a very important enumeration of the positions of the other Egyptian army units, clearly describing the situation at that moment, P §§ 56-64 (cf. 2.1.2). After this passage the story continues with the beginning of the Hittite attack (on the division Pre^c) and their subsequent discomfiture, P §§ 71-74 and B §§ 75-82 to which can be added R §§ 5-11, 17-19 (cf. 2.1.3 and 3.ff). The last indication of Ramesses II with his following must clearly reflect the farthest point of the march, i.e. the place where the camp was pitched; it is described as follows:

– to the north of the town of Qadesh, on the western side of the Orontes (P § 75)
– at the north-west of Qadesh (B § 29)
– to the north of Qadesh, on the western side of the Orontes (B § 32)

Although it is difficult to pinpoint this description to a location on the map, it seems most likely that the Egyptian camp was some 2 km to the north/north-west of Qadesh, at some distance from the river, more or less opposite the campsite of 'Qadesh the Old' at Sefinet Nuh (cf. Map 1). This location would enable the Egyptians to observe quite well the surrounding area and to be in a good defensive position if an enemy would approach the camp: enough space for their own chariotry to move around. It is furthermore indicated in the texts that Ramesses II takes his seat just outside the camp (cf.3.1.1).

These indications therefore make us familiar with the area, and we can mark the marching route of the vanguard units of the Egyptian army (Ramesses II with the Royal Guard and the division Amun). With the length of the marching route known, a calculation of the travelling speed of the army (basically determined by the infantry moving on foot) while allowing time for rest pauses and the crossing of the ford, and a calculation of the length of the army column, a hypothetical time-table can be reconstructed for this part of the day (de Bruyn forthcoming). The division Pre^c must have followed the same route until its march was interrupted by the Hittite attack (cf. 2.1.3).

2.1.2 *Positions of the Egyptian army units (still marching)*

One of the few crucial passages for the reconstruction of the events during this day is the enumeration of the positions of the Egyptian army units in P §§ 56-64 (cf. Fig. 1.1). From the immediately preceding text it is clear that these positions are described when Ramesses II is passing by the town of Qadesh (P § 40, cf. 2.1.1). The positions are described as follows:

§ 56: '(while) His Majesty was alone by himself with his followers,
§ 57: the army of Amun on the march behind him,
§ 58: the army of Pre^c crossing the ford
§ 59: in the neighbourhood south of the town of Shabtuna
§ 60: at a distance of 1 *jtr* from where His majesty was,
§ 61: the army of Ptah being to the south of the town Irnama,
§ 62: the army of Sutekh on the march on the road;
§ 63: and His Majesty had made a first battle-force from the foremost of his army,
§ 64: and they were upon the shore in the land Amurru.'

The importance for the reconstruction of the battle lies in the combination of three data (the earlier mentioned Who, When, and Where) which at this point can be perfectly related with one another:

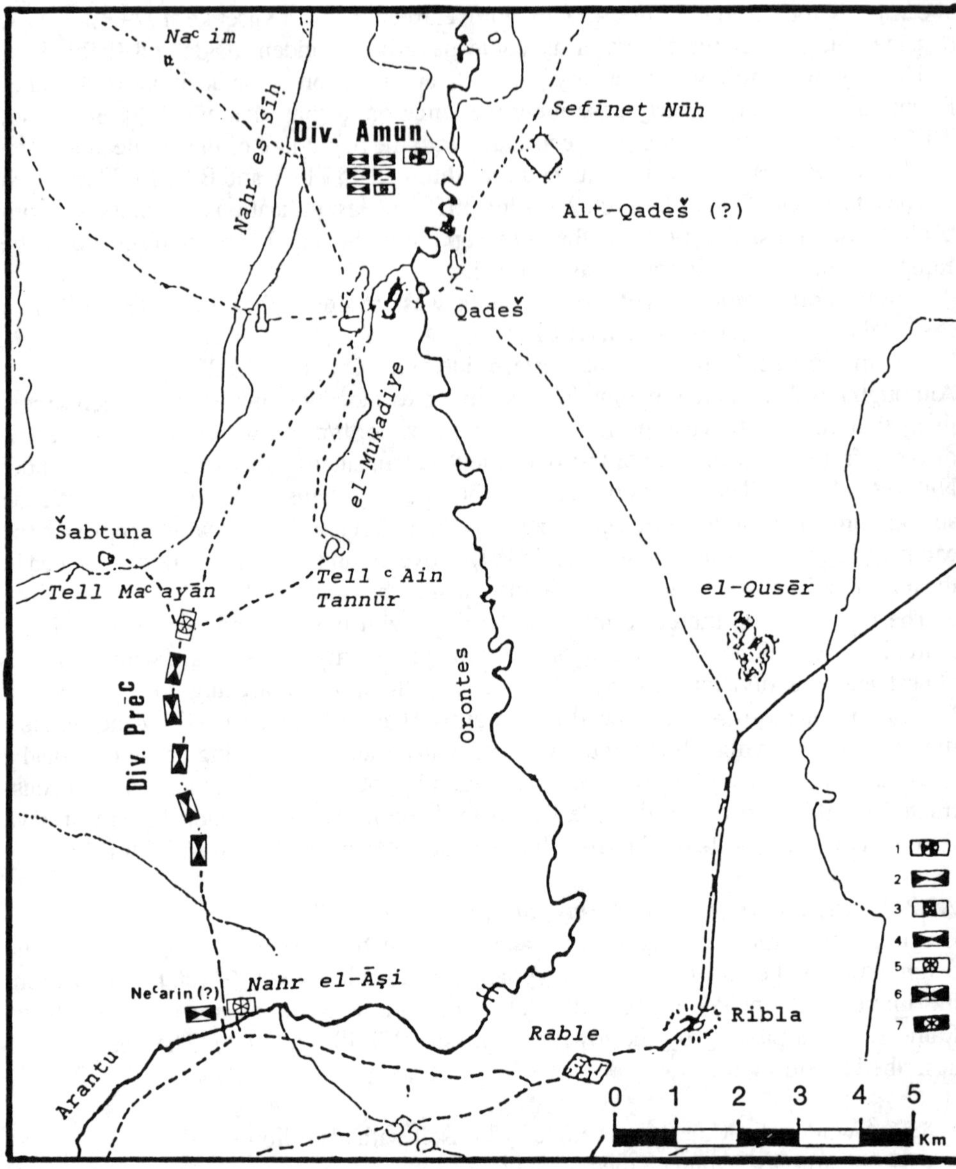

Map 3: Positions of Ramesses II/Div. Amun and Div. Pre' (B § 72- 74).

WHO: 1. Ramesses II (+ Royal Guard, + Division Amun)
2. Division Preᶜ
3. Division Ptah
4. Division Sutekh
5. 'First Battle-Force' (?); (further discussed below)

WHEN: ad 1. Ramesses II is still marching between Qadesh an the location of the later campsite;
ad 2-5. at the same moment in time as for 1), but with their own activities;

WHERE: ad 1. between Qadesh and later campsite;
ad 2) crossing the ford on the Orontes to the south of Shabtuna, at a distance of 1 *jtr* (ca. 10.5 km) cf. Schlott 1969:120-143) 'from where His Majesty was' (Ramesses II still marching);
ad 3. still marching to the south of the town Irnama;
ad 4. still marching 'on the road', further south than 3)
ad 5. not exactly clear, further discussed below.

Since the position of Ramesses II and the Division Amun and the position of the Division Pre^{c} are the most important for the subject of this paper, only these are indicated on Maps 2-5.

2.1.3 *The Hittite attack on the division Pre^{c}*

Apart from the mention in the enumeration above, the division Pre^{c} is referred to once again, slightly later on, when Ramesses II has already arrived at the location of the campsite and while he is conferring with his officers after they have been informed of the unexpected presence of the Hittite army (B §§ 52ff). As a result of this the Vizier is sent on a mission in order to hasten 'the army of His Majesty while they were still on the march to the south of Shabtuna' (B §§ 72-74). It will be clear that this can only be the division Pre^{c} which at this moment must be marching between the ford on the Orontes and Shabtuna (cf. map 3).

There is another reference to the division Pre^{c} in R § 11 from which it is clear that it has not yet been attacked while specific other events are taking place. From a chronological point of view this reference must be more or less simultaneous with B §§ 72-74. The importance of R § 11 lies in the fact that one of the other events is the attack on the Egyptian camp and Ramesses II. This presents a chronological difficulty since the attack on the division Pre^{c} is traditionally thought to have taken place before the attack on the Egyptian camp. It is therefore necessary to try to find out the exact order of events at the Egyptian camp and to try to relate them to the movements of the division Pre^{c} (cf. below).

The attack on the division Pre^{c} is only mentioned in P §§ 71-74 which can be juxtaposed to B §§ 78-81, the two passages are almost identical (cf. fig. 1.3 - 1.4). At the moment of the attack the division Pre^{c} must have been marching between Shabtuna and Qadesh. The Hittite chariotry is said to have come forth 'from the south side of Qadesh' and to have crossed the 'ford south of Qadesh'; it is not clear whether a ford on the Orontes is meant or one in the el-Muqadiye. In the traditionally accepted outline of the battle the Hittite chariotry first defeats the division Pre^{c}, puts it to flight, then veers northwards (while possibly pursuing fleeing soldiers of the division Pre^{c}) and attacks the Egyptian camp with Ramesses II (cf. Map 4).

3 CHRONOLOGY OF THE EVENTS AT THE EGYPTIAN CAMPSITE

3.1 *The movements of Ramesses II*

The texts of P, B, and R describe in full detail the movements of Ramesses II after the division Amun starts to pitch the camp to the north-west of Qadesh. The synchronization of the texts makes it possible to define the exact order and nature of his movements (cf.

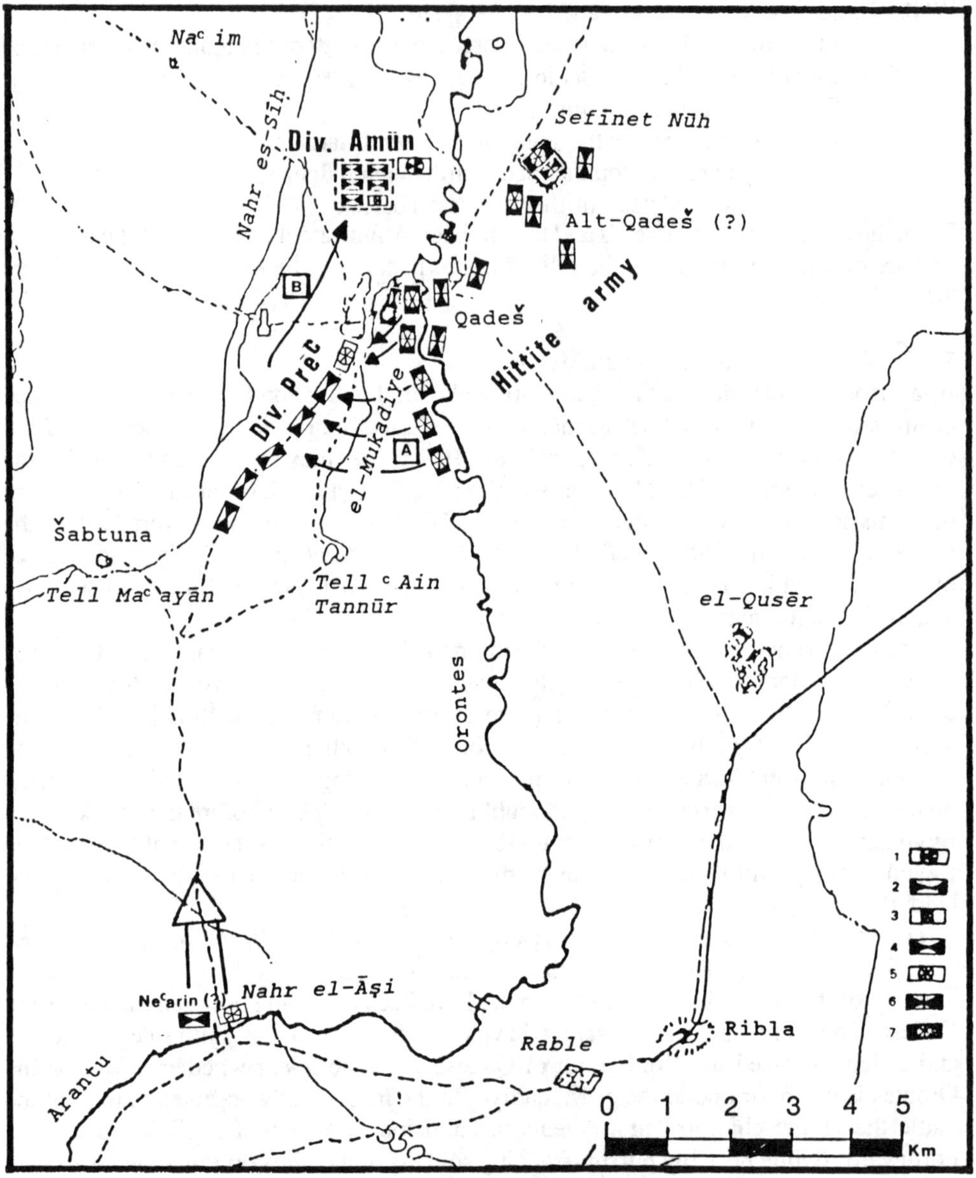

Map 4: Battle of Qadesh, traditional outline of the Hittite attack.

Fig. 1.2-1.5). Comparing the texts paragraph by paragraph reveals those places where they can be exactly juxtaposed and provides a chronological order for those events which are only told in either P, B, or R. The textual evidence renders the following movements or events which centre around the person of Ramesses II, summarized in Fig. 3 (a-o).

It will be observed that the total of the movements can be subdivided into three phases:

– at first Ramesses II remains seated (a-i)
– then Ramesses II arises; the enemy is near/at the camp (j-m)

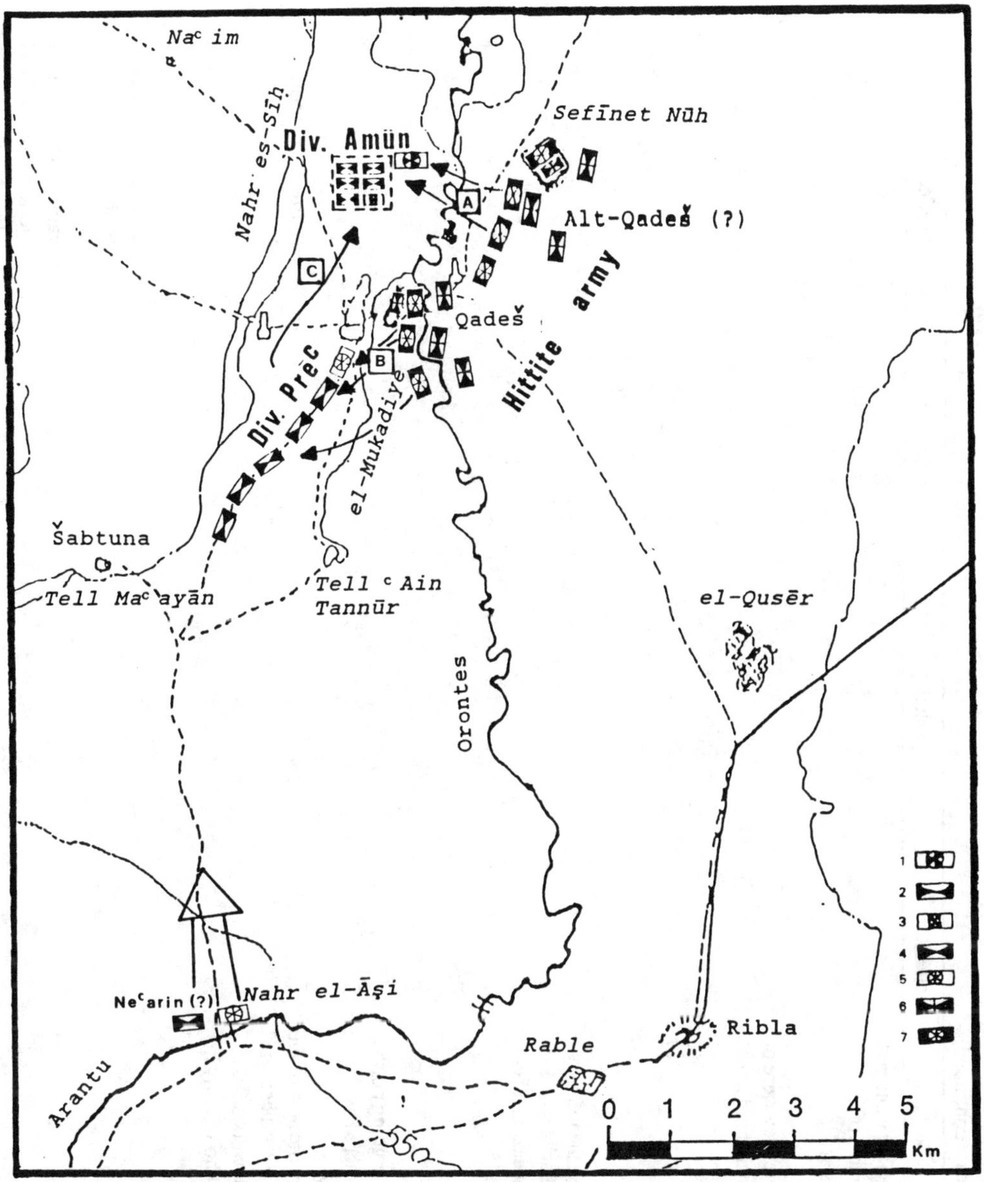

Map 5: Battle of Qadesh, revised outline of the Hittite attack.

– lastly Ramesses II mounts his chariot; drives outside camp (n-o)

At a first glance there seem to be no difficulties in interpreting this order of events but a closer look reveals a few unexpected details.

3.1.1 *Movements a-f*

These are very clear and do not need further explanation, only there where new information can be found, it will be explained.

Table 1.2 (Synchronized texts continued)

Poem	Bulletin	Reliefs	(Reliefs)	(Reliefs)	(Reliefs)
§ 72: and they broke into the army of Pre^{c} in its midst	§ 79: Then they entered into the midst of His Majesty's army				
§ 73: while they were on the march and did not know and were unprepared to fight.	§ 80: while they were on the march and did not know.				
§ 74: Thereupon the infantry and the chariotry of His Majesty weakened before them.	§ 81: Then the infantry and the chariotry of His Majesty weakened before them.				
	§ 82: while going northward to where His Majesty was.				
	§ 83: Then the host of the enemy of Khatti hemmed in the followers of His Majesty who were by his side.				(retrospective part of R § 11)
					§ 11: (... they found that) the host of the enemies of Khatti surrounded the camp of Pharaoh on its western side, while His Majesty sat alone, his army not with him, while

army, while the army of Amun in which Pharaoh LPH was had not yet finished the pitching of the camp, and while the army of Pre^{c} and Ptah were on the march and their ... (?) had not yet come out of the wood of Robawi.

§ 9: The coming of the Fan-Bearer of Pharaoh to say to the King's children together with those (?) of Mutnofre: 'Do not go out on the west side of the camp and keep clear of the battle.'

§ 10: The Fan-Bearer on the right of the King, King's scribe, great commander of the army, first charioteer of His Majesty, Pre^{c}hiwenamef.

§ 17: The goodly god smiting the Nine Bows and slaying a million in the completion of a moment. Making nonexistent the chiefs of foreign countries, prostrate in their blood, none other with him, the King of Upper and Lower Egypt, Usima^{c}re^{c}-Setpenre^{c}, given life, the son of Re^{c}, Ra^{c}messe-miamun.

§ 18: The goodly god, powerful of strength, great of victories, defeating all foreign countries, King who slays his enemies with his single arm, stout of heart, making them sprawl.

§ 19: The goodly god, powerful of strength, great of victories, defeating all foreign countries. the King of Upper and Lower Egypt, Usima^{c}re^{c}-Setpenre^{c}, the son of Re^{c}, Ra^{c}messe-miamun.

	§ 84: Then His Majesty caught sight of them.	His Majesty made ...(?) of the ....(?) followers of His Majesty as he sat to the north of Qadesh.	
	and thereupon he arose quickly	Then he arose quickly	The arising of His Majesty as he sat to the north-west of Qadesh
	§ 85: and was enraged against them like his father Monthu.		
§ 75: but His Majesty stood firm to the north of the town of Qadesh on the western side of the Orontes. § 76: Then they came to tell it to His Majesty, and then His Majesty appeared in glory like his father Monthu,			
§ 77: he assumed the accoutrements of battle and girded himself with his corselet,	§ 86: He assumed the accoutrements of battle and girded himself with his corselet.		
§ 78: he was like Ba^{c}al in his hour.	§ 87: he was like Sutekh at the moment of his power.		

Majesty being 'Victory-in-Thebes' of the great stable of Usima^{c}re^{c}-Setpenre^{c}, beloved of Amun.			
§ 80: Then His Majesty started forth at a gallop.	§ 89: and he started forth quickly alone by himself.		
	§ 90: His Majesty being powerful, his heart stout, and none could stand before him. § 91: All his territory was ablaze with fire and he burned every foreign country with his hot breath. § 92: his eyes savage when he saw them, and his might flared up like fire against them. § 93: He took no note of millions of foreigners, he regarded them as chaff.		
§ 81: and he entered into the host of the enemy of Khatti.	§ 94: Then His Majesty entered into the host of the enemy of Khatti	and he entered in among the Khatti enemies, being like a storm which goes forth from heaven and his might like fire in stubble, and he being powerful like his father Monthu.	and he entered into the host of the enemies of Khatti being along by himself, none other with him.
	§ 95: together with the		

	many foreign countries that were with him § 96: His Majesty being like Sutekh the great of strength and like Sakhmet at the moment of her raging	
§ 82: while he was alone by himself and none other with him. § 83: Then His Majesty went to look about him. § 84: and he found that 2500 chariots hemmed him in on his outer side. § 85: consisting of all the 'champions' of the enemy of Khatti with the many foreign countries which were with them. § 86: from Arzawa, Masa, and Pidasa, while they were three men on a cha-		He found 2500 chariots in four bodies hemming him in on every side of him.

Ad Movement c: The location where Ramesses II takes his seat is described as being 'to the north of Qadesh and to the west of the Orontes' (and 'to the north of Qadesh' in R § 18) instead of 'to the north-west of Qadesh' as the description for the camp in B § 29. Most authors have inferred from this that Ramesses II with his following separated himself from the camp (possibly while it was still being pitched).

3.1.2 *Movements g-h*

These are of crucial importance because it gives the exact moment when Ramesses II is attacked by the Hittite chariotry. There is a remark by Gardiner that there would be a contradiction in the texts concerning the 'hemming in'. It is worth to be repeated in full since it is indeed an important observation: 'B 83. This sentence in which the verb *inḥ* 'hem in' is used as in P 84, R 11, and R 19.6, suggests that when the Pharaoh was surrounded he had about him no more than his personal attendants. In R 11.1, however, it is 'the camp' which is surrounded, though this is said in the same passage to have been pitched by the army of Amun, but left unfinished by them. There seems to be somecontradiction here. (Gardiner 1960:33-34).

This contradiction does not exist if the hemming in of Ramesses II with his following is viewed as a separate event from the hemming in of the camp of the division Amun; both could of course have happened at the same time.

3.1.3 *Movement i*

This movement, derived from R § 18.5-6, is the last thing Ramesses II does before he indeed rises from his throne. The text of these lines is unfortunately slightly damaged (KRI II, 134.11):

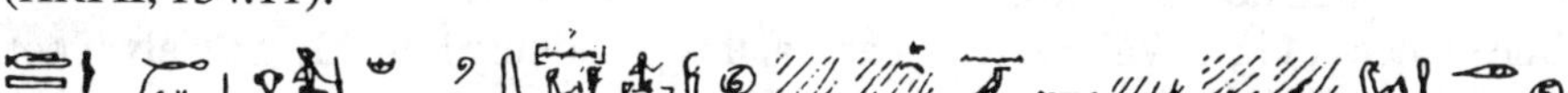

jr.n.f. ḥm.f [...] n n3 n [...] šmsw [n] ḥm.f jw.f ḥms (.w) ḥr mḥty Kdš

'His Majesty made [...] of the [...] the following of his Majesty as he was seated to the north of Qadesh'.

Any attempt to restore what has been damaged remains guesswork but the arrangement of the texts in Fig. 1.4 leaves little doubt as to what was probably stated. The Hittite attack comes probably so unexpectedly that the whole entourage – including non-combatants – has to take up arms in order to put up some defence. Since it is hard to imagine that Ramesses II would be without his Royal Guard, one must suppose that they were included into the 'following', or not mentioned specifically since they probably did not need to be organized into something.

A possible restoration of these lines could be made with a word like  *śkw* 'battle-force' in the first lacuna (cf. P § 63), and a word like 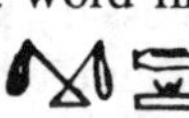 *dmḏ* 'total, all of'. Both suggestions would fit in the lacunae as they appear on the wall (cf. Desroches-Noblecourt et al. 1971, Pl. IV (top left), and Pl. 53). The lines would then have read:

jr.n.f ḥm.f [śkw] n n3 n [dmḏ] šmsw [n] ḥm.f jw.f ḥms(.w) ḥr mḥty Kdš

'His majesty made a [battle-force] (consisting) of [the whole of] the following of His majesty as he was seated to the north of Qadesh.'

3.1.4 *Movement j*

At this moment Ramesses II remains no longer seated. Exactly the same statement occurs in R § 18 and § 19 and provides one place where the texts can be synchronized. The texts of R § 18 and § 19 (together with R § 11 discussed below) belong to the longer ones among the Captions (or Legends). It is this moment which provides a clear chronological division for the course of events: certain events clearly take place before Ramesses II arises, others clearly afterwards.

3.1.5 *Movement k*

P § 75, synchronized with B § 85, can only mean that Ramesses II is at this moment really fighting the Hittite attackers which, after movements h, i, and j, is only logical[3].

3.1.6 *Movement l*

It is astonishing to find out how much can be hidden in a single, short sentence like the first line of P § 76: 'Then they came to tell it to His majesty,...'. Nowhere before or after this statement any other reference to this unknown message can be found in all the texts of P, B, and R, nor any reason why it had to be told. Yet it must have been a message of some importance or otherwise it would not have been included. It is surprising that most authors of studies on the Battle of Qadesh did not bother too much about this unknown message. In all there are three possibilities for the contents of this message:

1. Ramesses II is being told about the interrogation of the Hittite scouts (Gardiner 1960:18);
2. Ramesses II is being told about the attack on the camp of Amun (Gardiner 1960, text on frontispiece; Kitchen 1982:55);
3. Ramesses II is being told about the defeat of the division Pre[c] (Breasted 1903:11; Strum 1939:123; Faulkner 1958:58);

ad 1: One glance at Fig. 3 will make it clear that this possibility is highly unlikely if not impossible. It is more than clear that Ramesses II probably was present at the interrogation of the Hittite scouts, or at least already knows what they had to say. It is equally clear that he immediately acts upon this information (Movements e-f). Then follows the attack upon the camp, the arising of Ramesses II, and the first attempts to withstand the Hittite attack (Movements g-k). It only follows that there is no need whatsoever to inform Ramesses II at this moment that the Hittites stand behind Qadesh, ready to fight, when he himself is already fighting them! It also means that the scene in the reliefs with Ramesses II on his throne and his officers in front of him, cannot be the illustration to 'Then they came to tell it to His Majesty,...' (Gardiner 1960:18), because, when the message is told to Pharaoh, he is no longer seated but moving on foot.

ad 2: From the discussion of the first possibility it is already clear that the second possibility is as unlikely as the first. Not even the most servile follower of Pharaoh would have found it necessary to inform his lord about any attacking Hittites when his lord was already fighting them, while the follower himself might have had to fight his way through the turmoil of battle in order to reach his Pharaoh. One could argue that the attack upon the camp was different from the attack upon Ramesses II with his following since he was outside the camp. If so, then Pharaoh with his following would have been a primary target for the enemy and would undoubtedly have been attacked first. It seems more likely that both were attacked at the same time by the same group of Hittites, although fighting could have been going on at various places in or near the camp.

ad 3: This is no doubt the best option for the contents of the unknown message; not only does the synchronization of the texts suggest so, it also fits best in the course of events. It could very well refer back to an earlier event, namely when it was decided to send the Vizier to hasten up the division Pre[c] to the Egyptian camp. This was the best countermeasure Ramesses II could make at that moment: it would unite at least half of his army and he could more safely await whatever was going to happen next. It was the prospective arrival of the division Pre[c] upon which Ramesses II and those with him must have set their hopes when the camp was unexpectedly attacked, they were forced to hold out until it arrived. And then the events at the camp reach a dramatic climax when a certain messsage of crucial importance is delivered to Ramesses II: the division Pre[c] will not arrive and cannot relieve Ramesses II and his troops because it has been defeated by the Hittites. Those of the division Pre[c] who do arrive at the camp and who probably bring this news are fleeing in disorder and running for their life. Immediately prior to P § 76 the text reports on the attack upon the division Pre[c] and the fleeing of its infantry and chariotry to the north. This in itself is already a strong indication as to what could be referred to in P § 76.

3.1.7 *Movements m-o*

These are also very clear and do not need elaborate explanation. Apparently the Egyptians manage to put up some sort of defence and Ramesses II manages to prepare himself for further fighting by mounting his chariot.

3.2 *The chronological discrepancies*

Although the course of events resulting from the foregoing analysis looks fairly certain, there are a few points which need further consideration. Briefly summarized, the traditional general outline of the Battle of Qadesh is supposed to be as follows (cf. Map 4):

- Hittites attack the division Pre[c] to the south of Qadesh;
- Hittites then turn northwards, pursuing fleeing Egyptians;
- Hittites attack camp of division Amun and Ramesses II;

There is supposed to be only one group of attacking Hittites. It is clear that the reconstructed order of the events does not quite agree with this outline.

Firstly, in the reconstructed order of events Ramesses II is already fighting the enemy before the unknown message is delivered to him, i.e. before any of the fleeing soldiers of the division Pre[c] or any of the pursuing Hittites have reached the camp. This is certainly in disagreement with the generally supposed order of events. One could argue that the events in question happened so quickly and in so short a time that they seem to have happened more or less simultaneously, which is the impression one gets if one reads only the text of P, but this seems not very likely. The observation above remains as yet unexplainable.

Secondly, if the message delivered to Ramesses II is about the attack on the division Pre[c], than this attack itself must have taken place some time earlier. The fighting will have taken some time even if it was only a short fight and some time must have elapsed before the first persons (whether fleeing soldiers or attacking Hittites) moved from the place of the attack to the location of the camp (the distance can be estimated as at least 4-5 km, cf. Map 1).

It is therefore rather unlikely that part of the attacking Hittites could have appeared at

Table 2.1 Comparison of variants R § 11 (KRI II, 131-133)

Var1	VarL1	VarL3	VarR1	VarK2
(1) The coming of the Ne^{c}arin of Pharaoh LPH from the land of Amur	(1) The coming of the Ne^{c}arin of Pharaoh LPH from the land of Amur	(1) ... Pharaoh LPH from the land of Amur	(1)	(1)
... found ...(2) the camp of Pharaoh LPH on its western side	They found that the host (2) of the enemy of Khatti hemmed in the camp of Pharaoh LPH	They found that hemmed in (2) ... the camp of Pharaoh LPH	They found that ... enemy of Khatti hemmed in the camp of Pharaoh LPH on its western side	... the camp
	while His Majesty sat alone and his army was not with him	while His Majesty sat alone and his army was not with him		
while the host of chariots hemmed in ... had come (?)... his soldiers, (3) while the army of Amun in which Pharaoh LPH was had not			while the host of chariots hemmed in (2) ... His Majesty ...	... (3) His Majesty ...
yet completed the pitching of the camp, and while the army of Pre^{c} and the army (4) of Ptah were (still) on the march and their			... pitching of the camp, and while the army of Pre^{c} and of Ptah were (still) on the march and (3) ...	of Ptah were (still) on (4) ...
soldiers (?) had not yet come out of the wood of Robwai.			not ...	
And the Ne^{c}arin (5) broke into the host of the wretched enemy of Khatti as they were entering the camp	And the Ne^{c}arin (3) broke into them as they were entering the camp	(3) ... them as they were entering the camp	... broke into ... entering the camp of Pharoah LPH	... broke into (5)
and the servants of Pharaoh LPH (6) killed them They did not allow that even on of them escaped	and the servants of His Majesty killed them END	and the servants killed them END	and the servants of Pharaoh LPH killed them END	(6)

	[illegible] great strength of Pharaoh LPH (7) their goodly Lord, while he was behind them like a mountain of copper and like a wall of iron	
35	for ever and ever END	... ever ... END

Table 2.2 Chronological arrangement of R § 11 (combined variants)

Ne^{c}arin	Ramesses II	Amun Division	Pre^{c}/Ptah Division
		... the host of the enemy of Khatti hemmed in the camp of Pharaoh LPH on its western side,	
	while His Majesty sat alone, and his army was not with him, while the host of the chariots hemmed in ... ... had come (?) ... ... his soldiers.	while the army of Amun in which Pharaoh LPH was had not yet completed the pitching of the camp.	while the army of Pre^{c} and the army of Ptah were (still) on the march and their soldiers had not yet come out of the wood of Robawi.
The coming of the Ne^{c}arin of Pharaoh LPH from the land of Amur. They found that			
And the Ne^{c}arin broke into the host of the wretched enemy of Khatti as they were entering the camp, and the servants of Pharaoh killed them. They did not allow that even one of them escaped, their hearts filled with the great strength of Pharaoh LPH, their goodly Lord, while he was behind them like a mountain of copper and like a wall of iron for ever and ever.			

the campsite at the same time or so shortly after the attack upon the division Pre^{c}. This problem can be solved however with the analysis of the text of R §, which has been left out so far.

3.3 *The evidence of R § 11*

R § 11 has been preserved in five versions, one more complete than the other, some containing explicit statements that are absent in others (cf. KRI II, 131-133). These five versions can easily be dovetailed into one another, producing a combined version with all the information (cf. Fig. 2.1; Gardiner 1960:36-37)). The text of this combined version is chronologically arranged in Fig. 2.2 (cf. discussion below). The text of R § 11 belongs to the scene of the arrival of the N- ^{c}-rw-n3, Ne^{c}arin or Na^{c}runa, who are depicted marching in battle- order towards the camp of the division Amun. The text however contains much more than the announcement of their arrival: it also describes the situation in the camp as they find it, referring to what has happened earlier there and simultaneously to the situation of the other Egyptian units. Lastly, it also states briefly what happens after their arrival. In this paper only the evidence for the events at the campsite will be discussed, any problems concerning the Ne^{c}arin themselves are left out.

3.3.1 *The arrival of the Ne^{c}arin*

Ll. 1-3 (i.e. the line-numbers in the left margin of Fig. 2.1) are a factual statement announcing the arrival of the Ne^{c}arin. In l.4, 'They found that...', belongs to the same moment in time as their arrival, but the situation which is then described cannot be simultaneous with their arrival. If this situatition did not already exist, i.e. had begun earlier, then it could never have been stated that the Ne^{c}arin found the situation as such. From the context of R § 11 it is difficult to determine at what moment the Ne^{c}arin do exactly arrive at the campsite. It is clear that they must have arrived after the Hittites have already partly penetrated (or completely occupied?) the camp and when Ramesses II with his following is still being 'hemmed in'.

3.3.2 *R § 11 and the movements of Ramesses II*

R § 11.4-9 provides a first link with the so far synchronized texts, particularly with the movements of Ramesses II as discussed above (cf. Figs. 1.4, 3). From these lines it is clear that Ramesses II with his following is hemmed in while he is still seated. While the line in question clearly reads:  jw.f ḥm.f ḥms(.w) w^{c}(.w): 'while His Majesty *sat* alone', a number of authors have translated this differently: '(while) His Majesty had been camping alone' (Breasted 1903:116); 'while His Majesty was alone' (Schulman 1962:47). Others omit this line, occurring in R § 11 var L1/L3 only, while discussing the complete text of R § 11 Fecht argues that these variants 'sicher sekundär sind', and 'Hier versuchte man mit untauglichen Mitteln das Thema von der Einsamkeit des Königs nochmals herauszu-arbeiten, ...' (1984:28-1). As can be seen, var L1/L3 are shorter versions of R § 11, but whether one should look upon them as having no significance is another matter.

The line in question should be compared with a similar line like e.g., P § 56:
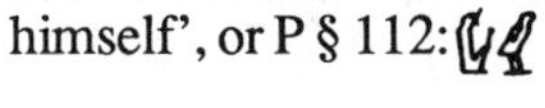
, jsṯ ḥm.f w^{c}.w ḥr tp.f 'while His Majesty was alone by himself', or P § 112:

, jw.j w^{c}.kwj 'while I was alone'. The Qadesh texts are

Table 3: Movements of Ramesses II at the Egyptian Camp

	Poem	Bulletin	Reliefs
		Ramesses II is seated	
a		- arrival n-w of Qadesh (§ 31)	
b			- in/near camp (§§ 1-3)
c		- takes seat on throne (§ 31)	
d		- interrogates Hittite scouts (§§ 35-51)	
e		- confers with his officers (§§ 52-71)	
f		- sends Vizier on mission (§§ 72-74)	
g		- being hemmed in with his following (§ 83)	
h		- catches sight of enemy (§ 84)	
i			- organizes his following into some sort of defence (?) (§ 18)
		Ramesses II moves on foot	
j		- arises quickly (§ 84)	- arises quickly (§ 18/§ 19)
k	- stands firm against enemy (§ 75)	- is enraged against enemy (§ 85)	
l	- is being told of ...(?) (§ 76)		
m	- assumes accoutrements of battle (§ 77)	- assumes accoutrements of battle (§ 86)	
		Ramesses II drives chariot	
n	- mounts his chariot (§ 79)	- mounts his chariot (§ 88)	- mounts his chariot (§ 18)
o	- enters host of enemy (§§ 80-81)	- enters host of enemy (§ 94)	- enters host of enemy (§ 18/19)

unambiguous in describing Ramesses' movements (see 3.1 above). If, therefore, it is explicitly stated that Ramesses II is seated, then there is no compelling argument to translate this otherwise; in doing so one misses an important detail, and, in this case, the confirmation of the arrangement of Ramesses' movements (see Fig. 3). It also lends further strength to the assumption that the 'hemming in' must have happened very unexpectedly, which would hardly be the case when Ramesses II had already known about the Hittite attack on the division Pre^{c}, or could have guessed that something was going on further to the south, or when the attacking Hittites could have been seen approaching from the south. The order in which these events are told in B §§ 83-84 might be significant: Ramesses II does not see the enemy approaching before he is being hemmed in, no, it happens the other way around! The Hittites are already upon him before he has had a real chance to see them coming. In other words, this would hardly have been the case when these Hittites had to drive from the south of Qadesh to the location of the camp, at least some 4 km to the north (cf. Maps 4/5), and after the attack on the division Pre^{c} had started or maybe was already over.

R § 11.10-12 is rather damaged but refers probably to the same attacking Hittite chariotry.

3.3.3 *R § 11 and the attack upon the division Pre^c*

From the whole of the foregoing discussion it has become clear that, if the traditionally supposed general outline of the Battle of Qadesh has any value, then the division Pre^c must be under attack before Ramesses II and the camp are attacked; it has become equally clear that this view can be seriously doubted. The next lines in R 11 give an enumeration of the positions of the other units of the Egyptian army, it can be considered as a parallel to P §§ 52-64 (cf. 2.1.2). Both enumerations can be connected with a specific moment in time: P §§ 56-64 at the moment when Ramesses II is passing by Qadesh, R § 11.13-21 at the moment when the Hittite attack upon Ramesses II and the camp begins. Important for this discussion are only the references to the divisions Amun and Pre^c:

R § 11.13: 'while the army of Amun in which Pharaoh LPH was had not yet completed the pitching of the camp, while the army of Pre^c and the army of Ptah were still on the march and their soldiers (?) had not yet come out of the wood of Robawi'.

There is no problem were the division Amun is concerned, but for the division Pre^c this statement seems to be false: it should be under attack or already be defeated by this time. It is surprising to find out that the line on the division Pre^c has been considered as false, or as a discrepancy, because it does not fit with a supposed course of events (Wrezinski 1935, Taf. 178; Fecht 1984:30). If it is clear from R § 11 that Ramesses II and the camp are attacked by the Hittites before the division Pre^c is attacked, then this should be accepted as a fact for any reconstruction of the events, even more so because the other texts do not contradict this. The implication is clear: the division Pre^c is attacked after Ramesses II, his following, and the camp, are already under attack, and not before. This is a confirmation of the earlier 'as yet unexplainable' observation that Ramesses II was fighting the Hittites before they could have been there! (cf. 3.2).

4 CONCLUSIONS

The analysis of the textual evidence reveals that the Hittites were at the camp before the division Pre^c was attacked. This can only mean that the latter was attacked by another part of the Hittite chariotry, i.e. the Hittites had divided their forces into two attack-forces, each with their own objective. The Hittite chariotry executing the attack upon the division Pre^c undoubtedly turned northwards after they had partly or wholly defeated it, but they did not start a new attack upon Ramesses II and/or the camp: they joined a northern attack-force which was already engaged in attacking the campsite.

But not only from the texts comes such an unambiguous statement: also the reliefs confirm the existence of two Hittite attack-forces. An analysis of the geographical situation in the reliefs (mainly the Battle Scene variants) reveals that this is correctly depicted (de Bruyn 1988). Unfortunately not all the reliefs have been completely preserved, but on two of them there are without doubt Hittite chariots crossing a river to the south of Qadesh as well as to the north of Qadesh. For what other purpose could this manoevre have been executed than for an attack upon the divsion Pre^c (to the south) and for an attack upon the Egyptian camp (to the north)? In one of the reliefs it even seems that the chariotry crossing to the north of Qadesh was accompanied by a small body of Hittite infantry.

If the analyses by the present author are correct, then the course of events during the Battle of Qadesh has to be revised as follows:

– Hittite chariotry (+ infantry?), a northern attack-force, attacks Ramesses II and the camp of the division Amun, to the north-west of Qadesh;

– simultaneously, or slightly later, Hittite chariotry, a southern attack-force, attacks the division Pre^{c}, marching between Shabtuna and Qadesh;

– after defeating the division Pre^{c} the southern attack-force turns northwards and joins in the attack upon the camp;

– at a critical point during the battle in/near the camp the Ne^{c}arin arrive and the Egyptians manage to break out.

– a Hittite reserve-force, also chariotry, joins the battle but cannot prevent the retreat of the Egyptian army.

This revised outline is shown in Map 5. In order to make any reconstruction as accurate as possible, the traditional general outline and the revised outline, in Maps 4/5 have been made with the help of the map prepared and published by Kuschke (Kuschke 1976, 1979). cf. also Goedicke (1985:85).

This still leaves us with the seemingly unavoidable question of who really won the battle. Seen from a strategical point of view, the Hittites won because the outcome of the battle enabled them to keep their sphere of influence in northern Syria, Qadesh and Amurru remained under Hittite control (Klengel 1969:32-33).

From a tactical point of view, one might argue that the Egyptians won because they managed to avert a total defeat under unfavourable conditions. The following quotation might elucidate the use of the terms 'strategical' and 'tactical': 'The terms 'strategy' and 'tactics' are so commonly used and misused that we should pause for a moment to consider their meaning before examining their application on the ancient field of battle. Military analysts know that the distinctions between strategy and tactics are often slippery, that they are sometimes so interrelated that they become blurred. Still, for reasons of clarity, the word 'strategy' is normally used to refer to the military plan for the conduct of a war or a campaign, while 'tactics'is reserved for battle plans.' (Ferril 1986:54).

There probably is no definite answer to the question of 'who really won', but a good description was given by Kitchen who called it a 'Triumph in Disaster' for the Egyptians, and might well have added a 'Disaster in Triumph' for the Hittites (Kitchen 1982: 53).

Finally, a few words may be said about Bronze Age warfare in general because it is an important period in the development of warfare, as expressed in the words of the late Yigal Yadin: 'It is equally evident that the art of warfare reached its highest standard in the first half of the second millennium. This was true of combat in open terrain, with the chariot and the composite bow as the principal battle instruments; of the structure of fortifications, with their glacis and moats; and of attack on a fortified city, in which the battering ram held a prized place. In no later period, until the evention of gunpowder, did the armies of the world succeed in introducing onto the battlefield new methods or weapons of war.' (Yadin 1963:75). To these might be added Ferrill's words from his discussion of two major 'Egyptian' battles during this period (battle of Megiddo, and battle of Qadesh): 'The strategic and tactical significance of these New Kingdom battles, especially if they are considered to be representative of the warfare of their period, is staggering. The size of the armies, their tactical organization, the use of chariots and other specialized units and the quality of generalship all show a degree of military sophistication that would be hard to match in many later historical periods. If one makes an allowance for the technological limitations of the Bronze Age, it is not difficult to conceive that the quality of generalship found in some of the Egyptian Pharaohs was competitive with the best generalship of any period down to modern times.' (Ferrill 1986: 60).

NOTES

1. Since the few short references from the Hittite sources do not tell us anything about the actual course of events during the battle, we are by necessity committed to the Egyptian sources comprising textual as well as pictorial evidence. As for the textual evidence, the so-called Qadesh texts, all references to them will be made with the help of their latest edition by Kitchen (1968; henceforth abbreviated KRI II). This edition contains a complete survey of all the various copies of the two versions of the Qadesh texts, the so-called 'Poem' and 'Bulletin', and also of the Legends or Captions accompanying the reliefs; henceforth abbreviated as P, B, R. As for the pictorial evidence, the Qadesh Reliefs, most references will be to Wreszinski 1935; henceforth abbreviated Wr II.
2. There are a number of good translations of the Qadesh texts and it seems unnecessary to present a complete new translation. The translations used in Figs. 1.1ff. for the texts of P, B, and R mainly follow those by Gardiner (1960:7-34). Other translations consulted (some with commentary) can be found in: Faulkner (1958:93-111); Lichtheim (1976:57-72); Fecht(1984:281-333); Von der Way (1984:279-366)[2].
3. Nevertheless, in Gardiner's dovetailing of the events the 'standing firm' is apparently something with only a symbolical meaning since Gardiner puts the 'seeing' of the enemy and the 'rising from his seat' *after* the arrival of the Hittites at the camp, but the 'standing firm' *before* their arrival of the Hittites! One can hardly believe that Gardiner thought Ramesses II to play the hero when the enemy was still conveniently out of sight (Gardiner 1960:49-50). That the meaning of the verb *smn* 'to stand firm' implies actual fighting is clear from the entries in *WB* IV:134, which clearly have a military context.
   The same verb is used by Ramesses II in his soliloquy when he is supposed to be fighting the enemy alone, saying: 'Not one among them (i.e. his own soldiers) stood firm (*smn*) to fight with them' (P § 91). That actual fighting is meant is also clear from the context of P §§ 74-75, where a comparison is made between the slackness of the army and the steadfastness of Pharaoh when confronted with the enemy. It would be ridiculous if the first part of this comparison was based upon reality, but the second part only a figure of speech. C.f. a few of the other less literal translations: 'Now, His Majesty had halted...', Breasted 1927[3]:140 § 311; 'Now His Majesty was stationed...', Lichtheim 1976:64; 'Seine-Majestät-aber-hatte-haltgemacht-im- Norden-von-Qidsa, ...', (Fecht 1984:290); 'SM hielt sich im Norden der Stadt Qades auf,...', Von der Way 1984:301.

REFERENCES

Abbreviations according to LÄ Bd 1, Lief. 1, V-X.

BREASTED J.H.
1903 *The Battle of Kadesh. A study in the earliest known military strategy*. Univ. of Chicago Pub. no. 5.
BREASTED J.H.
1927[3] *Ancient Records of Egypt III*. 294-351: The Asiatic War. Univ. of Chicago Press.
DE BRUYN M.J.
1988 *The Battle of Qadesh. A re-interpretation of its sources & reconstruction*. Amsterdam, forthcoming.
DESROCHES-NOBLECOURT CHR. ET AL.
1971 *Grand Temple d'Abou Simbel: La Bataille de Qadech*. CEDAE F28. Le Caire.
FAULKNER R.O.
1958 *The Battle of Kadesh*. in: MDAIK 16: 93- 111.
FECHT G.
1984a *Das 'Poème' über die Qades-Schlacht. in: SAK 11 (Fs Helck): 282-333.*
FECHT G.
1984b *Ramses II. und die Schlacht bei Qadesch (Qidsa). Ergänzende Überlegungen im Anschluss an meinem Aufsatz in der Fs Helck* (SAK). in: GM 80: 23-53.
FECHT G.
1984c *Nachträge zu meinem 'Das 'Poème' über die Qades- Schlacht (Fs Helck, SAK). in: GM 80, 55-57.*
FERRILL A.
1986 *The Origins of War. From the Stone Age to Alexander the Great*. London.
GARDINER A.H.
1960 *The Kadesh Inscriptions of Ramesses II*. Oxford Univ.Press.
GOEDICKE H.
1985 *The 'Battle of Kadesh': A reassessment*. in: Perspectives on the Battle of Kadesh. H. Goedicke (ed.)Baltimore.
HELCK W.
1972[2] *Die Beziehungen Ägyptens zu Vorderasien im 3. und 2.Jahrtausend v. Chr.* ÄA 5. Wiesbaden.
KITCHEN K.A.
1968 *Ramesside Inscriptions. Historical and Biographical.* (KRI) Vol. II, fasc. 2, 3. Oxford.
KITCHEN K.A.
1982 *Pharaoh Triumphant. The Life and Times of Ramesses II.* Warminster.
KLENGEL H.
1969 *Geschichte Syriens im 2. Jahrtausend v.u.Z. Teil 2: Mittel- und Südsyrien*. Deutsche Akademie der Wissenschaften zu Berlin, Inst. f. Orientforschung Veröffentlichung Nr. 70. Berlin.
KUSCHKE A.
1976 *Zur Siedlungskunde und Historischen Topographie der nördlichen Biqa*^{c}. in: TAVO Reihe B, Nr. 11: 103- 116.
KUSCHKE A.
1979 *Das Terrain der Schlacht bei Qades und die Anmarschwege Ramses' II.* in: ZDPV 95: 7-35
KUSCHKE A.
1983 *'Qadesch'; 'Qadesch-Schlacht', in: LÄ Bd. V, Lief. 1:27-37.*
LICHTHEIM M.
1976 *Ancient Egyptian Literature*. Vol. 2: The New Kingdom, 57-72.
SCHLOTT A.
1969 *Die Ausmasse Ägyptens nach Altägyptischen Texten*. Kap.III.A: Das Mass itrw. Tübingen.

STURM J.
1939 *Der Hethiterkrieg Ramesses II.* WZKM Bh. 4. Wien.
URK. IV
1955 *Urkunden der 18. Dynastie, IV.* Abt. Heft 17. bearbeitet von W. Helck. Berlin.
YADIN Y.
1963 *The Art of Warfare in Biblical Lands in the Light of Archaeological Study.* 2 vols.
VON DER WAY TH.
1984 *Die Überlieferung Ramses' II. zur Qades- Schlacht.Analyse und Struktur.* Kap. IV: Textmaterial. HÄS 22.

Ein neuer Typus halbmondförmiger Axtklingen aus Nordsyrien

FRIEDRICH LÜTH

Im 12. Jahrhundert errichtete Nur-ed-Din in Damaskus ein Krankenhaus mit angeschlossener ärztlicher Schule. In dem Gebäude, das heute zu den bedeutenden architektonischen Monumenten der Stadt zählt, arbeitete ein Stab von angesehenen Ärzten, der dem Krankenhaus bald zu internationalem Ruhme verhalf.

Heute ist das Maristan Nuri nach einiger Zeit der Restaurierung als Museum für Wissenschaft und Medizin für das Publikum zugänglich. Die dort gezeigte Ausstellung widmet sich vor allem der seit alters her im Hause angesiedelten Wissenschaften und führt den Besucher mit einigen schriftlichen und photographischen Dokumenten, vor allem aber über die detaillierte Präsentation des alten Instrumentariums in die Frühzeit chirurgisch-orthopädischer Medizin zurück.

In einem der Schaukästen finden sich allerdings einige Altsachen, namentlich Schmucknadeln, Waffen und Arbeitsgeräteaus Kupfer und Bronze, welche durchaus gängige Vertreter bekannter Formtypen und -varianten aus der frühen und mittleren Bronzezeit Syriens repräsentieren. Unter diesen Gegenständen fällt eine Axtklinge auf, die der Gattung der Klingen mit halbmondförmiger Gestalt zuzuweisen ist, jedoch in einigen Details von bisher bekannten Stücken abweicht[1]. Diese Axtklinge wurde im libanesischen Kunsthandel erworben, weshalb über ihre Provenienz auch keine Angaben vorliegen.

Die massive, aus Kupfer gegossene Klinge (Abb. 1) fügt sich in das Schema halbmondförmiger Axtklingen. An ihrem der Schneide gegenüberliegenden rückwärtigen Teile befinden sich noch zwei von ehemals drei Zungen, die einmal zur Befestigung um einen wohl hölzernen Griff gebogen wurden. Während eine Zunge zusammen mit dem Endstück der Axt rezent abgebrochen und nicht mehr erhalten ist, zeigt die einigermassen erhaltene Zunge am oberen Ende noch den Ansatz der zur Befestigung dienenden Biegung. Der Befestigungsteil an sich, wie auch die gesamte mittlere Zunge, sind bereits in alter Zeit abgebrochen.

Die Klinge ist auf der einen Seite flach und zieht auf der anderen Seite von der abgeschrägten Schneidenkante schwach gewölbt zur Hinterkante, an der sie im ganzen massiver ist. Da keine Schliff- oder Klopfspuren festzustellen sind, die auf eine die Form verändernde spätere Überarbeitung deuten, gibt der Querschnitt einen Hinweis auf die Beschaffenheit der zur Herstellung verwendeten Gussform: die Axt ist in einer Hohlform gegossen, die mit einem flachen Deckel abgedeckt war.

Die erhaltene Klingenlänge beträgt 27,2 cm. Ergänzt man sie symmetrisch um das fehlende Stück, so kann auf eine ehemalige Gesamtlänge von 31cm geschlossen werden.

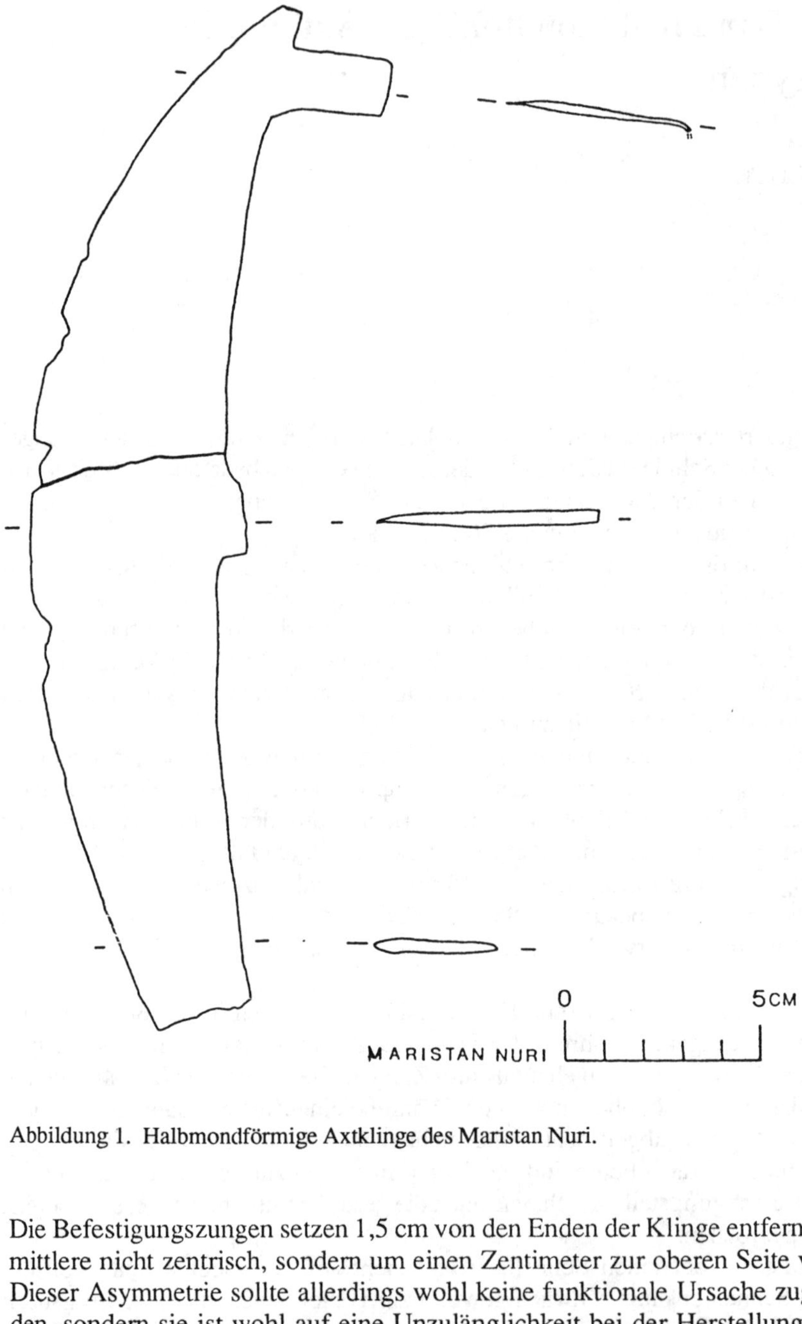

Abbildung 1. Halbmondförmige Axtklinge des Maristan Nuri.

Die Befestigungszungen setzen 1,5 cm von den Enden der Klinge entfernt an, wobei die mittlere nicht zentrisch, sondern um einen Zentimeter zur oberen Seite verschoben ist. Dieser Asymmetrie sollte allerdings wohl keine funktionale Ursache zugemessen werden, sondern sie ist wohl auf eine Unzulänglichkeit bei der Herstellung der Gussform zurückzuführen.

Entlang der Schneide finden sich einige rezente Beschädigungen. Abgesehen von der in alter Zeit bereits abgebrochenen mittleren Zunge muss die Axtklinge bei ihrer Auffindung also noch gut erhalten gewesen sein, so dass die Annahme naheliegt, sie stamme aus einem geplünderten Grabe.

Der durch die beschriebene Axt repräsentierte Typus lässt sich unmittelbar an die Äxte vom Amarna-Typus (Tubb 1982) anschliessen, unterscheidet sich von diesen jedoch in der Position der Befestigungszungen, weshalb es angebracht erscheint, diese Äxte als eigenständigen Typus neu zu definieren. Während für die Äxte des zuerstgenannten Typus eine Datierung in die Frühdynastische Zeit (ED II = EB III) nachgewiesen werden konnte (Tubb 1982:9 ff.), lässt sich eine chronologische Distanz des neuen Typus aus der typologischen Nähe zu den Amarna-Äxten nicht unmittelbar erschliessen. Ein überraschender Neufund kann aus diesem Dilemma hinausführen.

Vor kurzem fand sich auf dem im Gebiete des Assad-Stausees gelegenen Tell Halawa B eine Gussform aus Kalkstein, die einmal zur Herstellung halbmondförmiger Äxte genau jenes Typus gedient hat, der uns in dem Exemplar aus dem Maristan Nuri begegnet (Abb. 2). Die Gussform war auf dem offenen Hof eines Gebäudes liegengeblieben, als dieses einem zerstörenden Feuer zum Opfer fiel. Das Gebäude selbst befindet sich unmittelbar unter der Tempelterrasse des Bau II (Lüth 1981) und ist dadurch stratigraphisch abgesichert der Bauschicht 3 zuzuweisen. Für die jüngste Phase dieser Bauschicht ist eine Datierung in eine unmittelbar vor der Frühdynastischen Zeit gelegenen Periode vorgeschlagen worden, die auf einer komparativ-stratigraphischen Analyse verschiedener Kleinfunde und Keramikfazies gründet (Lüth, im Druck). Da der zur Diskussion stehende Axttypus zuerst auf dem Tell Halawa gefunden wurde, soll er als Typus Halawa dem Typus Amarna gegenübergestellt werden.

Gibt nun diese Gussform einen ersten Hinweis darauf, dass die Äxte vom Halawa-Typus älter als die des Amarna-Typus sein dürften, so lieferte eine weitere auf dem Tell Halawa B gefundene Axt, in diesem Falle vom Amarna-Typus, den endgültigen Beweis für eine chronologische Folge beider Typen. Auf der Oberfläche des bereits wieder unbesiedelten Hügels war einst ein Grab angelegt und darin ein Toter bestattet worden, dem ein Bronzedolch, mehrere Knochenringe (Gürtelbesatz ?) und eine Axtklinge vom Amarna-Typus beigegeben worden sind. Das Grab ist auf der bereits im Verfall befindlichen Stadtmauer angelegt worden, so dass ein echter *terminus post quem* für das Grab einerseits, respektive ein *terminus ad* oder *ante quem* für das Ende der Besiedlung des Tell Halawa B andererseits (Lüth, im Druck) besteht. Für die Äxte vom Halawa-Typus ist daraus eine gegenüber dem Amarna-Typus ältere Datierung zu folgern.

Äxte vom Amarna-Typus sind mit den Exemplaren aus Amarna (Tubb 1982:1, Fig. 1), Halawa (Lüth, im Druck) und Tawi (Kampschulte und Orthmann 1984:77-78, Taf. 41a) in einem recht engen Gebiet verbreitet. Dazu ist auch die aus dem Kunsthandel stammende Axt im Rijksmuseum, Leiden, zu zählen (Van den Boorn 1983:131-132, Taf. 5), für die Syrien/Palästina als Provenienz vermutet wird. Lediglich die Äxte aus Tell Agrab, die zu einem Hortfund von fünf unveröffentlichten Äxten desselben Typus (Hillen 1953, Fig. 1) gehören und aus einem Nebenraume des Shara-Tempels stammen (Delougaz und Lloyd 1942:268), fallen aus dem Rahmen der umrissenen Verbreitung. Die Äxte vom Halawa-Typus sind bislang erst in zwei Exemplaren bekannt, von denen das einzige mit bekannter Provenienz aus dem Euphrattal stammt. Nach der chronologischen Distanz ergibt sich also eine für beide Typen regionale entsprechende Verbreitung, woraus wiederum gefolgert werden darf, dass die Axtklingen vom Halawa-Typus die älteren Vorläufer zu denen vom Amarna-Typus bilden.

Auf einen Punkt sei hier noch hingewiesen: von den erwähnten Äxten des Halawa- und des Amarna-Typus ist lediglich die Axt aus Amarna aus getriebenem Blech gearbeitet, wohingegen alle anderen Äxte gegossen sind und damit aus massivem Kupfer

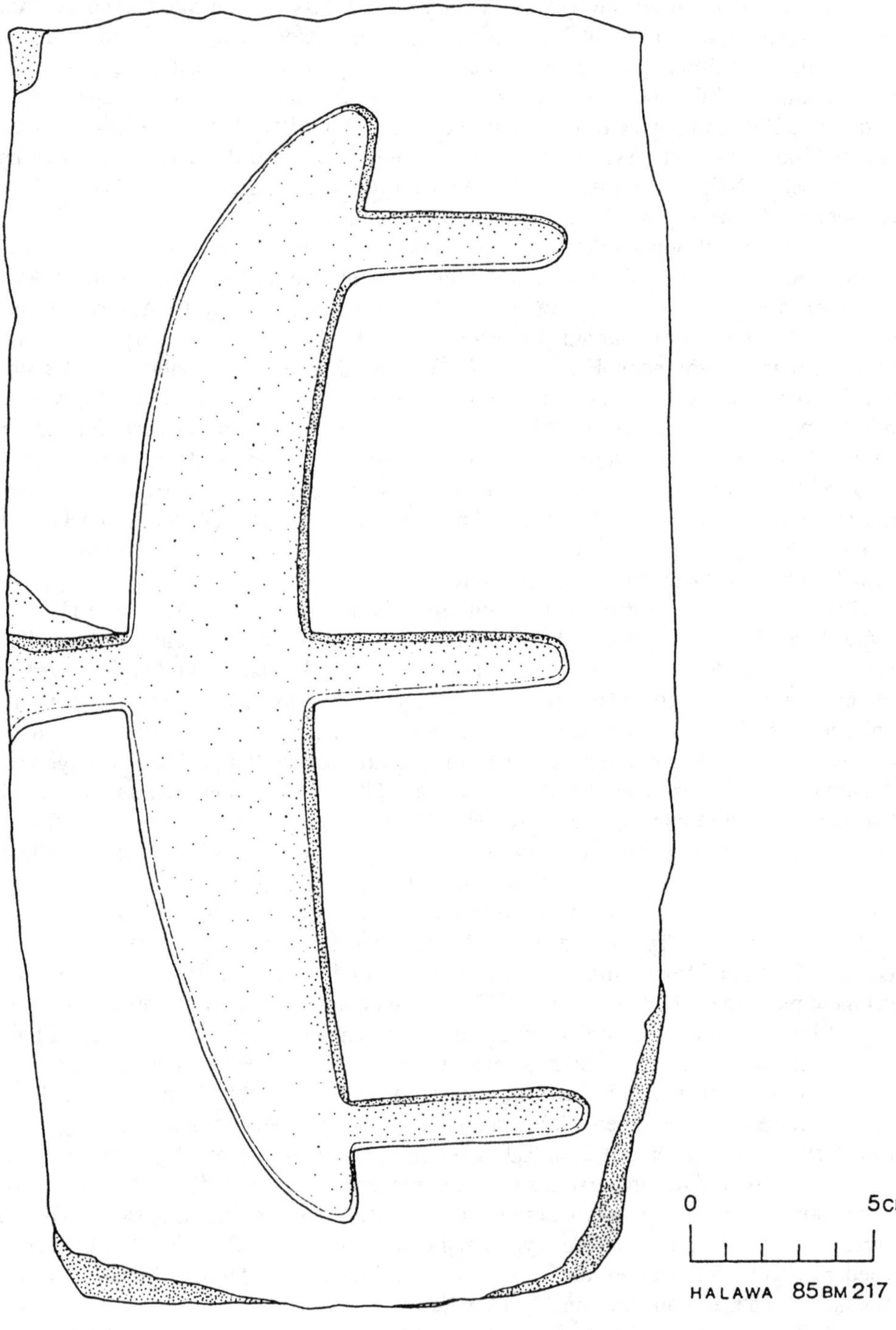

Abbildung 2. Gussform aus Kalkstein zur Herstellung halbmondförmige Äxte *(Typus Halawa)*.

bestehen. Beiden Exemplaren aus Tell Agrab und dem aus dem Rijksmuseum fehlen allerdings die entsprechenden Angaben.

Für die Frage nach der Funktion der Axtklingen kommt der Herstellungstechnik eine besondere Rolle zu. Die aus getriebenem Blech hergestellte Amarna-Axt kann kaum als Waffe benutzt worden sein, denn die mit 1 mm angegebene Blattstärke der Axt hätte bei der Verwendung als Schlagwaffe wie auch als Arbeitsgerät kaum ihre Form beibehalten. Die Axtklinge zeigt aber keine derartigen Benutzungsspuren. Für die gegossenen Axtklingen kann diese Annahme nicht ohne weiteres übernommen werden. Während die Axt aus dem Maristan-Nuri keine Benutzungsspuren aufweist, sind für die anderen Äxte die entsprechenden Beobachtungen nicht gemacht worden. Die Axt aus Tawi ist an zwei Stellen in alter Zeit beschädigt worden, doch handelt es sich auch hier wieder, wie bei der Axt aus dem Maristan Nuri, um Teile der Befestigung. Ob daraus abzuleiten sein darf, dass die Äxte während des Gebrauchs beschädigt wurden, erscheint zweifelhaft. Gleichwohl wäre an intentionelle Beschädigung als einem Schutz vor Grabräuberei zu denken. Für die einzigen Äxte, die keine Grabbeigabe darstellen, die Äxte aus dem Heiligtum von Tell Agrab, fehlt leider eine gute Beschreibung oder Abbildung. Für bestimmte anatolische Axtklingen, die keine geschärfte Klinge besitzen, hat Tezcan einmal die Vermutung geäussert, sie seien keine Gebrauchsgegenstände des täglichen Lebens, sondern Zeremonialgegenstände gewesen (Tezcan 1960:38). Tubb hat dagegen etwas ironisch eingewendet, der Gedanke sei zwar möglich, ebensogut sei aber denkbar, dass es sich um Gegenstände gehandelt haben könne, die von vorneherein als Grabbeigabe hergestellt wurden und daher gar nicht funktionsfähig gewesen sein müssen (Tubb 1982:7). Einmal abgesehen von der Frage, ob zwischen der Grabbeigabe und einer Funktion ausserhalb des alltäglichen Lebens ein ironiefähiger Unterschied besteht, geben die Fundumstände der Äxte doch einen gewissen Hinweis auf ihre Funktion. Die Äxte vom Amarna-Typus fanden sich in Gräbern (Amarna, Tawi, Halawa) und als Hortfund in einem Heiligtum (Tell Agrab). Eine Axt stammt aus dem Kunsthandel. Wenngleich es zu bedenken gilt, dass die Auffindungsmöglichkeit für Kupfergegenstände in Gräbern und nicht ausgeräumten Heiligtümern natürlich ungleich besser sind als in Siedlungsschichten, so ist die Ausschliesslichkeit, mit der die Äxte in eben den genannten Zusammenhängen auftreten, wohl am ehesten als Hinweis auf ihre Verwendung im Zeremonialwesen zu werten. Dass andere kupferne Arbeitsgeräte und Waffen durchaus in Siedlungsschichten gefunden werden, mag als hinlänglich bekanntes Faktum ein weiteres Indiz zur Stützung der Meinungs Tezcans bilden.

Zusammenfassend lässt sich festhalten, dass mit der Axtklinge aus dem Maristan Nuri zu Damaskus das erste Exemplar eines Axttypus (Halawa) vorliegt, der als Vorbild für die jüngeren Axtklingen vom Amarna-Typus in einer stratigraphischen Folge gesichert werden konnte. Während für die Axtklingen vom Amarna-Typus feststeht, dass sie in die Frühdynastische Zeit zu datieren sind, kann für die des Halawa-Typus nur konstatiert werden, dass sie zu eben dem Zeitpunkt bereits nicht mehr angefertigt wurden. Wann allerdings mit der Herstellung derartiger Äxte begonnen wurde, lässt sich beim derzeitigen Stande der Forschung noch nicht erschliessen. Halbmondförmige Axtklingen beider Typen dürften wohl am ehesten als Zeremonialgegenstände Verwendung gefunden haben, wenngleich eine profane Nutzung nicht grundsätzlich ausgeschlossen werden kann. Ob man nun die frühdynastischen Lokalkulturen als Teil der grossen Streitaxtkulturkomplexe (Zápotocky 1966) zu verstehen hat, bleibt noch zu untersuchen.

NOTEN

1. Für die Genehmigung zur Publikation dieser Axt bin ich dem Direktor der Generaldirektion der Antikenverwaltungen und der Museen in Syrien, Herrn Dr. A. Bahnassi, sowie dem Direktor des Maristan Nuri, Herrn A. al-Quaiem sehr dankbar. Die Vorzeichnung fertigte freundlicherweise Herr stud. phil. A. Pruss an.

BIBLIOGRAPHIE

DELOUGAZ P. AND LLOYD S.
1942 *Pre-Sargonic Temples in the Diyala Region*, OIP 58, Chicago.
HILLEN C.
1953 'A Note on Two Shaft-Hole Axes', *Bibliotheca Orientalis* 10: 211-215.
KAMPSCHULTE I. UND ORTHMANN W.
1984 *Gräber des 3. Jahrtausends im syrischen Euphrattal. 1. Ausgrabungen bei Tawi 1975 und 1978*, Saarbrücker Beiträge zur Altertumskunde 38, Bonn: Habelt.
LÜTH F.
1981 'Die Grabungen auf dem Tell B', in: W. Orthmann (Hrsg.), *Halawa 1977-1979*, Saarbrücker Beiträge zur Altertumskunde 31, Bonn: Habelt.
im Druck 'Grabungen auf dem Tell B', in: W. Orthmann (Hrsg.), *Halawa 1980-1986*, Saarbrücker Beiträge zur Altertumskunde.
TEZCAN B.
1960 'New Finds from Horoztepe', *Anatolia* 5: 29-46.
TUBB J.
1982 'A Crescentic Axehead from Amarna (Syria) and an Examination of similar Axeheads from the Near East', *Iraq* 64: 1-12.
VAN DEN BOORN G.P.F.
1983 'Recent Acquisitions from the Ancient Near East', *Oudheidkundige Mededelingen uit het Rijksmuseum van Oudheden te Leiden* 64: 123-141.
ZÁPOTOCKY M.
1966 'Streitäxte und Streitaxtkulturen', *Památky Archeologické a Místopisné (Prag)* 57: 172-209.

The beginning of the third millennium in Syria

HANS H. CURVERS

At this moment, paradoxically as it might seem, there is probably more argument about both details and broader aspects of Syrian chronology than there was 25 years ago, when the paucity of the data encouraged a more comfortable degree of certitude. At the very time when dating ranges had been narrowed and precision and complexity of argument and data had been refined, the unexpected reappearance of entire civilizations like Ebla forcefully reminded Syrian archeologists how much Syrian chronology exemplifies the paradox of increased knowledge that provides less information. The expeditions involved in salvage projects in the Tabqa Dam area and Middle Habur valley have added completely new and detailed data for chronological reconstructions.

SURVEY OF EARLY THIRD MILLENNIUM SITES

Hama. We shall begin our survey of the material evidence with the central Syrian site that has yielded one of the most complete sequences published so far: Hama. This site, of course, has its limitations. The early third millennium levels have been exposed on a considerably large scale, the stratigraphy is at present hard to check in detail and a quantified ceramic sequence has not yet been published. The finds, other than architecture have been published in minute line-drawings. Still, it is one of the few sites at which we can conveniently compare architecture as well as other finds retrieved from the early third millennium strata.

The appearrance of red-black burnished ware in the Hama levels K5-K1 dates these levels on acceptable grounds between 2700 and 2400 B.C. (Fugmann 1958:47). Level K1 is covered by an ash layer one meter thick and on top of this the first houses of phase J were built (Ingholt 1940:29). The latter phase has a new, highly distinctive material culture with few, if any, links to the preceding phase K. The street that ran north-south through the junction of the four excavated squares can be considered to perpetuate a configuration present in K1 (Fugmann 1958, figs. 53, 57). The general characteristics of the Hama K10-5 ceramics are large vessels with globular body and short flaring neck and plain or slightly thickened rims. Several rims of these large globular, round-based vessels are bevelled or have a wide groove on the outside. The occurrence of potters' marks is another common feature. The bowls are hemispherical or slightly carinated with plain rims. Several bowls have bevelled or slightly thickened rims. The goblets have round bases, globular bodies and plain everted rims.

Qal'at el-Mudiq/Apamea. The soundings at Qala'at el-Mudiq have yielded pottery which can be dated to the beginning, middle, and end of the third millennium B.C. The stratified sequence is not '..une succession chronologique bien définie.' (Collon et al. 1975:33). The graves uncovered in squares A2 and A3 have yielded inventories, which are parallel to the ceramics of Amuq G through H to I. 'C'est dans le contexte de la Phase I que Braidwood situe l'hypogée de Til Barsip, les tombes de Carchemish et la Tombe IV de Qatna. Il faut donc dater les tombes construites d'Apamée vers 2500 av. J.-C. et probablement entre 2500 et 2300.' (Collon et al. 1975:158). A radiocarbon date for a sample retrieved from 'couche IIIb, tombe no. 3' provides the following date: 4405 ± 245 B.P. (IRPA-170; Collon et al. 1975:191). The ceramic inventory of the graves is characterized by the cyma recta profiles, common in Amuq H (Braidwood and Braidwood 1960, figs. 269:9-12; 270:8, 9 and 271:2). Related vessels can be found at Ugarit on the coast (Courtois 1962, figs. 30:E, 33:E, 36:E). The distinctive small ring bases on the other ceramics associated with the cyma recta profiles can also be placed in a sequence, which shifts, changes and develops from the first half towards the second half of the third millennium B.C.

Ras Shamra/Ugarit. Ras Shamra is one of the largest Bronze Age sites located along the Syrian coast; yet the finds and occupation attributable to the third millennium B.C. are relatively sparse, although this may not be an accurate reflection of the extent of third millennium occupation. A useful summary for the evidence on the third millennium B.C. is provided by Henri de Contenson (1979, 1982). These publications make the evidence from Ras Shamra/Ugarit more accesible, because it is not easy to reconcile the statements dispersed throughout the various preliminary reports. Level IIIA1 represents the beginning of the third millennium B.C. at Ras Shamra/Ugarit. In his summary on third millennium chronology de Contenson (1979:857-8) indicates the existence of some parallels with Amuq G, but predominantly with Amuq H. He also lists some parallels with Hama K5-1 (de Contenson 1979:858). De Contenson dates Ras Shamra IIIA1 to EB III or 2900 - 2400 B.C.

In his later summary the absence of 'Chaff-Faced Ware', 'Multiple-Brush Painted Ware', and 'Reserved-Slip Ware', characteristic for Amuq G, may have strongly influenced the chronological date for the scarce material remains retrieved from Ras Shamra (de Contenson 1982). At the moment the evidence for a large hiatus at Ras Shamra, based predominantly as it seems on the absence of some ceramic groups, does not seem very convincing.

More Ras Shamra IIIA1 remains have been exposed on the northern slope of the acropolis. In a trench (5 x 7 m) the scant remains consist of at least two floor levels. Khirbet Kerak ware occurred in the upper level. The ceramics of the lower level are said to lack Khirbet Kerak ware (de Contenson 1969:76). If we accept a date for Khirbet Kerak contemporary with Amuq G and H (Braidwood and Braidwood 1960:292, 358 ff.), the lowest floor level(s) may belong to the earliest part of the third millennium B.C., in the period before the presence of Khirbet Kerak ware at Ras Shamra. The sounding in 1959, furthermore, has revealed the presence of Amuq F-G ceramics at Ras Shamra (Courtois 1962:452-3, fig. 43).

The beginning of the third millennium, or more specifically the hiatus between the late Chalcolithic period and the Bronze Age '.. n'est pas clairement attesté par des trouvailles caractéristiques.' (Courtois 1962:396). More deep soundings at Ras Shamra may result in a *local* early third-millennium ceramic assemblage, which seems not unlikely, because in the later periods of the third millennium B.C. the coastal region's isolation from inland Syria has been heavily stressed (de Contenson 1979:861-2; 1982:98).

Amuq Plain. The survey and soundings at Chatal Hüyük, Dhahab, Tell Ta'yinat, and Tell al-Judeideh have shown that Phases G and H can be assigned to the first half of the third millennium B.C. Remains of these phases have been exposed at the sites Tell al-Judeideh and Dhahab. The ceramic assemblage of Phase G is characterized by 49-54% wheelmade pottery including 'Plain-Simple Ware' (fine well-fired and sand-tempered), Syrian bottles and drooping spouts. The 'Reserved-Slip Ware' constitutes 3-8% of the assemblage; the 'Multiple-Brush Painted Ware' makes up 17-22% of the assemblage. Braidwood points out that the ceramic assemblage shows many novelties compared to the preceding phase F: '..the first really elaborate assemblage in a technological sense. Much of the pottery has an almost 'factory-made' look, ..' (Braidwood and Braidwood 1960:259).

The remains assigned to Phase H have been exposed at Chatal Hüyük, Tell al-Judeideh and Tell Ta'yinat. The pottery retrieved from the phase G remains also occurs in the Phase H remains. A new feature, however, is the occurrence of red-black burnished ware (50% and 75% at Judeideh and Ta'yinat respectively). This red-black burnished ware is grit- and straw-tempered, sometimes with a high luster and impressions or combinations of horizontals and zigzags. Another innovation in the ceramic assemblage is the 'Brittle Orange Ware', rare at Tell al-Judeideh, but constituting up to 10% of the assemblage at Tell Ta'yinat.

The information available in the publication on the results of the survey and sounding in the Amuq plain has provided a frame-work for Syrian archeologists to interprete their results. The challenge for future fieldwork and analyses is to refine the dating ranges with more precision, and to defy the paradox of increased knowledge providing less information.

Mardikh/Ebla. The brilliant civilization of Ebla in the second half of the third millennium B.C. has been treated at length in many publications (cf. Matthiae 1981). Since 1974 the excavation of Palace G has resulted in establishing identification of the site as Ebla, already known from many cuneiform documents. The results of the earlier activities at the site provide some information on the earliest periods. Some surface finds indicate that Mardikh was already occupied in the Mardikh I period (3500-2900 B.C.). The remains of Mardikh IIA have been exposed in the southern region of the acropolis. It seems that the settlement became increasingly more significant. The control and involvement of the city in the exchange of timber, copper, and silver from the Northern Libanon, the Amanus mountains, and the Taurus mountains may have caused this process of urbanization. The economic, political and cultural apogee of this development was reached in the Mardikh IIB period.

The ceramic assemblage of Mardikh IIA has been related chronologically to Amuq H, Hama K5-1, Ras Shamra IIIA1. The types which chartacterize the assemblage are goblets with green slip reserved in fine oblique radial bands, angular carinated bowls, high-shouldered jars with sharply flared and channeled rims, cooking pots with high shoulders and more markedly flaring collars (Mazzoni 1985:9, fig. 7). From the table which presents a relative chronology for Mardikh IIA, it seems that the Mardikh IIA assemblage could also have earlier chronological relations. Mazzoni describes the assemblage as '.. still in the post-Chalcolithic tradition, ...impure pastes ...; limited control over the firing; .. hand workmanship and artisan production.' (Mazzoni 1985:13). The transformation of ceramic production at Mardikh becomes extremely evident at the beginning of Mardikh IIB1. We must assume, however, that this development towards standardization and mass production had already started in the period preceding Mardikh IIB. The 1983 finds in the southern region of the acropolis seem to corroborate the idea of such a gradual development. The structure exposed immediately below the palace G seems to be of a palatial nature. The ceramic inventory can be related to late Hama K and Hama J8 (Pinnock 1985:89, note 21).

The River Qoueiq. During a survey in 1977-79 John Matthers and his colleagues collected archeological artefacts on the surface of eighty sites within the area of the river Qoueiq. Eight sites outside the area have been listed in their report on the survey as well (Matthers 1981). Taking the 'Reserved-Slip Ware' as a hallmark for the early third millennium, 21 sites in the area of the river Qoueiq seem to have been occupied in this period (Mellaart 1981, map XX). Other characteristics for Amuq G, or the beginning of the third millennium, like 'Multiple-Brush Painted Ware', spouts, and rims with triangular lugs bring the total of sites which seem to have been occupied in this period throughout the area of the river Qoueiq to 29. Khirbet Kerak ware has only been retrieved from the surface of two sites in the area. The information from the Quoeiq survey report fills in the gap between the Tabqa sites and the Amuq. The presence of Khirbet Kerak ware at 50 sites in the Amuq plain is in strong contrast to the two sites in the area of the Qoueiq river. The area of the Qoueiq river, however, seems to be part of the 'Reserved-Slip Ware' zone (Amuq 11 sites versus Qoueiq 29 sites (Mellaart 1981:154-155, Map XX; Braidwood 1937, Map XXIII). This zone of sites which yielded 'Reserved-Slip Ware' extends into northern direction along the Eufrates in the Adiyaman region, south of the Taurus, and even north of the Taurus the ware has been retrieved from strata with bevelled-rim bowls and four-lugged jars at Tepecik (Esin 1979: 108, 111, pl. 56, fig. 6; pl. 57, fig. 6). The later part of the first half of the third millennium B.C. seems to be represented by red burnished ware mostly from bowls with inverted rims. This type of ware is rare in the Amuq, but there are parallels with Ras Shamra III A1 and Hama K 5-1 (Mellaart 1981: 158-159, nos. 921-955).

Umm al-Marra. In 1978 and 1979 the Belgian team under the direction of Roland Tefnin, which has also excavated Abu Danneh, executed some soundings at Umm al-Marra. In stratum V stone walls with related plastered floors have been exposed. The pottery retrieved from this stratum is characterized by burnished ware, 'red burnished ware', in several shapes (Tefnin 1979:91). As no illustrations are available, we can only think of a relation to the above mentioned burnished wares in the area of the Qoueiq river. Goblets do not occur as frequently in stratum V as they do in the subsequent stratum IV, contemporary with Mardikh IIB1. Virgin soil has been approached, but not decisively reached, which does not exclude the presence of more occupation levels belonging to the early third millennium B.C. (Tefnin 1979:94; 1981/82:238).

THE TABQA DAM AREA (LAKE ASSAD)

In 1967 a new era in Syrian Archeology started with the general appeal launched by the

Directorate-General of Antiquities in Damascus for the rescue of the archeological monuments in the Euphrates valley threatened by the construction of a hydro-electric dam (Tabqa Reservoir Area). Maurits van Loon provided an accurate inventory of the sites in this area, which have witnessed an unprecendented concentration of international excavation teams since 1971.

Tell al-'Abd. Tell al-'Abd, located on the limestone cliff on the left bank of the Euphrates is a site of modest dimensions: 210 x 220 x 15 m. The excavations in 1971 and 1972 have revealed two levels which can be dated to the third millennium B.C (Bounni 1979:50-55). Level II yielded simple pottery and numerous clay figurines in great variety. Level II has been excavated in all investigated parts of the mound. Level III has been exposed in K8 and the sounding I11. The pottery is described as very simple; pots and jars were primarily handmade and bearing potters' marks. Some of the pots are said to be equipped with rough spouts. Few clay figurines have been retrieved from level III. The radiocarbon dates give an uncallibrated range of 2450 ± 60 to 2195 ± 60 B.C. from the deepest point reached to the upper deposits. The earliest date is clearly earlier than all radiocarbon dates from Selenkahiye, which corroborates a date for level III in the earlier part of the third millennium B.C (cf. Van Loon 1979:111). Callibration of the Tell al-'Abd radiocarbon dates provides a range in the first half of the third millennium B.C. (calibrated results in terms of 84% probability; Pearson et al. 1986). The evidence seems to point at an expansion of this site in the second part of the third millennium B.C.

Tell es-Sweyhat. Tell es-Sweyhat is located three km east of the Euphrates. The central mound measures 330 x 250 x 15 m; the outer town of 700 x 600 m is enclosed by an earth rampart. On the basis of the ceramic evidence the expansion of the site can be datad to the second half of the third millennium B.C. The operations in trench IIA in 1973-4 have yielded limited architectural remains which can be dated to the beginning of the third millennium B.C. The depth of trench IIA was 6.40 m which leaves several meters of deposit unexcavated. The top layers exposed in this trench included wash layers and pits dated to the Hellenistic period. The trench was laid on the west slope of the mound which leaves ca. two meters of deposit on top of the mound uninvestigated. This deposit could contain remains of the second half of the third millennium and some Hellenistic remains. The preliminary analysis of the ceramics retrieved from phases A-F in Trench IIA provided a date in the 'first half of the second [error: should be *third*] millennium B.C' (Holland 1976:48).

Halawa. Tell A, south of the wadi leading to the Euphrates (300 x 400 m), and Tell B, north of the wadi (100 x 100 m), are located on the east bank of the Euphrates. Operations on Tell A have revealed remains of late third and second millennium date. The remains belonging to the first half of the third milllennium B.C. were exposed in 1978-79 and during the more recent fieldwork (cf. Lüth, this volume). In Level I and II the architectural remains of a monumental building on a terrace have been exposed. The ceramics retrieved from these levels have been published as a complete assemblage. Establishing a quantified ceramic sequence has to wait till the final report. From the preliminary report we can conclude that this assemblage belongs to the first half of the third millennium B.C. (Lüth *apud* Orthmann 1981:44-48). Some general characteristics are the potters' marks on neck, shoulder, body or base of the vessels, the occurrence of few burnished sherds, and the majority of the bases being round, although some flat bases have been registered. The occurrence of sherds belonging to the 'Red-black burnished ware' (Orthmann 1981, Tf. 11:3), and sherds of large vessels with geometric seal impressions seem to corroborate a date in the first half of the third millennium B.C.

Habuba Kebira. The 1969-75 activities yielded the spectacular remains of a settlement interpreted as a colony of the Mesopotamian Uruk civilization, situated on the west bank of the Euphrates. The remains of this settlement are known as Habuba Kabira South and Tell Qannas. On the mound, Habuba Kabira North, a long sequence spanning the late fourth millennium through third millennium to the first half of the second millennium has been exposed on the southeastern slope. A thick deposit (8 m) comprised 20 building levels. The diameter of the mound is ca. 230 m which makes a maximum settlement size of 3 ha. Two preliminary reports present details on the architectural developments (Heusch 1980; Heinrich et al. 1971). The designation of the levels in both reports causes some confusion. It seems that level 6 (Heinrich et al. 1971) is equal to level 2 and/or 3 (Heusch 1980).

The earliest remains at the Habuba Kabira North (Heusch: level 1; Heinrich et al.: levels 7 and 8) have been related chronologically to the settlement of Habuba Kabira South and Qannas. The remains in the subsequent level 2/3 (Heusch) and level 6 (Heinrich et al.) have been interpreted as a settlement including a potters' workshop. The ceramic inventory of this level included coarse ware, 'bauchige Flasschen mit Wackelboden', bowls with convex profiles, and numerous sherds with incised potters' marks (Heusch 1980:164). In a description of the pottery from the potters' workshop the same pottery is described and related to the Hama K8-7 ceramics. Generally the pottery is dated to the earliest phases of the Early Dynastic period (Strommenger *apud* Heinrich et al. 1971:21-25). The excavators have supposed a gap between levels 8-7 and level 6 (Schmidt *apud* Heinrich et al. 1971:9-12). Details in support of such a chronological gap are absent.

These scant remarks on the deeper levels at Tell Habuba Kabira North indicate that the ceramic sequence of this site may add to our knowledge of the ceramic distribution and developments in the first half of the third millennium B.C. The architectural sequence indicates a gradual development in the levels 2-9 (Heusch 1980). Level 10 is said to display a structural change in lay-out and building techniques (Heusch 1980:168). Taking into account that the settlement at Habuba Kabira North was rather small and did not belong to the group of large urban settlements which displayed a considerable increase of settlement size in the second part of the third millennium (cf. Hadidi, Sweyhat), the structural change in level 10 seems to coincide with the developments in the larger urban centers in the Tabqa region. For the time being, however, the lack of a published ceramic sequence makes all correlations with other sequences highly speculative.

Tell Munbaqa. The site is located on the left bank of the Euphrates. Since 1969 the excavations have yielded important information for the Bronze Age in the Tabqa Dam area. Remains of the beginning of the third millennium have been exposed on a very limited scale (Orthman 1974, fig. 4-6). The level 4 remains suffered heavily from the building activities in the subsequent levels. The stone walls of *Steinbau* 1 have been dug into this level 4. A rather limited quantity of sherds has been retrieved from this level 4.

This assemblage is tentatively related to the Habuba North ceramics of level 6 (Orthmann 1974:65).

Tell Hadidi. Five seasons of excavation by the Milwaukee Public Museum have yielded considerable data for the Bronze Age occupation at this site, situated on the east bank of the Euphrates. Ceramics and architecture indicate that the assemblage of the beginning of the third millennium B.C. provide a vital link between the well-documented second part of the third millennium B.C. and the preceding Uruk, Protoliterate or Late Chalcolithic period. The sounding in Area R is of very limited size (1 x 1 x 1.5 m). This size is in considerable contrast with the remains of the second part of the third millennium excavated in 12 areas on the site. Stratum 1 included bevelled-rim bowls and red-black burnished wares. Stratum 2 yielded a similar assemblage without the bevelled-rim bowls. These ceramics have been related chronologically to the early third millennium sequences of the other Tabqa Dam sites (Sweyhat, Halawa, Munbaqa). A radiocarbon date from an early third millennium deposit suggests a possible coexistence of the settlement at Hadidi with the 'Late Uruk/Jemdet Nasr' settlement at Jebel Aruda (Hadidi: 4780 ± 60 (P-3199) (Dornemann 1984); Jebel Aruda: 4495 ± 35 (GrN-7989), 4490 ± 35 (GrN-8463) (Van Driel and Van Driel-Murray 1979:24). The evidence from the impressive fortified settlement at Habuba Kabira South/ Tell Qanas, the temple precinct and associated residences at Jebel Aruda and the occurrence of Uruk wares in the earliest stratum at Hadidi seem to stress the difficulties for the use of the term 'Late Uruk' for the material retrieved from these sites (cf. Sürenhagen 1986:32). The problem of period definition becomes very acute if we want to analyze the history of the alluvial Mesopotamian colonization of the Tabqa Dam area and its impact on the Syrian history.

For the moment we should think of a gradual decline of the Mesopotamian influence, after the abandonment of the fortified settlement at Habuba Kabira South/ Tell Qannas (Uruk VII-VI, cf. Sürenhagen 1986:32). The local cultures may have developped under the influence of the alluvial Mesopotamian control. Jebel Aruda may represent one of the latest strongholds of Mesopotamian presence in the area, whereas the finds of Uruk wares at the neighbouring sites suggest some kind of interaction. This interaction may have developed from the previous contacts during the period in which Habuba Kabira South/Tell Qannas heavily depended for its agricultural supplies on the countryside (Sürenhagen 1986:21-22, 23-24). A thorough analysis of this 'local' material at Habuba Kabira South and Jebel Aruda may provide more evidence for this suggestion of a more differentiated picture of the alluvial Mesopotamian presence in Syria.

THE BALIKH VALLEY

In 1938 Mallowan investigated the Balikh valley by soundings at various sites in the northern part of the valley. The interpretation of the results of this exploration of six weeks have heavily suffered from his findings in the Habur valley. Recently, however, the exploration of the valley was resumed by Jacques Cauvin, Lorraine Copeland and Maurits van Loon (cf. Van Loon 1981:31).

Hammam et-Turkman. The site is situated on the left bank of the Balikh and dominates its surroundings with its height and width of 40 and 500 m respectively. The excavations

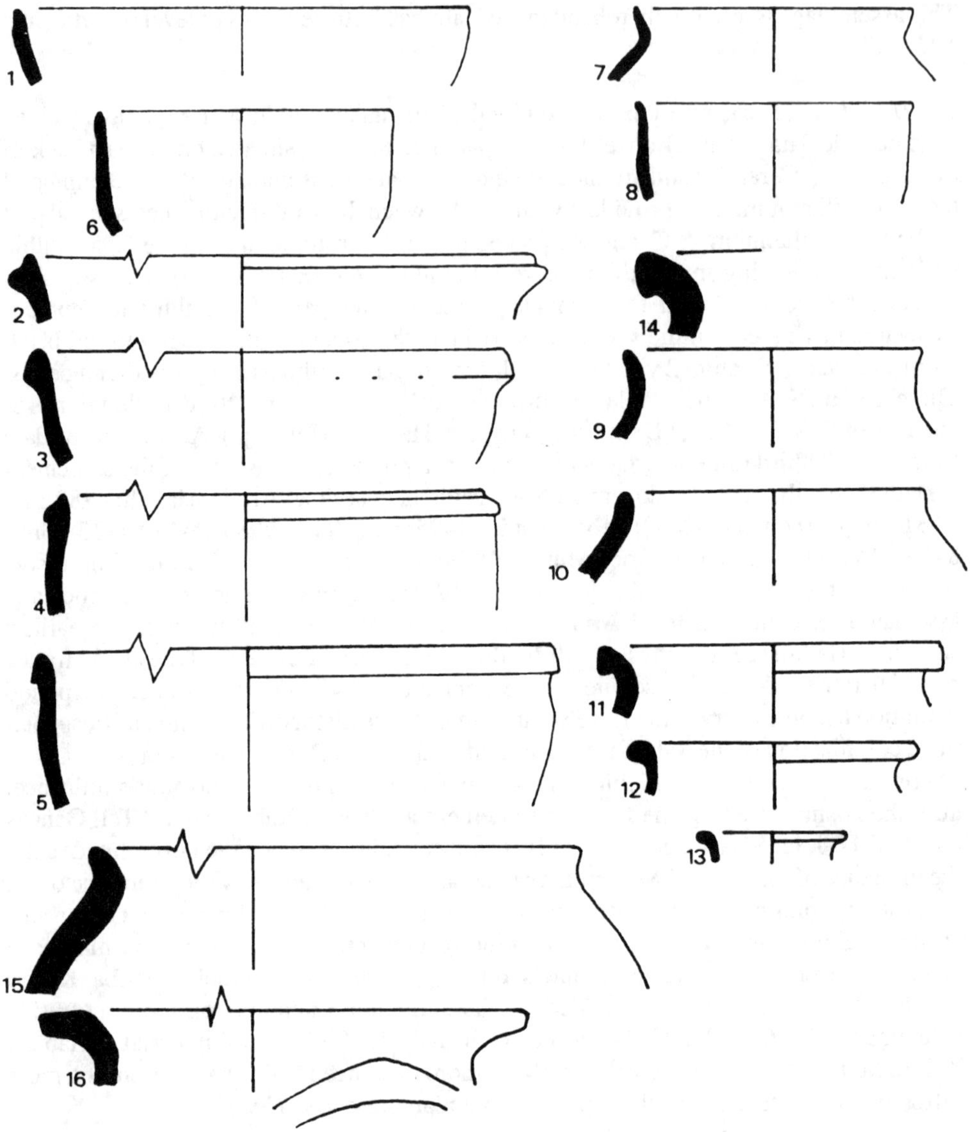

Fig. 1 Hammam VI east ceramics: *Hammam VI east early.* Scale 1:3.5

started in 1981 under the direction of Maurits N. van Loon. The ceramics retrieved from the 1981 and 1982 operations on the eastern slope of the mound have been designated as Hammam VI east ceramics. The analysis of a stratified and quantified preliminary sequence indicates that the sequence can be related to the Tabqa assemblages listed in this short survey (Curvers, in press). A break in the architectural stratigraphy is reflected in the ceramic changes. The early period comprises architecture of domestic nature which was built and rebuilt with only slight modifications in plan. The break in the architectural sequence is indicated by a change in orientation and other slightly different architectural

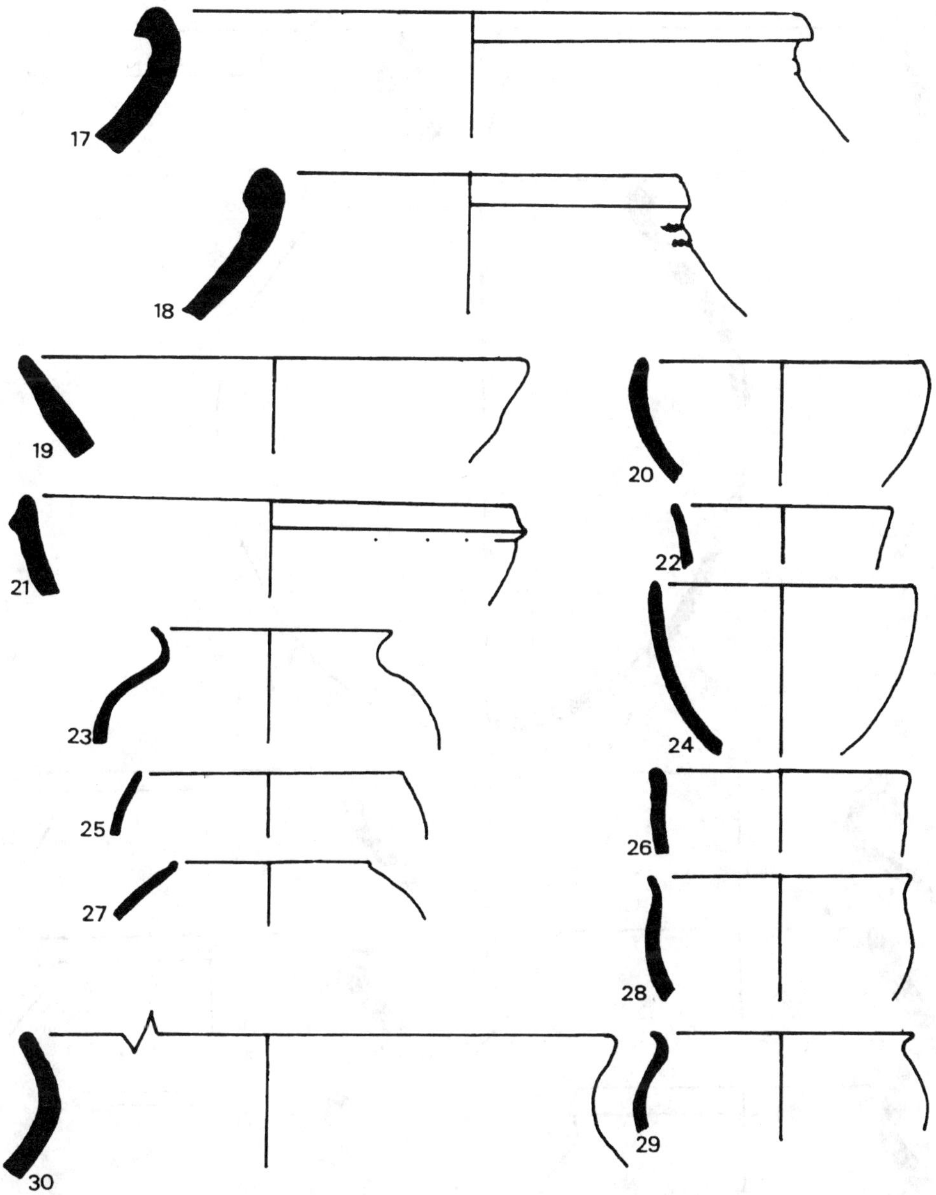

Fig.2 Hammam VI east ceramics: *Hammam VI east early* (17-18); *Hammam VI east common* (19-30). Scale 1:3.5

features (Meijer in press). On the basis of a presence/absence seriation matrix the assemblage has been divided in a group of early types (fig. 1:1-16; fig. 2:17-18), a group of common types (fig. 2:19-30; fig. 3:31-39), and a group of later types (fig. 4:40-51). The preliminary analysis has been limited to ceramics retrieved from floor deposits. In the debris of the later period two red-black burnished sherds (cf. Kelly-Buccellati *apud* Van

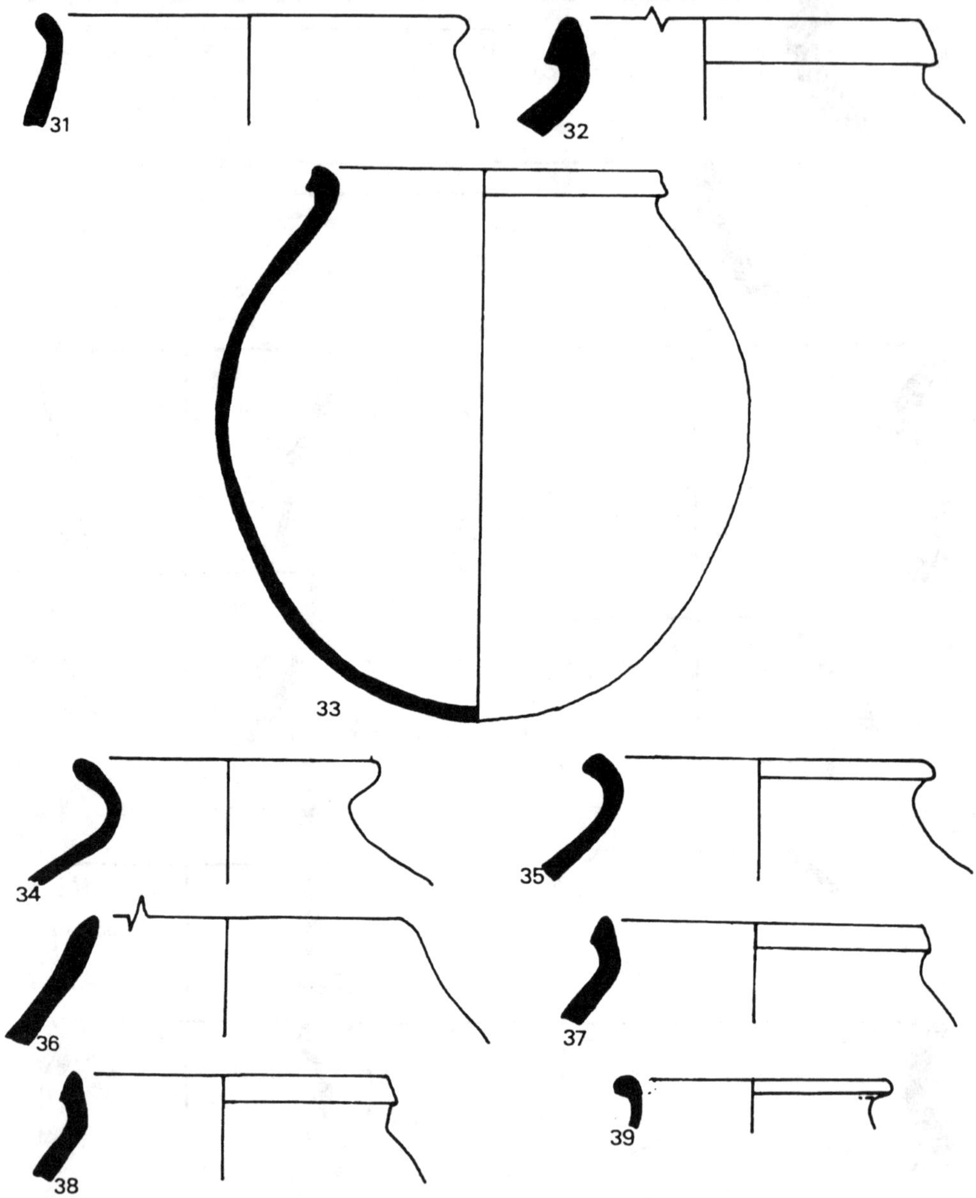

Fig. 3 Hammam VI east Ceramics: *Hammam VI east common*. Scale 1:3.5

Loon 1978, pl. 112N, pl. 113; Hauptmann 1979, pl.43:3), two stone ware body sherds, and a painted sherd (Van Loon 1982, fig. 4, note 36) provided evidence for a date towards the end of the first half of the third millennium B.C.

At the moment we lack a continuous sequence for the third millennium B.C. at Hammam et-Turkman. The Hammam VI west ceramics have been retrieved from the operations on the western slope of the mound. On the basis of the generally accepted

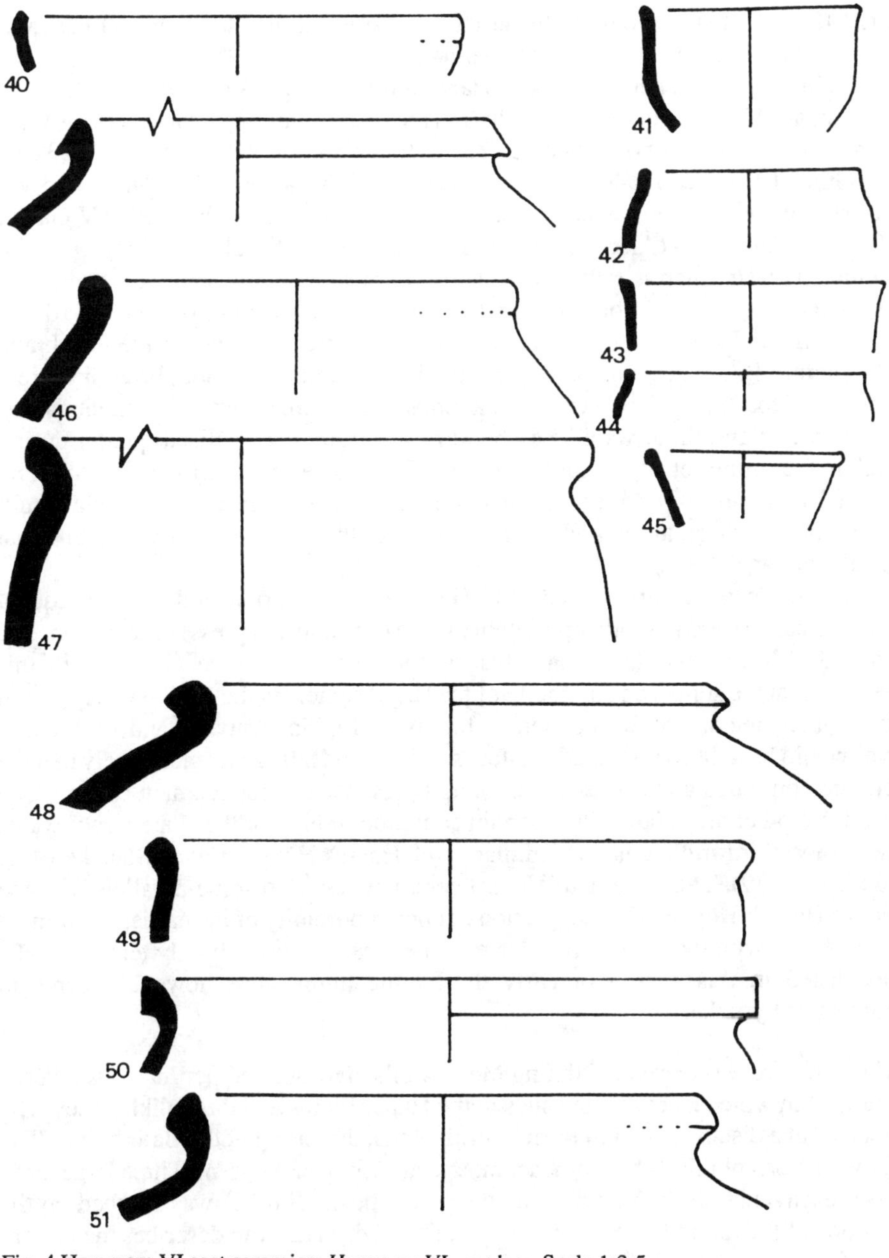

Fig. 4 Hammam VI east ceramics: *Hammam VI east late.* Scale 1:3.5

hallmarks for the second half of the third millennium, we have divided the ceramics in two subperiods: Hammam VI west early and Hammam VI west late. In the later period (EB IVB) collared-rim goblets (= sinuous rim 'dippers) and vertical-rim bowls (Curvers, in press, nos. 34, 8, 12-14) occur in large quantities, whereas these types are absent in the earlier period (EB IVA).

The EB III assemblage seems to fill in the gap between the Hammam VI east and Hammam VI west sequence (Thissen, this volume).

Stratigraphically the niched building of Hammam VB provides a *terminus post quem* for the Hammam VI east sequence. On a floor in the niched building charred beams were found which provide a series of radiocarbon dates for Hammam VB: (GrN11911) 3320 ± 25 B.C; (GrN-11912) 3285 ± 40 B.C.; (GrN-11913) 3285 ± 40 B.C. The Hammam VB ceramic assemblage has been related chronologically with Amuq F, late Leilan V, Gawra XI-X, Kurban Höyük VIB (Akkermans 1988). The absence of the classical Uruk wares at Tell Hammam et-Turkman is rather enigmatic. These wares, however, were present at several small mounds always located near the larger mounds within a distance of 10 km. These larger mounds seem to have a continuous occupation throughout the fourth and third millennium B.C. The reasons for this distributive pattern can only be conjectured. The absence of the 'Late Uruk' wares at Hammam et-Turkman may be a phenomenon caused by chance and these wares may be retrieved from a specialized quarter during fieldwork in the future (cf. Weiss and Young 1975). Another possibility is the structural absence at Hammam et-Turkman which implicates that the presence of 'Late Uruk' wares is limited to the small mounds in the vicinity of the larger centers with their local ceramic assemblage.

The assumption of a gap between the Hammam VB period and the subsequent Hammam VI east period is another possibility (cf. Akkermans, in press). The presence of the 'Late Uruk' ware would fit into this chronological gap. One objection to this hypothesis is that the supposed continuity of the larger centers in the Balikh valley would have been interrupted. The occupation in this period of interruption and of external character would have been restricted to a few small mounds. It seems more likely that the local culture continued and accepted the foreign presence outside its main centers. The evidence from other areas points to a certain comtemporeinity of the 'Late Uruk' wares with the earliest third-millennium assemblages (cf. Hassek Höyük: Behm-Blancke 1981; Tepecik: Esin 1979). The Hammam VI east ceramics revealed some parallels with the ceramics of Hassek Höyük. This suggestion of contemporeinity of the earliest Hammam VI east ceramics with the 'Late Uruk' ware of the smaller sites is highly tentative. The evidence listed in this survey of early third millennium sites, however, seems to corroborate this hypothesis.

Tell Jidle. This site was excavated during Mallowan's vigorous and pertinacious investigation project in which he excavated sites in the Habur Plains and the Balikh valley. The site is located at a distance of 4 km from 'Ain al-'Arus, the spring of the Balikh river. The overall dimensions of this rather compact mound are 150 x 120 x 15 m. Three large areas have been excavated on the mound. The deepest deposit (Jidle 8) was reached on the eastern slope of the mound; virgin soil was not reached. Mallowan describes the pottery of Jidle 7-8 as a mixture of handmade and wheelmade wares (Mallowan 1946:136). The 'burnished ware with brownish or brownish-black clay' may be similar to the small quantity of sherds in the Hammam VI east assemblage (fig. 1:10, 15, 16; fig. 2:30; fig. 3:31, 36; fig. 4:46). The few published sherds do not warrant more than a suggestion for a chronological correlation of Hammam VI east and Jidle 8-6.

The Balikh Survey. In 1983 Maurits van Loon has directed a survey in the Balikh valley (Akkermans 1984). The Bronze Age results of this survey are currently analyzed by the author. Some preliminary remarks will be presented here, as they seem to allude to the developments in the early third millennium in Syria. Within the Balikh valley surface evidence is available from the Palaeolithic period to the Middle Ages. The settlements from the Neolithic period to the Middle Ages favored a location along the Balikh river or its tributaries.

The fourth millennium occupation in the Balikh valley seems to be limited to the major sites in the valley. In the third millennium the number of settlements with Hammam VI east ceramics increased considerably. The major sites, already occupied in the fourth millennium B.C., yielded ceramics easily related with the Hammam VI east and Hammam VI west assemblages. This indicates that the major sites, such as Tell Hammam et-Turkman, Tell Sahlan, Tell es-Semen and Tell Abyad were occupied continuously during the third millennium B.C. There is no remarkable increase in the number of settlements with Hammam VI west ceramics, which indicates that the developments in settlement patterns of the first half of the third millennium developed gradually. Based on a preliminary analysis of the distributive patterns at the individual sites, it seems possible that there was an expansion of the major settlements in the second part of the third millennium B.C. This development seems to parallel developments noted for the Tabqa Dam region (Hadidi, Sweyhat) and the Habur Plains (cf. Weiss 1986).

The distribution of the classical Uruk wares reveals a pattern which may not be specific for the Balikh valley. The small sites with the classical Uruk wares are situated within a distance of 10 km from the major sites, which revealed the local fourth millennium pottery. This distribution pattern seems to occur in the Euphrates valley south of Der ez-Zor as well. The locations of Mari/Tell Ramadi and Terqa/Tell Qraya reveal a similar pattern. The explanation of this pattern can only be clarified on the basis of a better understanding of the Uruk expansion into the Syrian and Anatolian regions. For further research on the Uruk presence in Syria we should keep in mind Sürenhagen's warning not to use '.. the term 'Late Uruk' for the designation of northern sites like Habuba Kabira-South, Jebel Aruda, and others ..' (Sürenhagen 1986:32). Therefore, the problem of local period definition seems to become of major importance to deal with the developments of the alluvial presence in the northwestern periphery.

Tell Chuera. The investigations at this site started in the 1950s. The site has yielded evidence for an important third-millennium center with several noteworthy monuments. Among the monumental architecture there is a small temple, which yielded a ceramic sequence dated to the first half of the third millennium B.C. (Kühne 1976:30-33). Kühne based his chronology on the stratigraphic data collected until 1964. In 1973 the excavations at Tell Chuera were resumed. The large collection of bullae retrieved from the debris in the Antentempel is the only evidence so far for the existence of early third-millennium occupation at Tell Chuera (Orthmann 1986:63). On the basis of the excavated evidence from the lower town we may conclude that the settlement at Tell Chuera increased enormously in size in the middle of the third millennium B.C. (Orthmann et al. 1986).

THE HABUR PLAINS

Sir Austin Layard was one of the first explorers in this region described as 'a terrestrial paradise to which the wandering Arab eagerly turned his steps when he could lead his flocks thither in safety.' (Layard 1853:235). He started the exploration of the Habur Plains with an investigation of the Assyrian antiquities at Arban/Tell Ağağa.

Tell Leilan. Tell Leilan is certainly one of the more imposing sites in the Habur Plains. With its dimensions of 1100 x 850 x 15 m it belongs to the larger sites in Syria. In 1978 Yale University started the excavations at this site. Several seasons of excavations have yielded important remains spanning a period from the sixth millennium B.C. to the second millennium B.C. (Weiss 1985b).

The remains of the early third millennium have been retrieved from 20 building levels (Period III: levels 40-21) in Operation 1, a step trench on the northwestern slope (Schwartz 1982, Chapter 2). Period IV is characterized by bevelled-rim bowls which occur together with the first Ninevite V ware (Schwartz 1982:240). Period III has been dated into the Jemdet Nasr - Early Dynastic III periods (Schwartz 1982:245). This date indicates that in the definitions used throughout this survey the Period III ceramics belong to the early third millennium B.C. The calibrated results of radiocarbon samples listed in Schwartz's dissertation clearly indicate that all evidence seems to point to an earlier absolute date for the assemblages described in this survey of the 'early third millennium' sites (Schwartz 1982, Tables 10, 11, 13). The Period III ceramics are generally characterized as a Ninevite V assemblage. The earliest Ninevite V ceramics occurred together with bevelled-rim bowls in the preceding period IV (Schwartz 1982:240-241). The results of Operation 2 have clearly indicated that the Leilan III or Ninevite V settelement had not extended beyond the 15 ha area of the acropolis situated in the East. The sudden increase in settlement size seems to have begun with the construction of a defensive system. Due to the levelling of the previous deposits for the construction of the city wall an exact date for this development cannot be given. The excavators suggest a date in the middle of the third millennium B.C. (Weiss 1985b:24-25). The evidence from Tell Leilan points to the power of the locally developed culture which presumably did not accept any presence of the alluvial Mesopotamian culture.

Tell Barri. In 1980 the excavations at this site (400 x 300 x 30 m) on the Wadi Jaghjagh started under the direction of Paolo Pegorella. Area B on the western slope has yielded remains that have been dated to the first half of the third millennium B.C. (Pegcorella 1984:116; Biscione 1982:48-49). The lower horizon in the trench did not yield the incised/excised Ninevite V ware, although the other ceramics clearly belong to the Leilan III tradition. The upper horizon contained several incised/excised Ninevite V wares, which suggests the early third-millennium date for this sequence. The remains of the later third millennium are expected on a higher level, but have not been exposed so far (Pecorella 1984:116).

Tell Brak. In 1937 and 1938 Mallowan started the investigations on this huge mound in the southern portion of the Habur triangle. The excavations were resumed in 1976 by Oates. Mallowan exposed an important monumental building which revealed many characteristics of the monumental architecture exposed at Uruk-Warka in southern Mesopotamia. Another important find was the Naram-Sin palace (Mallowan 1947). One of the major goals for resumption of the investigations at Tell Brak was to obtain more stratigraphic evidence of the late fourth and third millennium B.C. (D. Oates 1985:159).

All efforts to obtain a continuous sequence for this period have so far been frustrated by the ancient building programs. To construct the Naram-Sin palace a large area of previous deposit had been scraped and levelled (D. Oates 1985:164). Till now three building phases have been exposed which are clearly earlier than the Naram-Sin palace (J. Oates 1985a; 1985b:137). A reliable, quantified and stratified ceramic sequence for the early third millennium is not available yet, which makes all correlations and historical reconstructions on the basis of the Tell Brak material highly tentative (cf. J. Oates 1985b).

THE MIDDLE HABUR SALVAGE AREA

A recent general appeal of the Directorate-General of the Syrian Antiquities Service has caused a concentration of archeological projects in this area south of Hasseke.

Tell Melebiya. This is the first site to be investigated in the Middle Habur Salvage area. The excavations started in 1985 under the direction of Mark Lebeau, who intended to expose a large area of third millennium architecture. The site is located on the west bank of the Habur and measures 300 x 150 m. The actual mound seems to represent less than half of thc original settelement (Lebeau et al. 1987, Pl. 2). This hypothysis implicates that the settlement at Tell Melebiya may be designated as a small urban center. The site yielded remains that have been dated to the early third millennium (Lebeau et al. 1985; 1986; 1987). The ceramics of the site have been well published in the preliminary reports. The large quantity of listed parallels clearly indicates that the settlement at Tell Melebiya came to be occupied in the early third millennium B.C. The ceramics of the upper levels indicate that the occupation of the site continued into the second half of the thid millennium B.C.

Tell Bderi. In 1985 the Free University of Berlin started excavations at this site. The results still have to be published, but on the basis of the kindly provided informal information of Peter Pfältzner during one of my visits to the site, I can provide a general description of the third-millennium results. Tell Bderi has yielded a long ceramic sequence, of which the oldest ceramics clearly can be dated into the first half of the third millennium B.C. The earliest architectural remains, part of a defense system and a gate, seem to indicate that Tell Bderi was one of the more important towns in the Middle Habur Salvage area.

Tell 'Atij. The Laval University (Québec, Canada) started the investigations at Tell 'Atij in 1986. The mound is located on the east bank of the Habur, and it comprises two mounds: the oval main mound (150 x 40 x 8 m) and the lower mound east of the main mound (200 x 40 x 2 m). The presence of several vaulted rooms, interpreted as silos, in a large building on top of the main mound indicates the importance of the settlement at Tell 'Atij (Fortin 1988: 155-162, fig. 18-20). The ceramics retrieved from one of the silos provide evidence for a date in the early third millennium B.C. (Fortin 1988, fig. 30). Besides the presence of storage facilities a cylinder seal stresses the importance of the settlement at Tell 'Atij (Fortin 1988, fig. 10).

Umm Qseir. This site was invesigated in 1986 under the direction of Frank Hole. Besides the remains of a Halafian tholos, this site also yielded 'classical Uruk' wares (pers. comm. Hole and Johnson) and third millennium wares. The presence of the classical Uruk wares seems to indicate that this small mound was inhabited by people, who were somehow related to the alluvial Mesopotamian core. The presence of early third millennium wares at Tell Bderi and its defensive system seem to provide a similar settlement pattern as observed for the Tabqa Dam area and the Balikh valley.

Tell Mashnaqa. In 1985 the investigations at Tell Mashnaqa were started under the direction of Jean-Yves Monchambert. The early third-millennium ceramics retrieved from strata 5-2 in area A reveal the general characteristics of a Ninevite V assemblage (Monchambert 1987:50-60, fig. 15). The expected Late Chalcolithic deposits have not been reached after two seasons of excavation (Monchambert 1985:223; 1987:60-62).

Tell al-Raqai. In 1986 the University of Amsterdam started its salvage project at Tell al-Raqai. Maurits van Loon, involved in the excavations at Tell Hammam et-Turkman, started the investigations at this site to expose a large part of third-millennium architecture. The size of the mound indicates that Tell al-Raqai had been a village between the larger centers such as Tell Melbiya, Tell Bderi and Tell 'Atij. Since 1987 project responsibilities are shared with Glenn Schwartz of the Johns Hopkins University. The results indicate that Tell al-Raqai came to be occupied in the early third millennium B.C. The final stage of occupation can be dated into the early second half of the third millennium B.C. (Curvers 1987). The ceramic assemblage retrieved from all exposed strata down to virgin soil displays the general characteristics of a Ninevite V assemblage.

THE MIDDLE EUPHRATES

Tell Ashara/Terqa. In 1975 investigations started under the direction of Giorgio Buccellati. The site is located on the east bank of the Euphrates and measures 450 x 300 m. The sharp escarpement in the East seems to be the result of erosion by the Euphrates, which has left probably less than half of the original settlement. The early third-millennium remains are sparse at Tell Ashara. The remains of a city wall were associated with ceramics, which clearly indicate relations with the ED I pottery of the Diyala region in Mesopotamia (Buccellati 1979:72, fig. 21). The ceramics have also been dated to the ED II period (Bucellati 1979:73, fig. 23). In general the assemblage associated with the city wall is rather homogeneous without any major discontinuities. In the historical interpretation there is again some confusion about a radiocarbon date which suggests a date in the fourth millennium B.C. without, however, the presence of archeological material to corroborate such a date. The fourth millennium material is present at nearby Tell Qraya which seems to reflect a distributive pattern already noted for the Tabqa Dam area and the Balikh valley.

Tell Hariri/Mari. A large fragment of a stone statue found by chance started the investigations on this large mound in 1933. Remains of a city wall and several temples and shrines have been exposed during the many years of excavation under the direction of

André Parrot and recently under the direction of Jean-Claude Margueron. The fiftieth anniversary of the project seems to have stimulated the analysis of the archeological and philological results collected at this important center in the third millennium B.C. Lebeau's preliminary ceramic sequence has given Mari its deserved prominent place in the research on the history of Syria in the third millennium B.C. The stratigraphic results of the excavations in 'chantier B' on the northern slope of the mound have yielded 18 strata with a continuous ceramic sequence for the third millennium B.C. (Lebeau 1985; 1987). This evidence for the foundation of the city of Mari in the early third millennium should again be related to the presence of 'earlier' or contemporary material collected at Tell Ramadi (Geyer and Monchambert 1987:336, fig. 8; Margueron 1987:498).

CONCLUSIONS

Whether completely representative or not, this short survey marks Syria as a region with a ceramic culture within the fragmented mosaic of early third millennium assemblages throughout the Near East. As mentioned in the introduction, the finds of the last three decades have revealed evidence for the existence of important cultural regions throughout Syria. On the Habur plains the archeological evidence might reasonably be considered to be an indication of an important, largely, Hurrian center (cf. Weiss 1983). The finds at Mardikh provide unparalleled evidence for the existence of a great Eblaite civilization (cf. Matthiae 1981).

The more recent developments of Syrian Archeology have been to some extent determined by practical perspectives such as the construction of hydro-electric dams in the Euphrates and Habur. The excavations have yielded ceramics that range from the Ninevite V assemblages in the Habur Plains and along the Upper Tigris and beyond, to the orange, brittle ware and red/black burnished tradition in the Amuq Plain. This fragmentation reflects a pattern of local interactions with the previous episode of alluvial Mesopotamian colonization of its northwestern periphery. The gradual collapse of this penetration from the south resulted in or was caused by the development of the local cultures. This development strongly contrasts the one sketched by Sürenhagen, who considers the urban growth along the Syrian Euphrates to be interrupted after the fall of the upper Euphratian colonies (1986:33). It now seems that throughout Syria indigenous developments after the Mesopotamian colonization finally culminated in the Eblaite and Hurrian culture of the later third millennium B.C. (cf. Weiss 1983).

Based on the evidence presented in the survey above, we should be aware of a possible overlap and contemporaneity of the assemblages retrieved from sites interpreted as 'southern colonies' and the local sites usually considered to be of a third millennium B.C. date. A similar overlap now seems to exist for the Late Uruk culture and the Ninevite V culture (Roaf 1983:71; Weiss 1985). Sürenhagen seems to agree with these suggestions for an overlap and adds further arguments (1986:33).

I would like to add similar arguments for the other Syrian regions based on the short survey of Syrian archaeological evidence for the beginning of the third millennium. The absolute date of the beginning of the third millennium may shift to a date before 3000 B.C. This depends on the use of callibrated radiocarbon dates or not (cf. Schwartz 1982).

Although systematic analyses of the pottery retrieved from the excavations and surveys listed above have not yet been completed, certain visual observations - of which the relative importance remains tentative - may help to situate the various sites in the network of local developments in northern Syria with respect to the stratified ceramics from other tentatively contemporary sites.

REFERENCES

AKKKERMANS P.M.M.G.

1984 'Archäologische Geländebegehung im Balih-Tal', *Archiv für Orientforschung* 31: 188-190.

1988 'An Updated Chronology for the Northern Ubaid and Late Chalcolithic Periods in Syria: New Evidence from Tell Hammam et-Turkman', *Iraq* 50.

in press 'The Period V Pottery', in: M.N. Van Loon (Ed.), *Hammam et-Turkman I, Report on the University of Amsterdam's 1981-84 Excavations in Syria*. Istanbul: Nederlands Historisch-Archaeologisch Instituut.

BEHM-BLANCKE ET AL.

1981 'Hassek Hüyük: vorläufiger Bericht über die Ausgrabungen der Jahre 1978-1980', *Istanbuler Mitteilungen* 31: 5-93.

BISCIONE R.

1982 'La Ceramica del III Millennio', in: P.E. Pecorella and M. Salvini (Eds.), *Tell Barri/Kahat 1, Relazione preliminare sulle Campagne 1980 e 1981 a Tell Barri/ Kahat nel bacino del Habur*. Roma: Servizio Pubblicazioni del Consiglio Nazionale delle Ricerce.

BOUNNI A.

1979 'Preliminary Report on the Archaeological Excavations at Tell al-'Abd and 'Anab al-Safinah (Euphrates) 1971-72', *Annual of American Society of Oriental Research (Archaeological Reports from the Tabqa Dam Project, Euphrates Valley, Syria)* 44: 49-61.

BRAIDWOOD R.J.

1937 *Mounds in the Plain of Antioch*. OIP 48. Chicago: The University of Cicago Press.

BRAIDWOOD R.J. AND BRAIDWOOD L.S.

1960 *Excavations in the Plain of Antioch I: The earlier Assemblages Phases A-J*. OIP 61. Chicago: The University of Chicago Press.

BUCCELLATI G. ET AL.

1979 *The Fourth Season: Introduction and the Stratigraphic Record, Terqa Preliminary Reports No. 11*. Malibu: Undena Publications.

COURTOIS J.-CL.

1962 'Contribution a l'étude des niveaux II et III de Ras Shamra', *Ugaritica* 4: 330-414.

CURVERS H.H.

1987 'The Middle Habur Salvage Operation: Excavation at Tell al-Raqai, 1986', *Akkadica* 55: 1-29.

in press 'The Period VI Pottery', in: M.N. Van Loon (Ed.), *Hammam et-Turkman I, Report on the University of Amsterdam's 1981-84 Excavations in Syria*. Istanbul: Nederlands Historisch-Archaeologisch Instituut.

DE CONTENSON H.

1969 'Les couches du Niveau III au Sud de l'Acropole de Ras Shamra', *Ugaritica* 6: 43-89.

1979 'Nouvelles données sur la chronologie du Bronze Ancien de Ras Shamra', *Ugarit-Forschungen, Internationales Jahrbuch für die Altertumskunde Syrien-Palestinas*, 11: 857-862.

1982 'Les phases préhistoriques de Ras Shamra et de l'Amuq', *Paléorient* 8: 95-98.

DORNEMANN R.H.

1979 'Tell Hadidi: A Millennium of Bronze Age City Occupation', *Annual of the American Schools of Oriental Research (Archaeological Reports from the Tabqa Dam Project, Euphrates Valley, Syria)* 44: 113-151.

1984 'Tell Hadidi: One Bronze Age Site Among Many in the Tabqa Dam Salvage Area', Unpublished Paper.

ESIN U.
1979 'Tepecik Excavations, 1973', in: *Keban Project 1973 Activities*. Ankara. Pp. 97-112.
FORTIN M.
1988 'Rapport Préliminaire sur la première campagne de fouilles (printemps 1986) à Tell 'Atij, sur le moyen Khabour', *Syria* 65: 139-171.
FUGMANN E.
1958 *Hama, Fouilles et rechereches, 1931-1938, II-1: L'architecture des périodes préhellénistiques*. Copenhagen.
GEYER B. AND MONCHAMBERT J.-Y.
1987 'Prospection de la moyenne vallé de l'Euphrate; rapport préliminaire 1982-1985', *M.A.R.I.* 5: 293-344.
HAUPTMANN H.
1979 'Die Grabungen auf dem Norşun-Tepe', in: *Keban Project 1973 Activities*. Ankara. Pp. 61-78.
HOLLAND T.A.
1976 'Preliminary Report on Excavations at Tell es-Sweyhat, Syria, 1973-4', *Levant* 8: 36-70.
1977 'Preliminary Report on Excavations at Tell es-Sweyhat, Syria, 1975', *Levant* 9: 36-65.
INGHOLT H.
1940 *Rapport préliminaire sur sept campagnes de fouilles à Hama en Syrie*. Kgl. Danske Videnskabernes Selskab, Archaeologisk-Kunsthistoriske Meddelelser III, 1. Copenhagen.
LAYARD A.H
1853 *Discoveries in the Ruins of Nineveh and Babylon; with travels in Armenia, Kurdistan and the Desert*. London: John Murray.
LEBEAU M.
1985 'Rapport préliminaire sur la sequence céramique du Chantier B de Mari (IIIe mill.)', *M.A.R.I.* 4: 93-126.
1987 'Rapport préliminaire sur la céramique des premiers niveaux de Mari (Chantier B-1984)', *M.A.R.I.* 5: 415-442.
LEBEAU M. ET AL.
1985 'Rapport préliminaire sur la première campagne de fouilles à Tell Melebiya (Moyen Khabour - printemps 1985)', *Akkadica* 45: 1-32.
1986 'Rapport préliminaire sur la deuxième campagne de fouilles à Tell Melebiya (Moyen Khabour - automne 1986)', *Akkadica* 46: 1-50.
1987 'Rapport préliminaire sur la troisième campagne de fouilles à Tell Melebiya (Moyen Khabour - automne 1986)', *Akkadica* 51: 1-74.
MALLOWAN M.E.L
1946 'Excavations in the Balih, 1938', *Iraq* 8: 111-159.
1947 'Excavations at Brak and Chagar Bazar', *Iraq* 9: 1-259.
MARGUERON J.
1987 'État présent des recherches sur l'urbanisme de Mari, I', *M.A.R.I.* 5: 483-498.
MATTHERS J. (ED).
1981 *The River Qoueiq, Northern Syria, and its Catchment. Studies arising from the Tell Rifa'at Survey 1977-79*. BAR International Series 98. Oxford.
MATTHIAE P.
1981 *Ebla, An Empire Rediscovered*. Garden City, New York: Doubleday & Company, Inc.
MAZZONI S.
1985 'Elements of the Ceramic Culture of Early Syrian Ebla in Comparison with Syro-Palestinian EB IV', *Bulletin of the America Schools of Oriental Research* 257: 1-18.
MEIJER D.J.W.
in press 'Tell Hammam: Architecture and Stratigraphy', in: M.N. Van Loon (Ed.), *Hammam et-Turkman I, Report on the University of Amsterdam's 1981-84 Excavations in Syria*. Istanbul: Nederlands Historisch-Archaeologisch Instituut.
MELLAART J.
1981 'The Prehistoric Pottery from the Neolithic to the Beginnings of E. B. IV', in:J. Matthers

(Ed.), *The River Qoueiq, Northern Syria, and its Catchment: Studies arising from the Tell Rifa'at Survey 1977-79*. BAR International Series 98. Oxford.

MONCHAMBERT J.-Y.

1985 'Mashnaqa 1985: Rapport préliminaire sur la première campagne de fouilles', *Syria* 62: 219-250.

1987 'Mashnaqa 1986: Rapport préliminaire sur la deuxième campagne de fouilles', *Syria* 64: 47-78.

OATES D.

1985 'Excavations at Tell Brak, 1983-84', *Iraq* 47: 159-174.

OATES J.

1985a 'Tell Brak: Uruk Pottery from the 1984 Season', *Iraq* 47: 175-186.

1985b 'Tell Brak and Chronology: The Third Millennium', *M.A.R.I.* 4: 137-144.

ORTHMANN W.

1981 *Halawa 1977-1979*. Saarbrücker Beiträge zur Altertumskunde 31. Bonn: Rudolf Habelt Verlag.

1986 The Origin of Tell Chuera', in: H. Weiss (Ed.),*The Origins of Cities in Dry-Farming Syria and Mesopotamia in the Third Millennium B.C.* Guilford, Connecticut: Four Quarters Publishing Co. Pp. 61-70.

ORTHMANN W. ET AL.

1974 Munbaqat 1973 ...', *Mitteilungen der Deutschen Orientgesellschaft* 106: 53- 97.

1986 *Tell Chuera in Nordost-Syrien, 1982-1983*. Vorläufiger Bericht über die 9. und 10. Grabungskampagne. Berlin: Gebr. Mann Verlag.

PECORELLA P.E.

1984 Tell Barri/Kahat', *Archiv für Orientforschung* 31:114-118.

PINNOCK F.

1985 'About the Trade of Early Syrian Ebla', *M.A.R.I.* 4: 85-92.

ROAF M.

1983 'A Report on the Work of the British Archaeological Expedition in the Eski Mosul Dam Salvage Project', *Sumer* 39: 68-82.

SCHWARTZ G.M.

1982 *From Prehistory to history on the Habur Plains: The Operation 1 Ceramic Periodization from Tell Leilan*, Ph.D. dissertation, Yale University. Ann Arbor: University Microfilms International.

SÜRENHAGEN D.

1986 'The Dry Farming Belt: The Uruk Period and Subsequent Developments', in: H. Weiss (Ed.), *The Origins of Cities in Dry-Farming Syria and Mesopotamia in the Third Millennium B.C.* Guilford, Connecticut: Four Quarters Publishing Co.

TEFNIN, R.

1979 'Exploration archéologique au nord du Lac Djaboul. Une campagne de sondage sur le site de Oumm el-Marra, 1978-79', *Annuaire de l'Institut de philologie et d'histoire Orientale et Slave* 23: 71- 94.

1981/82 'Tall Umm al-Marra', *Archiv für Orientforschung* 28: 235-239.

VAN DRIEL G. AND VAN DRIEL-MURRAY C.

1979 'Jebel Aruda 1977-78', *Akkadica* 12: 2-28.

VAN LOON M.N.

1979 '1974 and 1975 Preliminary Results of the Excavations at Selenkahiye Near Meskene, Syria', *Annual of the American Schools of Oriental Research (Archaeological Reports from the Tabqa Dam Project, Euphrates Valley, Syria)* 44: 97-112.

1982 'Hammam et-Turkman on the Balikh: Background and First Results of the University of Amsterdam's 1981 Excavation', *Akkadica* 27: 30-45.

VAN LOON M.N. (ED.)

1978 *Korucutepe, Vol. 2*. Amsterdam: North-Holland Publishing Company.

WEISS H.

1983 'Excavations at Tell Leilan and the Origins of North Mesopotamian cities in the third millennium B.C.', *Paléorient* 9/2: 39-52.

1985a Review of J. Curtis (Ed.), Fifty Years of Mesopotamian Discovery, in: *Journal of the American Oriental Society* 105: 327-330.

1985b 'Tell Leilan on the Habur Plains of Syria', *Biblical Archeologist* 48: 5-35.

1986 'The Origins of Tell Leilan and the Conquest of space in Third Millennium Mesopotamia', in: H. Weiss (Ed.), *The Origins of Cities in Dry-Farming Syria and Mesopotamia in the Third Millennium B.C.* Guilford, Connecticut: Four Quarters Publishing Co.

WEISS H. AND YOUNG T.C.

1975 'The Merchants of Susa: Godin V and Plateau-Lowland relations in the Late Fourth Millennium B.C.', *Iran* 13: 1-17.

An Early Bronze III pottery region between the Middle Euphrates and Habur: New evidence from Tell Hammam et-Turkman

LAURENS C. THISSEN

INTRODUCTION

The Early Bronze Age III pottery of Tell Hammam et-Turkman excavated in the 1986 season provides many morphological and technological parallels with Harran and Tell Chuera, allowing for the delineation of a pottery region connecting the Balikh valley and Chuera. This region is distinct from the contemporary assemblages west of the Middle Euphrates and from those in the Habur area to the east [1]. The extent of this ceramic region and its characteristic pottery types will be discussed in this paper.

Recently, Stefania Mazzoni has defined several ceramic provinces west of the Euphrates on the basis of the Mardikh IIB1-2 material (Mazzoni 1985a)[2]. One of her important conclusions was the fundamental role played by the Middle Euphrates as a dividing line between different ceramic cultures during the second half of the third millennium B.C. In her view, the left bank sites on the Euphrates stretch (Halawa, Munbaqa, Sweyhat), as well as the Balikh sites and Chuera belonged to an eastern oriented culture (Mazzoni 1985a:568-9).

The concept of ceramic cultures or pottery regions is not new (cf. Kühne 1976:106-108, Karte 4), but Mazzoni adds considerably to it by reassessing the nature of ceramic relations existing between sites. She rightly points out that a pottery region is not defined by the mere presence of certain shapes, types or decorations, but must be delineated by comparing the assemblages of different sites in their complex totality (Mazzoni 1985a:563). In my opinion, this 'complex totality' must be seen as the simultaneous occurrence in one local ceramic horizon of a standard repertoire of several ceramic categories which form the majority of the assemblage (Mazzoni 1985a:569), and other groups of pottery which are less frequent but constantly present. The technological and morphological variation from the standard repertoire of these groups indicates different manufacturing techniques, function, and import. In a given ceramic province there should be a standard repertoire with common shapes and techniques, together with a consistent preference for more specialized ceramics.

In this article I would like to take up Mazzoni's trail. It is now possible to trace the ceramic frontiers of the area beyond the Syrian Middle Euphrates more precisely, thus adding another pottery region to the ones Mazzoni already established for the second half of the third millennium B.C. (Mazzoni 1985a:562, here fig.1). This complex picture of diverse ceramic cultures corroborates the one painted in the Ebla texts of a series of local kingdoms in Syria, which were independent centers of political significance (Archi

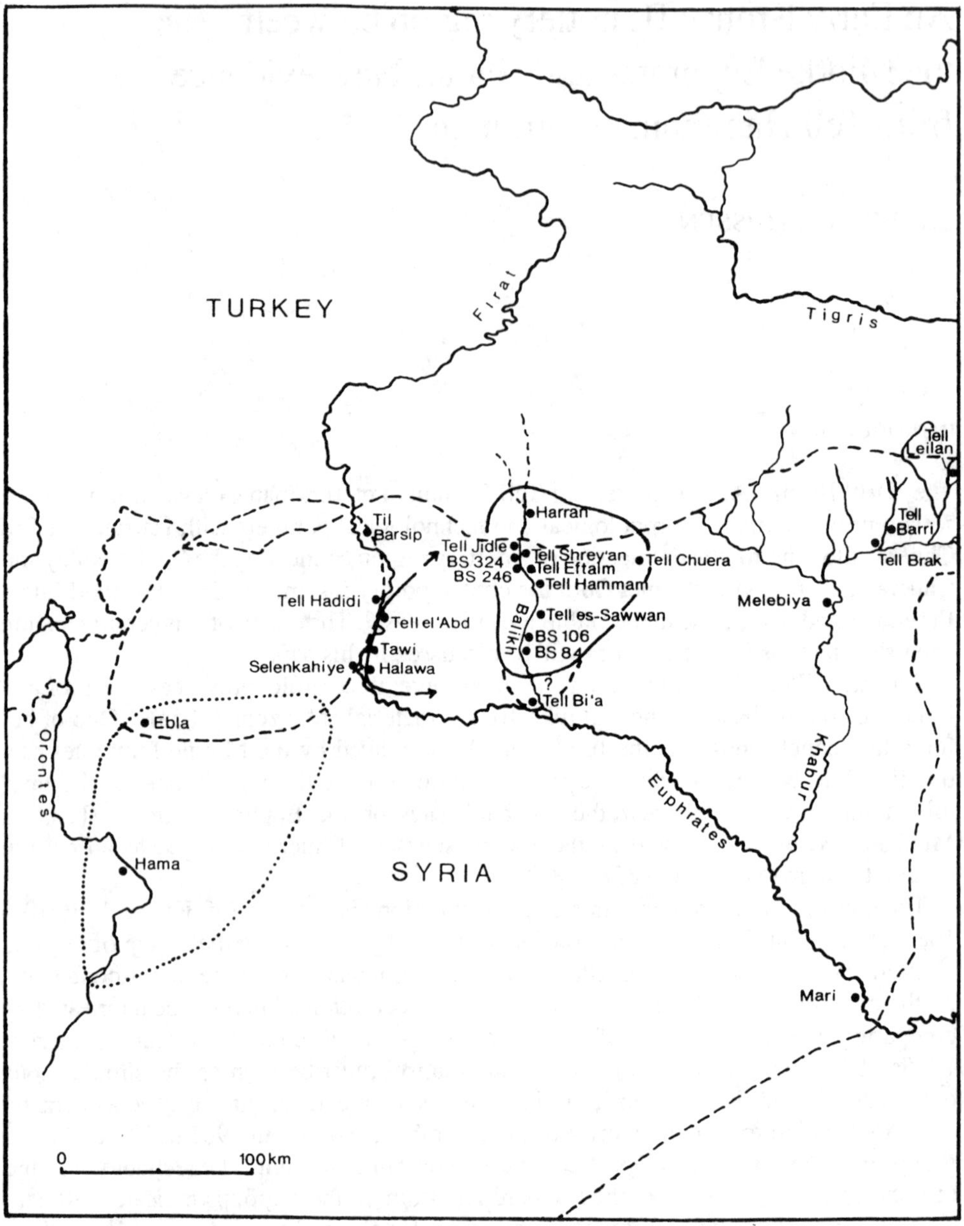

Fig. 1. The Balikh-Chuera EB III pottery region and sites mentioned in text. Also indicated are the ceramic regions west of the Euphrates (Mazzoni 1985a:562).

1980:3; Nissen 1983:200). Weiss and Algaze speak of great indigenous states and refer to the many fortified sites in the North during the EB III period such as Taya IX, Leilan II, Brak, Chuera, Mari, Ebla, and Khoshi (Weiss 1983:47, 49; Algaze 1986:146). Whether these large local states originated in the second half of the third millennium B.C. (2500 or 2400 B.C.), as Weiss suggests, referring to the sudden enlargement of Leilan from 15 ha in Period III to 90 ha in Period II (Weiss 1983:47,49), is not certain, in my opinion. The preceding Early Bronze II period is not well known in North Syria - only grasped at in Mardikh IIA, and elusive on Brak -, but the considerable amount of clay bullae datable to EDI-II found at Tell Chuera in 1976 indicate the presence of an important economic centre at the site in earlier periods (Moortgat and Moortgat-Correns 1978:16-33; Orthmann 1986:63).

At Tell Hammam et-Turkman, the 1986 excavation of a segment of a huge mud-brick city wall with carefully built and neatly plastered domestic architecture on top of it (Van Loon and Meijer 1987:1-2, fig.2) provides substantial evidence that this North Syrian site was fortified in EB III times and can be added to the list of the above-mentioned cities. The estimated settlement size of 3.75 ha for Tell Hammam et-Turkman pales before settlements such as Chuera (100 ha) or Leilan II (90 ha).

The rooms within the city wall were overlaid by several levels of domestic architecture aligned on a similar orientation. Two of these levels had burnt severely, probably without much of an interval between them[3]. Large masses of discarded pottery had accumulated outside the city wall, together with animal bones, rubbish and small finds. This pottery, in spite of its character of homogeneous debris, correlates exactly with the pottery retrieved from the city-wall rooms and the first burnt level.

The ceramics from the second burnt level and subsequent levels (VI west: 1-4) show important differences from the lower strata. The popular plain-rim, round-based goblets from the lower strata are absent in the later levels, replaced by goblets or bowls that are slightly larger in diameter, plain-rimmed but flat-based, with thicker, flaring walls[4]. Characteristic is a slightly convex exterior rim and an almost straight interior wall (fig. 2:8; Curvers, in press, Pl. 119:24-25). Another difference is that stone ware ('Metallische Ware') is absent in the assemblages retrieved from these later levels. There is, however, not a complete break between the first and second destruction levels: connective elements are cooking pots with triangular lugs and chaff-faced storage vessels, which remain in use (Curvers in press, Pl. 120:41; VI west:1). Combed-wash ware starts on a very limited scale in the first burnt level and is represented by a fine small bottle from a floor in the second burnt level (Curvers in press, Pl.122:73; VI west:1). The ware is conspicuous by its absence from the debris layers outside the city wall.

I will restrict myself here to the pottery from these debris layers, the city-wall rooms and the first burnt level, since it is completely homogeneous. An EB III date is suggested for this material, while the second burnt level and subsequent levels are considered to be part of EB IV (i.e. Akkad period). Radiocarbon dates from the first and second burnt levels point to a date between 2600 and 2300 B.C.[6], but I propose 2450-2300 B.C. for city wall - first burnt level[7].

THE EVIDENCE

Turning to the Tell Hammam pottery, I shall emphasize those ceramic categories and shapes that characterize the EB III assemblage by their quantity and shared technological aspects[8]. The standard ware at Hammam is a plain simple ware, in which the aforementioned plain rim, round based goblets dominate. They are wheelmade and have thin (4 mm) vertical or slightly incurving walls. The average diameter is 10 cm (fig.2:1-2). A variant is a more closed shape (no. 3).

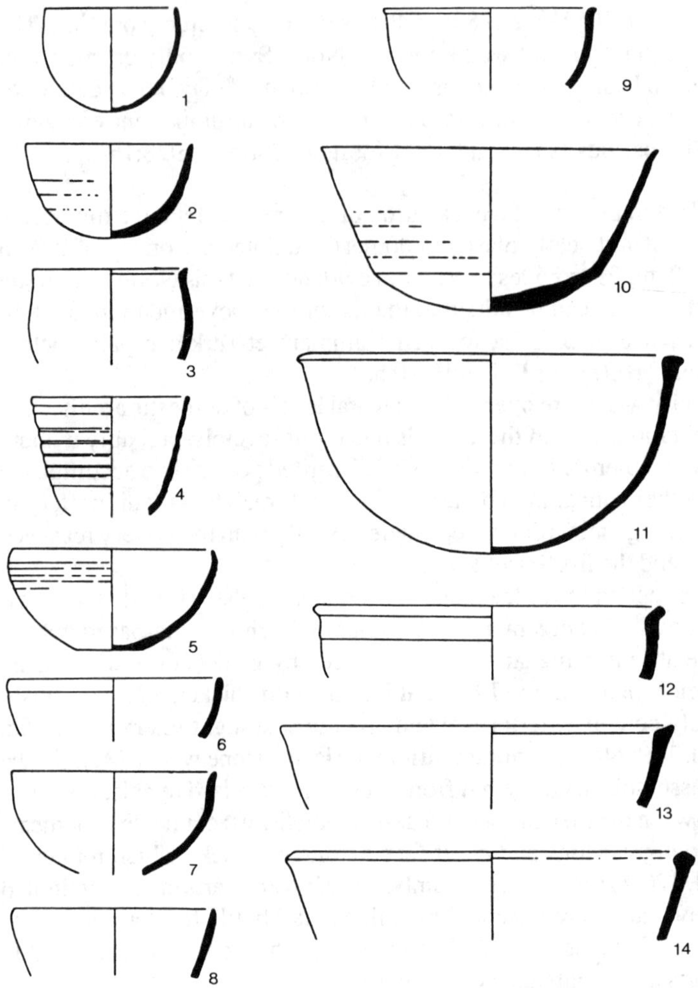

Fig. 2. Tell Hammam et-Turkman EB III Pottery. Plain simple ware (nos. 1-3, 7-9, 11-14); stone ware (nos. 4-6, 10). Scale 1:4.7.

Wheelmade large open bowls (diameters between 18 and 26 cm) form a second distinctive category. Lips are thickened in a variety of ways (nos. 11, 14), everted (no. 12) or bevelled inward (no. 13). All lip varieties have invariably round bases.

A third characteristic group consists of small pots (average diameter 22 cm) with rather vertical walls, creating a bag-shaped aspect. They often possess abruptly bent outward rims with vertical-faced exterior lips (nos. 15, 18), or else rounded (no.16) or bevelled outward lips (no. 17). A more globular body with similar lips is often preferred above the bag shape (no. 18).The pots have round bases.

A fourth group consists of small S-shaped jars with round bases. Diameters vary between 10-14 cm. Lips are plain (no. 23) or folded over and slightly thickened (no. 24).

Larger jars form a fifth group. Variability in rim treatment is higher in this group than in the other four groups. A basic characteristic, however, is a round-based, globular body with an often abrupt transition into a funnel-shaped neck (nos. 25-27). The rims have outrolled lips (no. 27), everted lips (no. 26), or lips with triangular section (no. 25).

From a technological point of view, the homogeneity of the material is even more obvious than the shapes indicate. The same clay is used for the manufacture of all categories, firing to a distinctive pale orange colour. Fine to medium sand, often with mica inclusions, are the non-plastics. A standard surface treatment was the careful scraping in oblique directions of the lower and middle parts of the interior and exterior walls while the vessel was in a leather-hard state. This technique was primarily used on the larger pieces: large bowls, bag-shaped pots and larger jars. The small plain rim goblets sometimes show a similar treatment. Rims show fine, horizontal striations. The goblets and bowls were made in one piece, while the pots and larger jars often had their rims made separately and joined to the body of the vessel afterwards. This method of manufacture accounts for the often abruptly bent outward rims and a considerable difference in wall thickness of body and neck (cf. nos. 15, 18).

The plain simple ware of Tell Hammam et-Turkman clearly forms one body of material with the same firing conditions, the same methods of manufacture and surface treatment, and a limited standard repertoire for finishing-off rims and lips. These characteristics point to a powerful local production, able to manufacture most of the necessary ceramic categories. It is important to note, however, that the production techniques were still rather laborious in contrast to the fast-wheel mass production of the late third millennium Hammam VI west period (Curvers in press).

There are also less frequent but consistently present wares and types in the Hammam EB III ceramic assemblage. These are: chaff-faced vessels used for storage, which are handmade, coarsely chaff-tempered and have rounded or flat (slightly concave) bases (nos. 21-22); grit or calcite-tempered cooking pots with triangular lugs, handmade, with pebble-burnished exterior (no. 19); stone ware, predominantly corrugated goblets (no. 4) and small bowls (nos. 5-6). Other stone ware shapes are the miniature (no. 30) and thick-walled body sherds (not ill.). Unique, but important for its chronological implications[7] is a complete green stone ware bowl (no. 10) from a floor of the first burnt level. Eggshell ware, grey-black or orange-brown, with 1 mm wide spiral burnish strokes (not ill.), red-banded jars, also with spiral burnishing and eggshell-thin walls (no. 31), and combed-wash ware (not ill.) complete the picture.

The EB III pottery of Tell Hammam et-Turkman corresponds perfectly with the Early Bronze material from Harran, phases IIa and IIb (Prag 1970). Several Hammam plain simple ware categories are represented in the Harran PSW also (often providing exact parallels): plain rim, round-based goblets (Prag 1970, fig. 7:14), large open bowls (Prag 1970, fig. 7:2, 7, 8), vertical-walled pots (Prag 1970, fig. 7:3, 6).The Harran assemblage also includes the more specialized wares, such as stone ware (nos. 4-6; cf. Prag 1970, fig. 7:36, 37, 27, 15), eggshell ware (Prag 1970, fig. 8:43-44), and triangular lugged cooking pots (Prag 1970, fig. 8:54-55). On the basis of a parallel standard repertoire and an identical preference for specialized ceramics, Harran seems to belong to the same ceramic culture as Tell Hammam EB III. This allows us to equate Harran IIa-b with the Hammam city wall and associated architecture. Furthermore, the stone ware in Harran precedes the combed-wash ware, as in Tell Hammam (Prag 1970:78). Combed-wash ware starts to appear in Harran in phase IIIa, separated by destruction level IIIb from phases IIIc and IV, the final Early Bronze phase where combed-wash ware is still present. Phase IIIc was destroyed as well (Prag 1970:72). Considering the appearance of

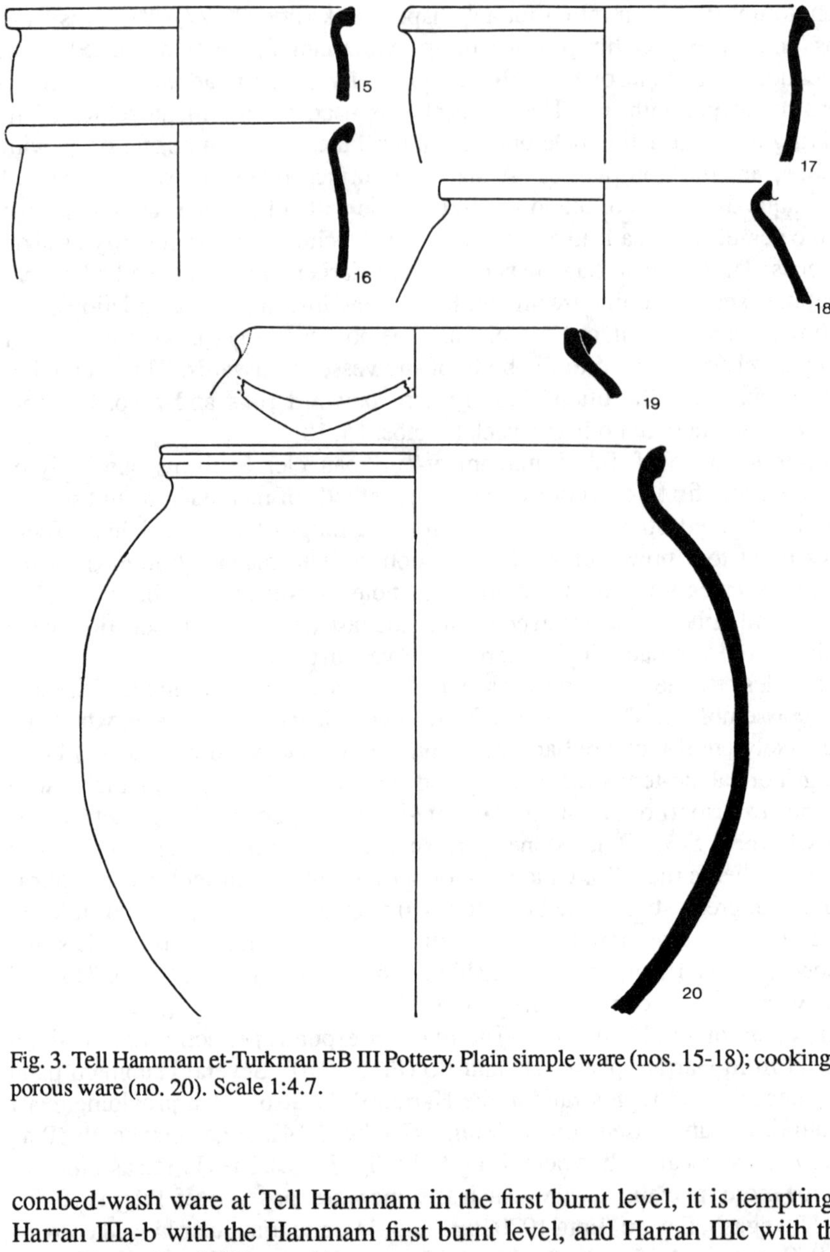

Fig. 3. Tell Hammam et-Turkman EB III Pottery. Plain simple ware (nos. 15-18); cooking ware (no. 19); porous ware (no. 20). Scale 1:4.7.

combed-wash ware at Tell Hammam in the first burnt level, it is tempting to correlate Harran IIIa-b with the Hammam first burnt level, and Harran IIIc with the Hammam second burnt level(VI west:1).

One would expect the other excavated Early Bronze Balikh site, Tell Jidle, in a central position between Harran and Hammam et-Turkman, to have had a similar ceramic assemblage. Although the published evidence is of limited use, I think that Mallowan's Sargonid date for level 5, perhaps contemporary with Naramsin's reign, is entirely justifiable, rather than the earlier date (ED III) preferred by Kühne (Kühne 1976:66, 70,114 table 2; Mallowan 1946:134). Indeed, many vessels are definitely later than Hammam EB III and have exact parallels in the later Hammam VI west pottery (EB IV):

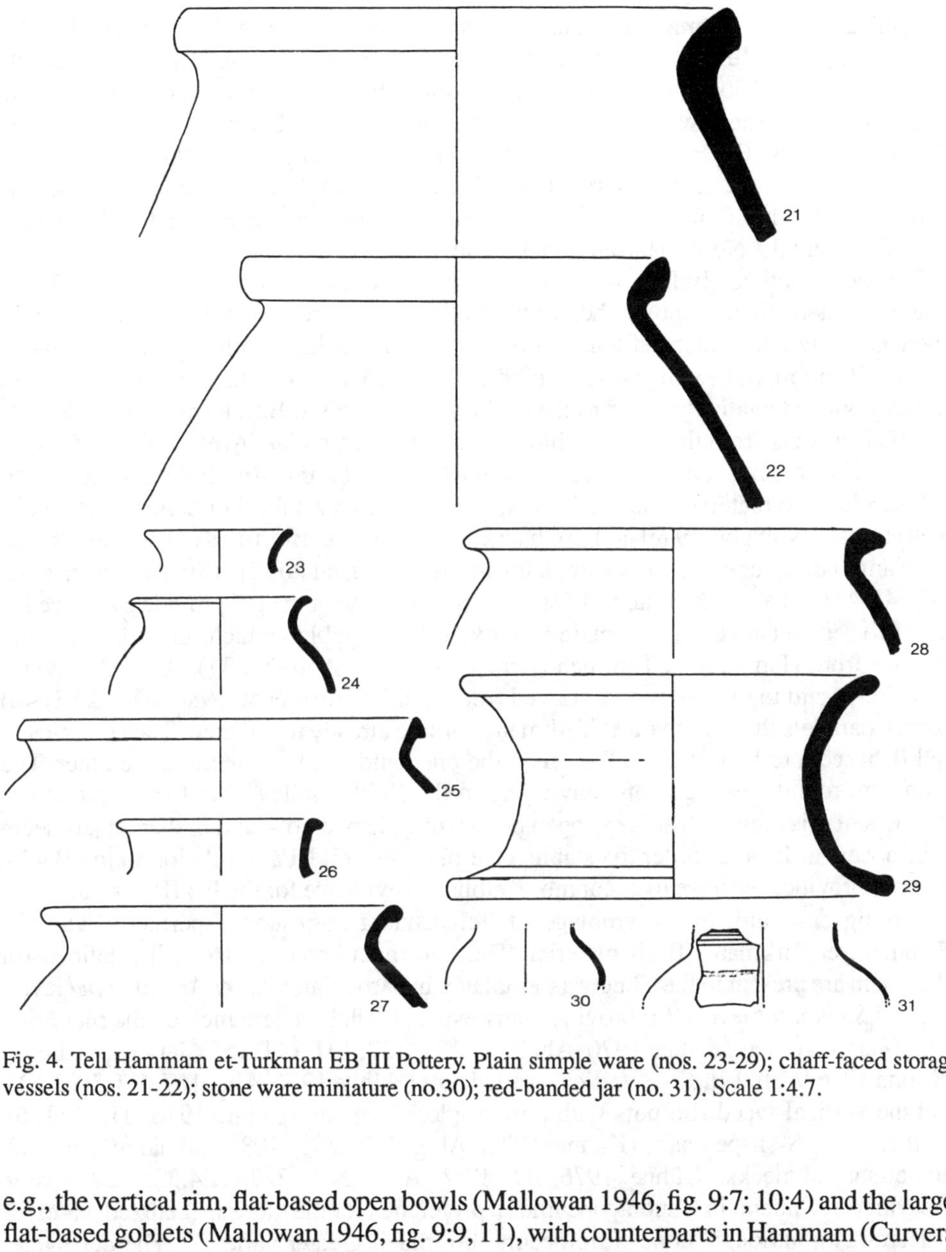

Fig. 4. Tell Hammam et-Turkman EB III Pottery. Plain simple ware (nos. 23-29); chaff-faced storage vessels (nos. 21-22); stone ware miniature (no.30); red-banded jar (no. 31). Scale 1:4.7.

e.g., the vertical rim, flat-based open bowls (Mallowan 1946, fig. 9:7; 10:4) and the large, flat-based goblets (Mallowan 1946, fig. 9:9, 11), with counterparts in Hammam (Curvers, in press, Pl. 117:13 and 120:39 respectively). Several rim-profiles from level 6 which fit more easily into the Hammam EB III assemblage include large open bowls, triangular lug pots, and a jar with funnel-shaped neck (Mallowan 1946, fig. 12:22, 29, 21, 27, 24 resp.). In spite of the small pottery sample and insecure stratigraphic position of level 6 (Mallowan 1946:136), Tell Jidle is here tentatively incorporated in the Balikh EB III pottery region.

More evidence for this Balikh ceramic region is provided by surface finds collected during the 1983 Balikh survey (Akkermans 1985). Perfect parallels with the EB III assemblage of Tell Hammam et-Turkman occur among the sherds from site BS 84, situated opposite Tell es-Semen on the left bank of the Balikh. This huge EB III settlement (380 x 550 x 3.6 m) had been devastated by a severe fire. Many sherds show scrape marks on the lower and middle body parts comparable to those noted on the Hammam sherds. Other Balikh sites with similar material are BS 106, BS 147 (Tell es-Sawwan), BS 212 (Tell Eftaim), BS 246, BS 264 (Tell Shrey'an) and BS 324. The distribution of these sites demarcates a Balikh pottery region in the EB III period stretching from the North (Harran) to the South (BS 84).

The most southern Balikh valley site is Tell Bi'a, under excavation since 1980. It would be interesting to know whether the Balikh pottery region extends this far south. There is substantial evidence for an EB III settlement at Bi'a. On hill C, massive remains of a fortification wall with rooms behind it were discovered. The complex had burnt severely and is tentatively dated to the mid-third millennium B.C. (Arne et al. 1984:44-48). Four vessels from this context have been published so far (Arne et al. 1984, Abb. 21m-p). They are not comparable to material from Tell Hammam et-Turkman but instead can be related to material from Tell Barsip. In another area a third millennium level has been exposed with pottery related to Chuera. The main characteristics of this assemblage are triangular lug cooking pots and plain-rim, round-based goblets (Strommenger et al. 1987:41-43). Finally, the contents of several burials north of the tell provide evidence for an EB III cemetery at Bi'a, with plain-rim, round-based goblets which correspond readily to those from Hammam et-Turkman (Arne et al. 1984, Abb. 34, 37). A combed-wash ware bottle said to to have been retrieved from a burial (Arns et al. 1984, Abb. 35:U 40) exactly parallels the Hammam et-Turkman example already mentioned. The ceramics of Tell Bi'a seem to be related to Barsip on the one hand and to Chuera on the other. The plain-rim, round-based goblets may relate to the Balikh valley[9], but the characteristic products of this area such as large open bowls, bag-shaped pots, and S-shaped jars seem to be absent at Bi'a. In order to establish the place of Tell Bi'a in relation to the Balikh ceramic province we have to await more published evidence for the EB III period.

Turning eastward, the assemblage of Tell Chuera corresponds perfectly with the Hammam et-Turkman EB III material. The five main ceramic categories defined for Hammam are present in the Chuera assemblage. In particular, *Kleiner Antentempel* levels 4-2 and *Steinbau I* levels 7-6 provide many exact parallels. These include the plain rim, round-based goblets (Kühne 1976, Abb. 133-134, 137, 141-144, 155), large open bowls (Kühne 1976, Abb. 171, 172), vertical-walled pots (Kühne 1976, Abb. 167, 170, Tf.12:1), and the vertical-faced rim pots with a more globular body (Kühne 1976, Tf. 17:3, 6). Furthermore, S-shaped jars (Kühne 1976, Abb. 202, 206, 208) and large jars with funnel-shaped necks (Kühne 1976, Tf. 17:2, Abb. 223, 225, 224,228, 220) were excavated at Chuera. In Chuera, one finds a preference for the more specialized wares, as at Hammam, although stone ware occurs in a much greater variety[10]. The corrugated goblets, small bowls and the miniature in the Tell Hammam et-Turkman assemblage have exact parallels in Chuera (Kühne 1976, Abb. 2, 10, 12, 13, 24, 29, 43, 46, 47, 56). Chaff-faced storage vessels are also present at Chuera (Kühne 1976, Tf. 30:1-2, Abb. 349), although a chaff temper for these pots is not mentioned by Kühne. Cooking pots with triangular lugs, eggshell ware with spiral burnish (Kühne 1976, Tf. 2:3,6), and red-banded jars with spiral burnish (Kühne 1976, Abb.88) complete the Chuera assemblage. On the basis of these morphological correlations, a similar category repertoire, and a comparable preference for the smaller ware groups, I propose to include Tell Chuera in the ceramic region established for the Balikh valley. Several plain simple ware vessels from Chuera, furthermore, show the same surface treatment, a careful scraping of the vessel walls, as similarly shaped vessels at Tell Hammam et-Turkman (vaguely visible on Kühne 1976, Tf. 12:1-2; 17:2,6). It may be that this surface treatment - not mentioned by Kühne in his technical description of the 'Einfache Ware' (Kühne1976:73) - was more common in the Chuera assemblage than these few examples indicate. Obviously, the Chuera potters shared the same approach towards certain details of pottery production as their colleagues at Tell Hammam et-Turkman. On the basis of the noted correlations, I propose to equate Hammam EB III (city wall and first burnt level) with Chuera *Kleiner Antentempel* 4-2, *Steinbau I* 7-6 and possibly the deeper levels in the *Häuser* area (i.e. below -3.80 m, Kühne 1976:28-29, Liste 5). The excavated material of Tell Chuera is therefore not earlier than EB III (contra: Kühne 1976:31, 114). A similar late date for Chuera has been proposed by Weiss (1983:47) and Schwartz (1985:57-58).

Further to the east, the Habur sites appear to belong to another ceramic culture, although they have several contacts with the Balikh-Chuera pottery province.

The recent excavations at Tell Brak have established a reliable sequence for the EB III - EB IV periods. Three levels (8-6) seem to belong to the Early Bronze III period (termed 'Late ED III' by D. and J. Oates). Levels 7 and 6 have been heavily destroyed, deeds imputed to kings Lugalzaggesi and Sargon respectively (Oates 1985:143). These two levels are separated by a transitional level 5 from level 4, associated with the Naramsin palace. The composition of the Brak level 6 ceramic assemblage is considerably different from the assemblages of the Balikh valley and Chuera[11]. At Brak, the ceramic assemblage of level 6 is characterized by flat-based, sloping sided plain simple ware bowls (Oates 1982, fig. 2:20-23, 26-30), which are completely absent at Hammam et-Turkman and Chuera. The large open bowls and bag-shaped pots do not occur in the assemblage of Tell Brak. Connecting elements between Hammam/Chuera and Brak are perhaps the S-shaped jars (Oates 1982, fig. 4:59, 62; 5:74, 75) and jars with funnel-shaped necks (Oates 1982, fig. 4:64, 65). An important difference is provided by the small, flat-based stone ware bowls with plain rim and sloping sides (Oates 1982, fig. 1:5-10), which do not occur either on Hammam or Chuera. The larger stone ware bowls with rounded bases seem to be characteristic for the Brak assemblage (Oates 1982, fig. 1:14-15), although a similar bowl was found at Hammam (no. 10) on a floor in the first burnt level. The corrugated goblets, common at Chuera and Hammam (no. 4) do not occur at Brak, whereas the small, globular stone ware jars (Oates 1982, fig. 1:2-4) at Brak correspond to similar jars at Chuera (Kühne 1976, Abb. 30-32, 45-49), although they are not present at Hammam et-Turkman. Further connective elements between Brak and the west are the cooking pots with triangular lugs (Oates 1982:205) and chaff-faced storage vessels (Oates 1982, fig. 5:77). Similar vessels occurred in 'phase 2' at Brak (Fielden 1977, Pl. 13:15-16, note that no. 16 has scrape marks on the exterior wall), thus confirming the continued occurrence of these vessels from EB III into the EB IV period, as noted in the Hammam et-Turkman sequence. Finally, the storage vessel from Brak (Oates 1982, fig. 3:48) supplies another possible parallel with Tell Hammam et-Turkman (no. 20).

One can conclude that the EB III assemblage of Tell Brak belongs to a pottery region outside the Balikh-Chuera province, despite the occasional parallels. There seem to be more connective elements with Chuera than with the Balikh valley, such as jars with long necks (Oates 1982, fig.4:61, 63; Kühne 1976, Abb. 265, 266) and small, spouted pots (Oates 1982, fig. 4:70-73; Kühne 1976, Abb. 256-261). The location of Tell Chuera between the Balikh and the Habur allows for such parallels. However, the parallels between Chuera and the Balikh are much stronger than those between Chuera and Brak.

The ceramics of another EB III site in the Habur triangle, Tell Leilan, very much resemble the Brak level 6 assemblage (Weiss 1983:47). The published pottery shows no affinities with Hammam or Chuera. There is one exception, a straw-tempered jar (Weiss 1983, fig. 10:19) which corresponds with the Hammam, Chuera and Brak chaff-faced storage vessels.

Downstream, in the Middle Habur salvage area south of Hasseke, Tell Melebiya yielded remains dated from ED I to Akkad period. The plain rim, round-based goblets, characteristic of the EB III period in the Balikh-Chuera region, are already present in the ED I contexts and continue into the ED II period, but they are gradually replaced by goblets/bowls with plain rims, sloping sides and flat bases in ED III, which I propose to be contemporary with the EB III period (Lebeau et al. 1987, Pl. XI:1, 2, 5; XII:1; XIII:1-3). At the end of the ED III period at Melebiya these goblets/bowls have become the standard type (Lebeau 1987, Pl. XV:1-3). The ED III pottery at Melebiya has many affinities with the level 6 pottery at Tell Brak, while there are few relations with the Balikh valley and Tell Chuera. In spite of the many parallels Lebeau draws with Chuera (Lebeau 1987:11-12), the characteristic Hammam/Chuera EB III types, plain-rim, round-based goblets, large open bowls, vertical walled pots, and S-shaped jars, are absent at Melebiya. There are, however, several correspondences in the specialized wares: cooking pots with triangular lugs (Lebeau et al. 1987, Pl. XII:21), a chaff and mineral tempered pot with folded-over lip (Lebeau et al. 1987, Pl. XII:23), and a stone ware miniature (Lebeau et al. 1987, Pl. XII:5). Clearly, Melebiya belongs to a separate pottery region related to Tell Brak and Tell Leilan and centered in the Habur Triangle.

The EB III assemblage of Mari is completely different from the Balikh-Chuera region. Although several red-banded, spirally burnished jars occur in the Ishtar temple assemblage (Parrot 1956, fig. 107:1548-49), they may be interpreted as imports. The deep sounding at Mari provides a better picture of the Mari pottery (Lebeau 1985; 1987). Levels 6-3 are dated to the ED III period (LeBeau 1985:126). The essential difference with the Balikh-Chuera pottery province is again illustrated by the Mari ED III goblets/bowls. They usually have a flat base, plain rim and sloping sides and are generally larger in diameter (larger than 12 cm) than the goblets from the western region. The ceramic evidence from Mari compares better to the Habur EB III assemblages than to the contemporary Balikh-Chuera assemblages. The main ceramic categories from the Balikh-Chuera region are absent in Mari.

In order to examine the western extent of the Balikh-Chuera EB III pottery region, we will now turn to the Middle Euphrates. Mazzoni already noted the difference in the ceramics of the right-bank and left-bank sites. The only unequivocal evidence for the EB III period is provided by a few graves from Halawa and Tawi. While Sweyhat and Munbaqa are later, Tell el-'Abd does have contemporary levels, characterized by plain-rim, round-based goblets and small jars with a carination halfway up the body (Bounni 1979:53). Without more details on this assemblage, however, its relations with the east cannot readily be assessed.

At Halawa, settlement remains of the EB III period are not yet excavated, but the EB III graves H-70 and H-123 suggest the existence of a contemporary settlement on tell A (Orthmann 1981:57). The presence of a double tubular lug pot and an alabastron-shaped bottle with recessed neck suggests a slightly later date for grave H-64 (Orthmann 1981, Tf.

58:10; 59:28). The contents of grave H-70 include plain-rim, round-based goblets, large round-based open bowls, and S-shaped jars. The open bowls are not exactly parallel to the examples in the Hammam et-Turkman assemblage, but they show affinities in proportion as well as in lip treatment (Orthmann 1981, Tf. 60:2-4, 13; cf. also lip treatment of Tf. 60:8, 9, though to be completed with a pedestal base). Notable also is a specific surface treatment for these bowls: '*... die Schalenwandungen [wurden] nach dem Drehprozess häufig abgezogen oder beschnitten.*' (Orthmann 1981:54-5). This surface treatment, though not precisely paralleled at Hammam et-Turkman, certainly indicates a similar laborious approach towards finishing off these large bowls. The round-based S-shaped jars from Halawa (Orthmann 1981, Tf. 61) also have parallels at Hammam et-Turkman, but the characteristic vertical-walled pots do not occur at Halawa. A further connection with the Balikh-Chuera region is a red-banded spiral burnished jar (Orthmann 1981, Tf. 62:33). The second EB III grave H-123 contained the same vessel types as H-70, but in smaller quantities. The plain-rim, round-based goblet, S-shaped jar and red-banded jar have parallels at Hammam et-Turkman (Orthmann 1981, Tf. 71: 1, 3, 6).

At Tawi, two graves yielded a ceramic assemblage similar to the EB III Halawa graves: Tawi T-70 and T-71 (Kampschulte and Orthmann 1984, Tf. 33, 34).

One must conclude that the evidence at present is too scarce (burial contexts only) to situate Halawa and Tawi in the Balikh-Chuera pottery region. There seems to be, however, a similar preference for several vessel categories, with a conspicuous absence of vertical-walled pots. There is a shared group of specialized ware (the red-banded jars), and there are certain similarities in surface treatment. For the time being, it seems safer to consider the Halawa/Tawi EB III material as part of a pottery region which is a variant of the Balikh/Chuera region, with its own characteristics and solutions for technical problems. It certainly has no connections with the west-bank Euphrates sites, apart from the red-banded burnished jars. These EB III west-bank sites, Hadidi and Selenkahiye, show more parallels with the west than with the east, as Mazzoni has already pointed out (Mazzoni 1985a:568). An important connection with the Balikh valley and Chuera is provided by the red-banded, burnished jars at Hadidi, which occur in large quantities in several tomb inventories(*1972 tomb*: Dornemann 1979, fig. 13:1-5, 11-15, 17, 20-22, 26; *tomb LI*: Dornemann 1984, fig. 13:19, 20; *tomb EI*: Dornemann 1984, fig. 20:36). These jars seem to occur at Hadidi in greater quantity than elsewhere (Dornemann 1984:9). In addition to the well-known jar-shape with ring base, several of these jars occur with a pedestal base; a bowl on a pedestal base also has the red bands and burnish (Dornemann 1984, fig. 13:16-18). At Hadidi, change in the shape of these jars through time has been observed (Dornemann 1979:132, and fig. 16:16). Mazzoni's suggestion that these red-banded jars probably belong to a local, northern tradition (Mazzoni 1985a:568) can perhaps now be emended with the suggestion that Hadidi was the main production centre for these vessels. The tombs LI and EI seem to have been used over several centuries of the second half of the third millennium B.C. The earliest vessels, datable to the EB III period, clearly fit into a western oriented pottery culture. Corrugated goblets, thickened rim bowls, spouted jars and a corrugated jar (Dornemann 1984, fig. 13:14, 15, 2, 6; fig. 14:7, 8; fig. 17:12 resp.) have exact parallels in the Mardikh IIB1 assemblage. Only a few vessels relate to the east, such as plain rim, round-based goblets and a combed-wash jar with a funnel-shaped neck and round base (Dornemann 1984, fig. 15:24-25; fig. 19:5; fig. 20:17 resp.). The Hadidi EB III assemblage clearly belongs to the Ebla-Hama pottery region (Mazzoni 1985a, 1985b), and the red-banded jars represent the only link to the Balikh-Chuera region. These may be considered as imports from Hadidi into the Balikh-Chuera region.

The Early Selenkahiye assemblage is contemporary with Hama J8-6 and Mardikh IIB1 (Van Loon, n.d.). Selenkahiye is included in the Hama-Ebla pottery region. Connections with the east include several red-banded, spirally burnished jars (Schwartz in press, Pl. 26a-c), a combed-wash jar (Schwartz in press, Pl. 26b), and cooking pots with triangular lugs (Schwartz in press, Pl. 28a-b).

Finally I would like to suggest a possible parallel between Tell Hammam et-Turkman and Mardikh IIB1. An almost complete storage vessel retrieved from a room on top of the city wall at Hammam et-Turkman had a plant-and-sand temper, causing a porous effect (no. 20). The shape and the technological aspects of this vessel seem comparable to the large pots with flat bases and grooved rims interpreted as water jars at Mardikh (Mazzoni 1982:173 Classe C, Tipo 1; cf. fig. XXV:1, 2). These water jars represent the only parallel between the contemporary assemblages of Ebla IIB1 and Tell Hammam et-Turkman city wall (EB III). A similar vessel from Brak level 6 is the only parallel between Brak and Ebla (Oates 1985:143).

CONCLUSIONS

I have tried to demonstrate the remarkable homogeneity of the ceramic assemblage of the Early Bronze Age III period of Tell Hammam, Harran and Tell Chuera. These sites have the same standard repertoire in plain simple ware, and they share a preference for more specialized wares such as stone ware and calcite-tempered cooking ware.

The northern border of the region must be situated somewhere between Harran and the Turkish Euphrates: the Karababa sites have a different, though related assemblage.

The different goblet/bowl type with flat base illustrates the essential difference between the Habur Triangle and the Balikh-Chuera region. In the Habur, the evolution from older, round-based goblets to later, flat-based ones developed at a quicker pace than in the Balikh-Chuera region. In the EB III period, the flat-based goblet had superseded the round-based goblet in the Habur, whereas the latter goblet still occured in the EB III assemblages of the Balikh-Chuera region. The stone ware in the Habur Triangle also had its own peculiar characteristics: flat-based bowls with sloping sides. This shape is not attested in the Balikh-Chuera region, except as a unique import at Tell Hammam.

Certain shapes and ware groups, however, link both areas, such as S-shaped jars in plain simple ware, calcite-tempered cooking pots with triangular lugs, and chaff-faced storage vessels. The high quality and extreme functionally of the cooking pots may have caused their wide geographical distribution. For the chaff-faced storage vessels such qualities cannot immediately be discerned. It is unknown, at present, whether their occurrence in both areas is to be attributed to trade or to independent development.

The southern border of the Balikh-Chuera pottery region should be situated somewhere between the Balikh site BS 84 and Tell Bi'a.

To the west, the left-bank Euphrates sites seem to belong to a strongly related pottery region with characteristics of its own. Completely different are the assemblages of the right-bank settlements and of the Ebla-Hama region. Here, the EB III ceramics display advanced technological methods of manufacture. The emphasis on fast-wheel, mass production (flat bases) sharply contrasts with the production methods of the Balikh-Chuera region (round bases, separate manufacture of rim and body). The absence of mass production should not be interpreted as the absence of full-time potting specialists, however, since the homogeneity visible in the plain simple ware assemblage of Tell Hammam et-Turkman indicates a well-trained and qualified potter's class. From the ceramic point of view, the Balikh-Chuera region seems to lag behind contemporary developments west of the Euphrates and in the Habur Triangle. But this technological backwardness need not, in my opinion, be taken as a cultural backwardness. It could, instead, be explained by a more traditional approach towards pottery-making. The mercantile contacts of Harran with Mardikh IIB1, documented by the Ebla texts, clearly indicate that the Balikh valley was not an isolated region (Archi 1980:3-4). The presence of imports at Hammam et-Turkman also attest to inter-regional contact: red-banded jars came from the west, stone ware from the east. The Balikh-Chuera region had its own interests and was a part of the patchwork ensemble of EB III kingdoms in North Syria.

NOTES

1. *Early Bronze III* is regarded here as being roughly equivalent to the third quarter of the third millennium B.C. For the beginning of EB III no clear criteria exist yet. Good, continuous ceramic sequences for the third millennium in northern Syria are very scarce. For the sake of convenience a date of ca. 2500 B.C. is proposed here. The end of EB III is put around 2300 B.C. and coincides with the end of the Early Dynastic III period. At this time (the second half of Sargon's reign, 2334-2279 B.C.) the mature Akkad period seems to start, with its floruit during the reign of Naramsin (2254-2218 B.C.). For this mature Akkad period and later I propose to reserve the Early Bronze IV designation (cf. Lebeau 1986:3, 12ff.). This nomenclature deviates from the one used by Matthiae and Mazzoni, among others, who regard pre-Akkad Mardikh IIB1 as part of EB IV.
   It is important to stress that the Tell Hammam et-Turkman EB III assemblage discussed in this article does not cover the entire EB III time-span and that earlier EB III levels are still to be excavated. At the moment Tell Hammam et-Turkman has yielded a continuous ceramic sequence which starts somewhere in EB III. The transition from EB III into EB IV is situated in the second burnt level, i.e. Stratum VI West:1 (cf. note 3).
2. In fact, several of her ideas were already put forward in Mazzoni 1982.
3. For a report on the architecture and pottery of the second burnt levels and subsequent strata see Meijer *apud* Van Loon (Ed.), in press, Chapter 4 and Curvers *apud* Van Loon (Ed.), in press, Chapter 9. The second burnt level was termed Stratum VI west:1a-b.
4. The same evolution from older, round bases to later, flat bases was already observed at Tell Chuera (Kühne 1976:23, 27, 28, 74).
5. Others prefer to call it smeared-wash ware (J. Oates, Algaze), *Wellenware* or *mit Streifen bemalte Ware* (Kühne).
6. City-wall architecture:
   (GrN-14649) 4125 ± 35 BP (charred wood).
   First burnt level:
   (GrN-14650) 3860 ± 30 BP (grain);
   (GrN-14644) 3930 ± 40 BP (charcoal);
   (GrN-14646) 3975 ± 30 BP (charcoal);
   (GrN-14645) 4050 ± 30 BP (charcoal).
   Second burnt level:
   (GrN-11150) 4170 ± 60 BP (charred beam);
   (GrN-11154) 3905 ± 35 BP (charred wood).
7. The proposed date is based on the following considerations:

- the green stone ware bowls from the first burnt level at Tell Hammam et-Turkman (fig. 2:10) and Brak level 6 indicate a possible synchronism;
- Joan Oates has imputed the destruction of Brak level 6 to Sargon (1985:143). This suggestion perhaps allows us to ascribe the destruction of the first burnt level to the same king (and the destruction of the second burnt level (Hammam VI West:1) to Naramsin);
- my proposal that the first and second burnt levels represent the transition from EB III to EB IV seems to be corroborated by the fact that the combed-wash ware appears on a very small scale in the first destruction level. Subsequently this ware occurs in slightly larger quantities in the secondary debris between the first and second burnt levels. The second burnt level yielded a complete piece. This agrees with the evidence from other sites, where combed-wash ware is mainly of Akkadian date. The ware, however, starts to appear late in the EB III period: Mardikh IIB1 and 2 (Mazzoni 1982:172), Harran III, IV (Prag 1970:72, 85), Chuera upper levels (Kühne 1976:25, 27, 30), but on a small scale in *Kleiner Antentempel* 3/2 (Kühne 1976:111, note 938), Tell Brak levels 5 and 4 (Oates 1985:143).

8. Final analysis must await the results of the 1988 excavations of the deeper Early Bronze levels at Tell Hammam et-Turkman.
9. Plain rim, round-based goblets clearly have their limitations when used as chronological indicators, and they may similarly be of limited use for the recognnition of the diagnostic characteristics of a ceramic region. At Tell Hammam et-Turkman, these goblets have a long tradition and are attested as early as the beginning of the third millennium B.C. (Curvers, in press, Pl. 111:3, 4; 112:14-16). At Hadidi, plain rim, round-based goblets are said to continue from the EB I to the EB III/IV period (Dornemann 1984:8). In the Habur Triangle and in Mari they occur frequently in the earlier periods (ED I) and are replaced by flat-based goblets in the ED III period.
10. It is important to note, however, that Kühne's publication of the Chuera pottery gives an extremely biased picture of the composition of the assemblage. Plain simple ware is the largest ware group, whereas the stone ware at Chuera occurs in limited quantities (cf. Orthmann 1986:68-69, figs. 5-7).
11. Thus far only the pottery of level 6 has been published.

CATALOGUE

1. HMM 86.G5. Floor in room associated with city wall. wheelmade plain rim, round-based goblet. Medium sand temper with lime inclusions. Wet-smoothed, green surface. H. 6,5 cm, D. 7.5 cm.
2. HMM 86.B22. Floor in room associated with city wall. Wheelmade plain rim, round-based goblet. Wheelmade, medium sand temper. Wet-smoothed greyish cream surface, scraped around base. H. 5.6 cm, D. 10.0 cm.
3. HMM 86. G 23 lot 54:12. City-wall debris. Wheelmade goblet with bevelled inward rim. Fine sand temper. Wet-smoothed cream surface, interior buff, and orange paste. D. 10.0 cm.
4. HMM 86. G 23 lot 56:13. City-wall debris. Wheelmade corrugated stone ware goblet. No visible temper. Wet-smoothed surface, sharp striations near base. Rim orange, body grey. D. 10.0 cm.
5. HMM 86.B27. City-wall debris. Stone ware, plain rim bowl with string-cut base, wheelmade. Upper part corrugated. No visible temper. Slightly burnished surface except for corrugated, sharp parallel striations. Brown-greenish surface, corrugated part alternating purple grey and orange. H. 6.5 cm, D. rim 13.0 cm, D. base 4.0 cm.
6. HMM 86. G 23 lot 46:13. City-wall debris. Wheelmade stone ware bowl with folded-over lip. No visible temper. Wet-smoothed orange surface, rim (purple-black) slightly burnished. Orange paste. D. 14.0 cm.
7. HMM 86. G 23 lot 47:1. City-wall debris. Wheelmade plain rim bowl. Sand temper. Wet-smoothed, light green surface. Some faint scrape marks on exterior below carination. D. 13.0 cm.
8. HMM 86. I 23 lot 1:2. Debris just below second burnt level (=Stratum VI:1). Wheelmade plain rim bowl. Fine sand temper. Wet-smoothed, greenish cream surface. D. 15.0 cm.
9. HMM 86. G 23 lot 45:8. City-wall debris. Small wheelmade bowl with everted lip. Fine sand temper. Slightly wet-smoothed, cream surface. Orange paste. D. 13.0 cm.

10. HMM 86.B5. Floor of first burnt level. Wheelmade stone ware, plain rim bowl. Fine sand temper. Carefully wet- smoothed green interior and exterior. Base pared down. H. 10.3 cm, D. 21.0 cm.
11. HMM 86.B6. Floor in room associated with city wall. Wheelmade round-based bowl with thickened rim. Sand temper with very fine lime inclusions. Upper part of cream exterior and interior scraped, lower parts slightly wet-smoothed. H. 13.0 cm, D. 24.0 cm.
12. HMM 86. G 23 lot 39:8. From city-wall debris. Large wheelmáde bowl with everted rim. Fine sand temper with mica inclussions. Wet-smoothed pale brown surface, heavily scraped below bulge. D. 22.0 cm.
13. HMM 86. G 23 lot 50:8. City-wall debris. Large wheelmade bowl with rim bevelled inward. Fine sand temper with mica inclusions. Wet-smoothed cream surface, interior lightly scraped in oblique directions. D. 23.0 cm.
14. HMM 86. G 23 lot 38:1. City-wall debris. Large wheelmade bowl with thickened rim. Sand temper with mica and very fine lime inclusions. Wet-smoothed, cream surface, ochre interior and orange paste. D. 25cm.
15. HMM 86. G 23 lot 38:7. City-wall debris. Vertical-walled pot. Wheelmade rim (faint horizontal striations), body separately made and joined: scrape marks near join on shoulder.Sand temper with mica and very fine lime inclusions. Wet-smoothed, cream surface, creamy orange interior and orange paste. D. 19.0 cm.
16. HMM 86. G 23 lot 39:10. City-wall debris. Vertical-walled pot. Wheelmade rim, body made separately and joined (Cf. no. 15). Heavy scraping on exterior below rim. Sand temper with mica inclusions. Wet-smoothed, orange-buff surface, orange paste. D. 22.0 cm.
17. HMM 86. G 23 lot 52:25. City-wall debris. Wheelmade vertical-walled pot. Fine sand temper with mica inclusions. Slightly wet-smoothed and scraped cream surface and orange paste. D. 25.0 cm.
18. HMM 86. G 23 lot 44:16. City-wall debris. Wheelmade pot with globular body. Rim like no. 15. Body made separately. Sand and mica temper. Wet-smoothed, cream surface, buff interior and orange paste. Oblique scrape marks on exterior. D. 21.0 cm.
19. HMM 86. G 23 lot 39:21. City-wall debris. Handmade cooking pot with triangular lugs. Calcite temper. Orange brown surface. Exterior and interior rim pebble-burnished. D. 20.0 cm.
20. HMM 86. H 23 lot 90. Room on top of city wall. Storage vessel with porous paste. Rim wheelmade, body handmade. Fine chaff and sand temper. Exterior wet-smoothed and very faint traces of a reserved wash (in horizontal, parallel strokes). Buff colour, light-cream wash. D. 32.0 cm.
21. HMM 86. I 23 lot 21:1. Floor of first burnt level. Handmade (coiled) storage vessel. Coarse chaff temper, with some very large pebble inclusions. Wet- smoothed, buff surface, grey interior grey, and black core. D. 35.0 cm.
22. HMM 86. H 23 lot 90. Room on top of city wall. Handmade (coiled) storage vessel. Coarse chaff and sand temper. Wet-smoothed, brown surface, brownish grey inmterior, black core. D. 27.0 cm.
23. HMM 86. G 23 lot 45:23. City-wall debris. Wheelmade S-shaped plain rim jar. Fine sand temper. Wet-smoothed, cream interior and exterior. D. 9.0 cm.
24. HMM 86. G 23 lot 45:19. City-wall debris. Wheelmade S-shaped jar with slightly bevelled outward lip. Fine sand temper. Wet-smoothed cream surface, oblique scrape marks on lower part. D. 11.0 cm.
25. HMM 86. G 23 lot 39:12. City-wall debris. Large wheelmade jar with sharply folded-over lip. Fine sand temper. Wet-smoothed, cream surface. D. 25.0 cm.
26. HMM 86. G 23 lot 39:14. City-wall debris. Small wheelmade jar with everted lip. Fine sand temper with mica inclusions. Cream surface, orange paste. D. 12.0 cm.
27. HMM 86. G 23 lot 38:8. City-wall debris. Large S-shaped jar with rolled-out lip. Rim wheelmade, body attached separately (at interior joint scrape marks and applied clay pellets, below exterior joint oblique scrape marks. Sand temper. Wet-smoothed, greenish surface (slightly vitrified). D. 22.0 cm.
28. HMM 86. G 23 lot 42:12. City-wall debris. Large wheelmade jar with everted rim. Fine sand temper. Slightly wet-smoothed, buff surface. D. 24.0 cm.
29. HMM 86. H 23 lot 90. Room on top of city wall. Handmade (coiled) potstand. Irregular surface, incidently scraped. Medium sand temper. Slightly wet-smoothed, greenish cream surface. D. top 26.0 cm, D. base 28.0 cm.
30. HMM 86. G 23 lot 46:28. City-wall debris. Fragment of a wheelmade stone ware miniature jar. No visible temper. Wet-smoothed, grey surface. D. neck 6.0 cm.

31. HMM 86. G 23 lot 49:25. City-wall debris. Fragment a red-banded jar. Eggshell ware. Wheelmade, fine sand temper with medium lime inclusions. Exterior spirally burnished, thin red painted band on shoulder. Neck corrugated. Ochre surface colour. D. neck 9.0 cm.

REFERENCES

AKKERMANS P.M.M.G.

1984 'Archäologische Geländebegehung im Balih-Tal', *Archiv für Orientforschung* 31: 188-190.

ARCHI A.

1980 'Notes on Eblaite Geography', *Studi Eblaiti* II/1: 1-16.

ARNE R. ET AL.

1984 'Ausgrabungen in Tall Bi'a 1982 und 1983', *Mitteilungen der Deutschen Orientgesellschaft* 116: 15-63.

ALGAZE G.

1986 *Mesopotamian Expansion and its Consequences: Informal Empire in the Late Fourth Millennium B.C.* (Ph. D. dissertation), Chicago.

BOUNNI A.

1979 'Preliminary Report on the Archeological Excavations at Tell al-'Abd and 'Anab al-Safinah (Euphrates) 1971-72', *Annual of the American Schools of Oriental Research* 44: 49-62.

CURVERS H.H.

in press 'The Period VI Pottery', in: M.N. Van Loon (Ed.), *Hammam et-Turkman I, Report on the University of Amsterdam's 1981-84 Excavations in Syria.* Istanbul.

DORNEMANN R.H.

1979 'Tell Hadidi: A Millennium of Bronze Age Occupation', *Annual of the American Schools of Oriental Research* 44: 113-151.

1984 'Tell Hadidi: One Bronze Age Site Among Many in the Tabqa Dam Salvage Area'. Unpublished paper.

FIELDEN K.

1977 'Tell Brak 1976: the Pottery', *Iraq* 39: 245-255.

KAMPSCHULTE I. AND ORTHMANN W.

1984 *Gräber des 3. Jahrtausends v. Chr. im Syrischen Euphrattal. 1. Ausgrabungen bei Tawi 1975 und 1978.* Bonn.

KÜHNE H.

1976 *Die Keramik vom Tell Chuera.* Berlin.

LEBEAU M.

1985 'Rapport préliminaire sur la sequence céramique du Chantier B de Mari (IIIe mill.)', *MARI* 4: 93-126.

1987 'Rapport préliminaire sur la céramique des premiers niveaux de Mari (Chantier B-1984)', *MARI* 5: 415-442.

LEBEAU M. ET AL.

1986 'Rapport préliminaire sur la deuxième campagne de fouilles à Tell Melebiya (Moyen Khabour - automne 1986)', *Akkadica* 46: 1-50.

1987 'Rapport préliminaire sur la troisième campagne de fouilles à Tell Melebiya (Moyen Khabour - automne 1986)', *Akkadica* 51: 1-74.

MALLOWAN M.E.L.

1946 'Excavations in the Balih Valley, 1938', *Iraq* 18: 111-162.

MAZZONI S.

1982 'La produzione ceramica del palazzo G di Ebla e la sua posizione Storica nell'orizzonte Siro-Mesopotamico del III millennio A.C.', *Studi Eblaiti* V: 145-199.

1985a 'Frontières céramiques et le Haut Euphrate au Bronze Ancien IV', *MARI* 4: 561-577.

1985b 'Elements of the Ceramic Culture of Early Syrian Ebla in Comparison with Syro-Palestinian EB IV', *Bulletin of the American Schools of Oriental Research* 257: 1-18.

MEIJER D.J.W.

in press 'Tell Hammam: Architecture and Stratigraphy', in: M.N. Van Loon (Ed.), *Hammam et-Turkman I, Report on the University of Amsterdam's 1981-84 Excavations in Syria*. Istanbul.

MOORTGAT A. AND U. MOORTGAT-CORRENS

1978 *Tell Chuera in Nordost-Syrien*. Berlin.

NISSEN H.J.

1983 *Grundzüge einer Geschichte der Frühzeit des Vorderen Orients*, Darmstadt.

OATES J.

1982 'Some Late Early Dynastic III Pottery from Tell Brak', *Iraq* 44: 205-219.

1985 'Tell Brak and Chronology: The Third Millennium', *MARI* 4: 137-144.

ORTHMANN W.

1981 *Halawa 1977 bis 1979*. Bonn.

1986 'The Origin of Tell Chuera', in: H. Weiss (Ed.), *The Origins of Cities in Dry-Farming Syria and Mesopotamia in the Third Millennium B.C.*. Guilford. Pp. 61-70.

PARROT A.

1956 *Mission archéologique de Mari 1. Temple d'Ishtar*. Paris.

PRAG K.

1970 'The 1959 Deep Sounding at Harran in Turkey', *Levant* 2: 63-94.

SCHWARTZ G.M.

1985 'The Ninevite V Period and Current Research', *Paléorient* 11/1: 53-70.

in press 'The Pottery from Selenkahiye: 1972, 1974, 1975', in: M.N. Van Loon (Ed.), *The Excavations at Tell Selenkahiye*. Leiden.

STROMMENGER E. ET AL.

1987 'Ausgrabungen in Tall Bi'a 1985', *Mitteilungen der Deutschen Orientgesellschaft* 119: 7-49.

VAN LOON, M.N.

in press 'The Final Stage of Early Syrian Culture and the Beginning of the Amorite Period', in: P. Matthiae (Ed.), *L'Archéologie et l'Histoire de la Syrie, Tome I: De la préhistoire aux royaumes araméens*.

VAN LOON M.N. (ED.)

in press *Hammam et-Turkman I, Report on the University of Amsterdam 1981-84 Excavations in Syria*. Istanbul.

VAN LOON M.N. AND MEIJER D.J.W.

1987 'Hammam et-Turkman on the Balikh: First Results of the University of Amsterdam's 1986 Excavation', *Akkadica* 52: 1-9.

WEISS H.

1983 'Excavations at Tell Leilan and the Origins of North Mesopotamian cities in the third millennium B.C.', *Paléorient* 9/2: 39-52.

Metal animal headed cups at Mari

SALLY DUNHAM

As has long been noted, among the numerous gold and silver vessels mentioned in the texts from Old Babylonian Mari are a group that were probably animal headed cups and perhaps represent distant ancestors of later Persian and Greek rhytons. These are the vessels described as GAL.SAG +(animal's name) = cup[1], head of (animal). They have been noted and/or discussed by several scholars and their probable archaeological analogs in the terracotta animal headed cups from Kültepe and Alishar have been pointed out[2]. With the recent publication of many more of the Mari texts[3] the repertoire of references concerning these vessels has increased many fold. This brief essay is intended to bring these references to the attention of other scholars and to point out some interesting details and the possible implication for the archaeological record.

All the references found so far are listed in Table I, where they are grouped according to the animal represented. In the first column are the ARM/ARMT volume numbers, text numbers, and line numbers. The next three columns give the number of vessels listed for the entry, the material they are made of, and the total weight of the vessels mentioned (if two vessels are mentioned in the entry, the weight is of both vessels together). Table 1 shows six different kinds of cups: bull's head cup, calf's head cup, gazelle's head cup, deer's head cup, ibex's head cup, and lion's head cup [4].

They were made of either silver or gold and weighed from about 1/2 to 1 mina (= ca. 250-500 grams; Neugebauer and Sachs 1945:4). This weight range suggests comparison to the well known silver 'Stag Rhyton' of Hittite date in the Schimmel collection (Muscarella 1974, no. 123). However, the weight given for this vessel is said to be 'compromised by wax used in restoration.' This later vessel is probably more elaborate, since the stag's whole front end - head, neck, shoulders, front legs - are represented. At Mari the descriptions refer only to the head of the animal, so this was probably all that was depicted, as is the case with terracotta bull's head cups from Kültepe.

The Mari texts reveal that these cups were made at Mari and that sometimes they were melted down to make other objects, including other animal headed cups. ARMT XXV:217 records the receipt from Mukannišum of a deer's head cup to be melted down to make a block of silver. ARMT XXII:251 states that one silver lion's head cup and one silver deer's head cup, which together weighed 5/6 minas 4 1/3 shekels and which Yar'ip-Abba had made, are given to Yar'ip-Abba for the making of two gazelle's head cups[5]. ARMT XXI:227 is dated to the same day as ARMT XXII:251 (Month 12 (Eburim), day 16, Zimri-Lim year 1') and records a small quantity of silver given to Yar'ip-Abba for the ears and horns of two gazelle's head cups, presumably the same ones. ARMT XXV:730 is a record of eleven shekels of copper given to make certain small objects (ŠU-ZI-GA TUR) needed for the making of 2 gold bull's head cups by Rimši-ili. Unclear is whether the ŠU-ZI-GA are to be part of the vessels or are some kind of tool used in making the vessels. In a letter to Zimri-Lim Mukannišum says he is sending two silver bull's head cups, eight silver calf's head cups, one silver deer's head cup, one silver gazelle's head cup and gold *ṭapīḫu*-vessel which Yašūb-Ašar made[6]. Yašūb-Ašar is well known as a goldsmith at Mari from many other texts.

Four texts list animal headed cups as diplomatic gifts (*šûbultum*; cf. Durand, ARMT XXI:512-515) sent to other rulers. ARMT XXV:3 says two gold gazelle's head cups were sent to Sheplarpak, king of Anshan in Iran. ARMT XXII:253 and ARMT XXV:13 record that a silver calf's head cup and a silver bull's head cup were sent to Sharriya, king of Razama, a place in northeastern Syria, perhaps in the Sinjar plain (Dalley 1984:xix, 41, note 21; Groneberg 1980:250). These two texts are dated only two days apart. ARMT XXV:18 lists one silver gazelle's head vessel and one silver bull's head vessel as a diplomatic gift sent to Ibal-pī-El, king of Eshnunna.

These last three texts (ARMT XXII:253; XXV:13 and 18) have the phrase 'ša ÁŠ.KÀR PN' after the description of the vessel. The three people named, Ana'iš (ARMT XXII:253; XXV:13), Yašūb-Ašar (XXV:18, for gazelle's head cup) and Idin-^{d}IGI.KUR (XXV:18, for bull's head cup) are all well known as goldsmiths at Mari from many other texts in which they receive supplies, make objects, and deliver various items. The word ÁŠ.KÀR (= *iškaru*) is interpreted by Limet to refer to materials which a craftsman has at his disposal or to finished objects he keeps in his workshop (ARMT XXV:271, l. 13; Charpin and Durand 1983:96, l. 254). Since the item referred to is a finished product that a goldsmith would make, one could perhaps ask if the phrase here implies that mentioned animal headed cup is the product of the goldsmith named (or of his workshop?)[7].If so, then the Mari texts supply the names of five different goldsmiths who made animal headed cups at Mari: Yar'ip-Abba, Rimši-ili, Yašūb-Ašar, Ana'iš and Idin- ^{d}IGI.KUR.

Animal headed cups also came to Mari from other places. ARMT 25:48, ll. 9-12 lists the delivery from Šennam, king of Uršu of a silver ibex head cup whose horns and beard were plated with gold. Uršu is not definitely identified yet, but probably was located in the area of Gaziantepe or Urfa in southern Anatolia (Groneberg 1980:250). A gold bull's head cup is one of the objects sent to Mari by Yarim-Lim, king of Yamhad (ARMT XXV:82, ll. 1-3). Another place associated with the animal headed cups is Tukrish. No cups are actually said to be delivered from there, but 6 entries are described as 'Tukrish-type of bull's head cup' (Table 1:14-17). While this description does not necessarily mean that the cups mentioned were made in Tukrish, since they could have been made at Mari in imitation of a type of cup made in Tukrish, it does imply that Tukrish was a place where a certain type of bull's head cup was made. Unfortunately, the location of Tukrish is still unsure. Shamshi-Adad claimed to have received the tribute of 'the kings of Tukrish and of the king of the Upper Land' (Grayson 1972:20). An inventory of precious objects from Mari dating to the rule of Yasmah-Addu includes *kannu*-stands of ivory that are said to be the 'work of Tukrish'[8]. Other references to Tukrish in historical and literary texts suggest it was northeast of Mesopotamia, but more precision is not possible with the available evidence (Gelb 1944:41,55,57; Stol 1972:41; Komoróczy 1972:113-123; Steinkeller 1982:248, note 37). Bottéro (ARMT VII:312) suggested it might be Luristan, but Moorey suggests a location further north in Kurdistan or Azerbaijan (Moorey 1985a, xviii; Moorey 1985b:34). Since in the Mari texts Tukrish is associated with animal headed cups and with ivory objects, it is perhaps interesting to note that both are found in Azerbaijan in the early first millennium B.C[9].

The Mari texts show that metal animal headed cups were certainly well known and produced in north Syria in the early second millennium B.C. and must have been known in Mesopotamia and Elamite Iran. Moorey tentatively suggested that the references to Tukrish-type bull's head cups in the Mari texts might point to an ultimate Iranian origin for the taste for such metal vessels (Moorey 1985b:34), but the Mari texts indicate that they were produced in the west, in Syria (Mari, Yamhad) and southern Anatolia (Uršu) at the same time, so one cannot definitely claim an Iranian origin. Terracotta animal headed cups are known from level II of the Karum-Kanish at Kültepe, which makes them earlier than any of the Old Babylonian references at Mari. Whether these earlier cups were imitating contemporary metal ones cannot be known, but is certainly very possible (and, indeed, probable). While the Mari texts cannot tell us exactly what these cups looked like, or how the 'Tukrish'-ones differed from the others, they do serve to illustrate how much more widespread an artifact or motif could have been than is known from the archaeological record alone.

NOTES

1. The translation of GAL as 'cup' perhaps needs some explanation. While Dietrich and Loretz (1964:232-239) make a convincing case for GAL = *kāsu* = goblet, cup in the later second millennium, the CAD (s.v. *kāsu*) and Durand (ARMT XXI:350-351) note that for the Old Babylonian period the Akkadian equivalent of GAL is still uncertain. However, the vessels under discussion all seem to be in a weight (and, hence, size) range that is suitable to hold in the hand (see discussion). Sometimes animal headed cups are referred to as 'rhytons'. However, the word 'rhyton' comes from the ancient Greeks who defined it as a drinking vessel in the shape of a curved horn ending in an animal head or protome and having a second opening or spout in the lower part from which the liquid could flow out (Yon 1981:208-209). Since we do not know if the vessels of the Mari texts had such second holes, and indeed, the terracotta animal headed cups from Kültepe and Alishar, which they may have resembled did not, the word 'cup' seems preferable to 'rhyton'.
2. ARMT VII:308-314; Limet 1985:518-519; Deller 1985:337-340; Deller 1987:219-220; Moorey 1985b:34; Dalley 1984:59-62. For the terracotta animal headed cups found at Kültepe and Alishar in levels dating to the early second millennium B.C., see Özguç and Özguç 1953:223f., Abb. 281, 283; Özguç 1959:112f., pls.XLV-XLVII; Özguç 1986:65-67, pls. 115-117; von der Osten 1937:165, 190-191, fig. 208-214.
3. J. M. Durand, *Textes Administratives des Salles 134 et 160 du Palais de Mari* (ARMT XXI, Paris, 1983); J. R. Kupper, *Documents administratifs de la salle 135 du palais de Mari* (ARMT XXII, 2 vol., Paris, 1983); Philippe Talon, *Textes administratifs des salles Y et Z du palais de Mari* (ARM XXIV and ARMT XXIV, Paris, 1985); and H. Limet, *Textes administratifs relatifs aux Metaux* (ARMT XXV. Paris, 1986).
4. Limet (1985:519) suggests some vessels were made in the shape of cat's heads and bird's heads as well. However, the references he cites do not say head (SAG) but rather list the animal's name right after GAL (GAL *šu-ra-nim* ARMT 25:538, rev. 3; GAL UZ ^{MUŠEN} ARMT 25:527:line 1), suggesting that the whole animal was represented. One can note, however, that at Alishar a cup in the shape of a cat's head was found (von der Osten 1937, fig. 208) and that at Kültepe there were vessels in the shape of eagles' heads as well as whole birds (Özguç 1986, pl. 114-115).
5. The text is slightly damaged where *'a-na'* would be expected, ARMT 22, p. 392, note 1 and pl. III, no. 251.
6. ARMT XIII:8: lines 8-16. See collations of this text by Durand (1983:142). However, I am somewhat uncertain if the phrase *'ša Yašūb-Ašar īpušu'* refers only to the '*tapīḫu*'-vessel which precedes it or to the animal headed cups as well.

7. This would seem to be the way Kupper interprets it in ARMT XXII:253, since he translates 'de la production d'Ana'iš.' A broken text, ARMT XXV:12, recording another *šubultum* explicitly says Yašub-Ašar made the object (line 4: '*ša Yašūb-Ašar īpušu*'), but the name of the object is not preserved. The CAD (s.v. *iškaru*) lists a meaning '(3) finished products, or materials to be delivered.' If a goldsmith is delivering a finished gold or silver cup, it seems reasonable to assume that he made it, especially at Mari where all the quantities of gold and silver entrusted to craftsmen were very tightly controlled (Durand 1983:123-139).
8. ARMT XXI:222, l. 56 '*š[ip]ir Tukriš*[ki]'. Note also line 28 which lists two other vessels described as 'Tukrish-type'. Charpin shows that it was part of a general inventory of the palace treasures made on the occasion of the Assyrian take-over of Mari (Charpin 1983:211-214). Whether the word *kannu* here denotes a pot stand or a kind of vessel, is uncertain (CAD s.v. *kannu*; Durand, ARMT XXI:361-364).
9. E.g., at Hasanlu (Porada 1965:121; Muscarella 1980). For a recent survey of animal headed vessels in neo-Assyrian times, see Calmeyer *apud* Kleiss 1979:194-202. To this corpus may now be added a terracotta ram's head cup found in neo-Assyrian levels at Khirbet Khatuniyeh in the Eski Mosul region of northern Iraq (Curtis and Green 1986:76, pl. 4).

Table 1. Animal headed cups at Mari.

	No. of cups	Material	Weight
I. GAL SAG GU_4 (bull's head cup)[1]			
A. GAL SAG GU_4			
1. ARM VII:218, ll. 1-2	2		
l. 6	2		
ll. 9-10	4		
2. ARMT XIII:8, ll. 9-10	2	silver	1 mina 17⅔ shekels
3. ARMT XXII:256, l. 1	8[2]		
4. ARM XXIV:103,l. 4'	1	silver or gold[3]	
5. ARMT XXV:13, l. 4	1	silver	⅔ mina 4⅓ shekels 14 grains
6. ARMT XXV:18, l. 6	1	silver	½ mina 8⅚ shekels
7. ARMT XXV:82, l. 1	1	gold	½ mina 4 shekels
8. ARMT XXV:393, rev. 4	2	silver or gold[4]	
9. ARMT XXV:502, l. 4	5[5]	[..]	
10. ARMT XXV:510, l. 3	1	[..]	[..]
11. ARMT XXV:519, rev. 15	3	silver	2⅓ mina
12. ARMT XXV:525, l.7	1	[..]	[..]
13. ARMT XXV:527, l.4	1	gold	
14. ARMT XXV:529, rev. 2'	2	silver or gold[6]	
B. GAL SAG GU_4 *tukrišû* (Tukrish-type of bull's head cup)			
15. ARM VII:239, l. 12'[7]	1	silver	
l. 18'[8]	1	silver	
16. ARM XXIV:91, l. 9[9]	1	silver	
17. ARMT XXV:347, l. 22	1		
l. 23[10]	2		

Table 1. (continued)

	No. of cups	Material	Weight
18. ARMT XXV:499, rev. 5-6[11]	1	silver	⅔ mina
C. GAL SAG GU_4 *tu-tu-da-*..[ ] (a ... bull's head cup)			
19. ARMT XXV:393, l. 2	2	silver or gold[4]	
II. GAL SAG AMAR (calf's head cup)			
20. ARMT XIII:8, l. 12	8	silver	ca. 5⅙ mina 8 shekels[12]
21. ARMT XXII:253, ll. 1-4	1	silver	⅔ mina 4⅓ shekels 15 grains
22. ARMT XXV:347, l. 5	2	gold	
l. 24[13]	2		
23. ARMT XXV:499, rev. 10'	2	gold	
24. ARMT XXV:511, l. 4'	1	gold	½ mina 3 shekels
25. ARMT XXV:524, rev. 1-2	1	gold	½ mina 2⅔ shekels
III. GAL SAG MAŠ.$DÀ_3$ (gazelle's head cup)			
26. ARM VII:239 l.13	10	silver	
27. ARMT XIII:8, ll. 13-14	1	silver	⅓ mina 7⅔ shekels[14]
28. ARMT XXI:227, l. 8	2[15]	at least part silver[16]	
29. ARMT XXII:251, ll. 7-8	2	silver	⅚ mina 4⅓ shekels
30. ARMT XXII:256, l. 2	1		
31. ARMT XXIV:91, l. 8	6	silver	
32. ARMT XXV:3, ll. 1-2	2	gold	1 mina 1 shekel
33. ARMT XXV:18, l. 3	1	silver	½ mina 8⅚ shekels
34. ARMT XXV:347, l. 21	10[17]		
35. ARMT XXV:499, rev. 3-4	6	silver	3½ mina
rev. 9	1	gold	
36. ARMT XXV:501, l. 6	1	silver	
37. ARMT XXV:502, l. 5	5	[..][5]	
38. ARMT XXV:510. l. 2	2	[..]	[..]
39. ARMT XXV:518, rev. 4	1	gold	⅔ mina 6 shekels
40. ARMT XXV:519, rev. 14	1	[..]	[..]
41. ARMT XXV:524, edge 3-4	1	gold	½ mina 3 shekels
42. ARMT XXV:534, l. 2	1	[..]	[..]
43. ARMT XXV:538, l. 11-rev.1	1	silver	1 mina 2⅔ shekels

Table 1. (continued)

	No. of cups	Material	Weight
44. ARMT XXV:600, l. 4’	2	silver	[..]
IV. GAL SAG *nālim* (deer’s head cup)[18]			
45. ARMT XIII:8, ll. 11-12	1	silver	ca. ⅔ mina 1 shekel[12]
46. ARMT XXII:251, l. 4	1	silver	ca. 5/12 mina 2⅙ shekels[19]
47. ARMT XXII:256, l. 3	1		
48. ARMT XXV:217, l. 1	1	silver	
49. ARMT XXV:501, l. 5	1	[..]	[..]
50. ARMT XXV:519, rev. 13	1	[..]	[..]
V. GAL SAG *turāḫim* (ibex’ head cup)			
51. ARMT XXV:48, ll. 9-11	1	silver	½ mina 2 shekels
VI. GAL SAG UR.MAḪ (lion’s head cup)			
52. ARM IX:84, l. 7[20]	[..]	gold	[..] mina 5 shekels
53. ARMT XXII:251, l. 3	1	silver	ca. 5/12 mina 2⅙ shekels[19]
54. ARMT XXV:393, l. 1	2	silver or gold[4]	
55. ARMT XXV:499, rev. 11	2	gold	
56. ARMT XXV:507, l. 3	1	gold	
57. ARMT XXV:510, l. 4	1	[..]	
58. ARMT XXV:511, l. 3’	1	gold	⅓ mina 6½ shekels
59. ARMT XXV:524, edge 1-2[21]	1	gold	½ mina 5 shekels
60. ARMT XXV:528, ll. 1-2[21]	1	gold	⅚ mina 6 shekels

NOTES TO TABLE 1

1. GU$_4$ = *alpu* can refer to a bull or an ox (CAD). One cannot really know, then, if SAG GU$_4$ refers to an ox or a bull.
2. ARMT XXII:256, l. 1: 8 GAL SAG.ḪI.A GU$_4$.
3. ARM XXIV:103, l. 4: 1 SAG GU$_4$ KÙ.[...]. Although GAL is not written, the text is a list of vessels, so this is most likely a vessel as well.
4. ARMT XXV:393: The end totals of the text indicate that all the vessels listed are either silver or gold (seven gold, 25 silver), but the material of each vessel is not always indicated.
5. ARMT XXV:502: This is a list of GAL vessels, but the reverse is gone, so we do not know whether they are silver or gold.
6. ARMT XXV:529: rev. 2’: 2 GAL SAG GU$_4$ KÙ.[...].
7. ARM VII:239, l. 12’: Qualified as ‘big’. The entry reads: 1 GAL SAG GU$_4$ GAL *tu-uk-ri-šu-ú* KÙ.BABBAR.

8. ARM VII:239, l. 18': according to ARMT XXI, p. 350, note 1, this line reads '1 GAL SAG GU_4 *tu-uk-ri-š[u-ú* KÙ].BABBAR'.
9. ARM XXIV:91, l. 9: *tukrišû* spelled *'tu-uk-ri-šu-ú'*.
10. ARMT XXV:347, ll. 22-23 read: '1 SAG GU_4 *tu-uk-ra-šu-ú* GAL; 2 SAG GU tu-uk-ra-šu-ú TUR' = 1 large Tukrish-type bull's head cup, 2 small Tukrish-type bull's head cup. As in ARM XXIV:103 (above note 3) GAL is left out before some of the vessel names.
11. ARMT XXV:499: *tukrišû* spelled *'tu-uk-ri-šu-ú'*.
12. ARMT XIII:8, ll. 11-12 give 5 5/6 mina 9 shekels as the weight of 8 calf's head cups and one deer's head cup. Therefore, each weighs ca. 2/3 mina 1 shekel, so 8 would weigh ca. 5 1/6 mina 8 shekels (assuming all the vessels were the same weight). For the reading of this line as '8 GAL SAG AMAR 1 GAL SAG *na-lim* KÙ.BABBAR', see Durand 1983:142.
13. ARMT XXV:347 leaves out GAL before SAG AMAR in lines 5 and 24. See note 10 above.
14. ARMT XIII:8, l. 14: For this reading of the weight see Durand 1983:142.
15. ARMT XXI:227, l. 8: 2 GAL SAG *ṣa-bi-tim*.
16. See discussion of this text above.
17. ARMT XXV:347, l. 21: 10 SAG *MAŠ.DÀ*. See above notes 10 and 13.
18. *Nālu* (*najalu*) is 'roe deer' according to the CAD. Two other Akkadian words for deer are *ajālu* and *lulīmu* ('red deer, stag'--CAD). That *nālu* and *ajālu* were distinct species of horned animals is clear from the occurrence of both words in ARMT XIII:55. This is a letter from Iasīm-Sūmû to Mukannišum concerning the preparation of a certain object (or objects) described as (lines 4-7): '*enūt kanni qaqqadāt nēšim qarnat nāli u aiālī*'. Birot (ARMTXIII:78) translated this as 'ornements des *kannu* (savoir) les têtes de lions, les cornes de cerfs et de chevreuils (?)'. *Kannu* has been suggested to be some type of potstand or vessel (CAD s.v. *kannu*; Durand ARMT XXI:361-364). In the phrase É *kannim* it refers to a storeroom and at Mari *kannum* alone sometimes has this sense (e.g. ARMT XXII, 256:4). Deller (1985:339) has suggested this is the meaning of *kannum* here and that *qaqqadāt nēši* and *qarnat nālī u ajālī* refer to separate vessels and not ornaments. However, one should note that *qaran nāli* (horn of a roebuck) and *qaran ajāli* (horn of a stag) never seem to occur at Mari as descriptions of vessels.
19. ARMT XXII:251, ll. 1-4 say 5/6 mina 4 1/3 shekels of silver is the weight of 1 lion's head cup and 1 deer's head cup. So each should weigh half of this amount --assuming they are of equal size.
20. ARM IX:84, l. 7 actually has [x GA] L SAG MAḪ KÙ.GI. The emendation SAG ‹UR›MAḪ has been suggested by Durand (ARMT XXI:351, note 6).
21. ARMT XXV:524, edge 1 and 528, l. 1 leave out GAL before UR.MAḪ.

REFERENCES

BIROT M.
1964 'Lettres de Iasîm-Umû',in: *Textes divers à l'occasion du XXXe anniversaire de la découverte de Mari* (ARMT XIII). Pp. 49-79.

CHARPIN D.
1983 'Un Inventaire général des trésors du palais de Mari, *M.A.R.I.* 2: 211-214.

CHARPIN D. AND DURAND J.M.
1983 'Relectures d'A.R.M. VII', *M.A.R.I.* 2: 75-115.

CURTIS J.E. AND GREEN A.R
1986 'Preliminary Report on Excavations at Khirbet Khatuniyeh,1985,' *Researches on the Antiquities of Saddam Dam Basim Salvage and Other Researches*. Baghdad.

DALLEY S.
1984 *Mari and Karana, Two Old Babylonian Cities*, London and New York.

DELLER K.
1985 'SAG.DU UR.MAḪ, 'Löwenkopfsitula, Löwenkopfbecher'', *Baghdader Mitteillungen* 16: 337-340.
1986 'SAG.DU UR.MAḪ Eine Nachlese', *Baghdader Mitteillungen* 18: 219-220.

DIETRICH M. AND LORETZ O.

1964 'Der Vertrag zwischen Šuppiluliuma und Niqmandu, eine philologische und kulturhistorische Studie', *Welt des Orients* 3: 206-245.

DURAND J.M.

1983 *Textes Administratives des Salles 134 et 160 du Palais de Mari* (ARMT XXI). Paris.

1983 'Relectures d'ARMT VIII, II: ARMT VIII, 89 et le travail du métal à Mari', *M.A.R.I.* 2: 123-139.

GELB I.J

1944 *Hurrians and Subarians* (SAOC 22). Chicago.

GRAYSON A.K.

1972 *Assyrian Royal Inscriptions I*, Wiesbaden.

GRONEBERG B.

1980 *Répertoire géographique des textes cuneiformes III: die Orts- und Gewässernamen der altbabylonischen Zeit.* Wiesbaden.

KLEISS W.

1979 *Bastam I.* Berlin.

KOMORÓCZY G.

1972 'Das Mythische Goldland Haralu im Alten Vorderasien,' *Acta Orientalia* 26: 113-123.

KUPPER J.-R.

1983 *Documents administratifs de la salle 135 du palais de Mari* (ARMT XXII, 2 vol.). Paris.

LIMET H.

1985 'Bijouterie et orfèvrerie à Mari', *M.A.R.I.* 4: 509-521.

MOOREY P.R.S.

1985 'The Iranian Contribution to Achaemenid Material Culture', *Iran* 23: 21-38.

1985 *Materials and Manufacture in Ancient Mesopotamia: the evidence of Archaeology and Art* (B.A.R. International Series 237). Oxford.

MUSCARELLA O.W.

1974 *Ancient Art: the Norbert Schimmel Collection.* Mainz.

1980 *The Catalogue of Ivories from Hasanlu, Iran.* Philadelphia.

NEUGEBAUER O. AND SACHS A.

1945 *Mathematical Cuneiform Texts*, New Haven.

ÖZGUÇ T.

1959 *Kültepe-Kaniš.* Ankara.

1986 *Kültepe-Kaniš II.* Ankara.

ÖZGUÇ T. AND ÖZGUÇ N.

1953 *Ausgrabungen in Kültepe.* Ankara.

YON M.

1981 *Dictionnaire illustré multilingue de la céramique du Proche Orient Ancien.* Lyon.

PORADA E.

1965 *The Art of Ancient Iran.* New York.

STEINKELLER P.

1982 'The Question of Marḫaši: a Contribution to the Historical Geography of Iran in the Third Millennium B.C.', *Zeitschrift für Assyriologie* 72: 237-265.

STOL M.

1972 *Studies in Old Babylonian History.* Leiden.

TALON P.

1985 *Textes administratifs des salles Y et Z du palais de Mari* (ARM XXIV). Paris.

TALON P. AND LIMET H.

1986 *Textes administratifs relatifs aux Metaux* (ARMT XXV). Paris.

VON DER OSTEN H.H.

1937 *The Alishar Hüyük Seasons of 1930-32*, OIP 29. Chicago.

Ground plans and archaeologists: On similarities and comparisons

D.J.W. MEIJER

INTRODUCTION

It is well known that the ancient Near Eastern architects left us no theoretical treatises on building or on buildings. The ensuing difficulties for archaeologists attempting a reconstruction of the built-up surroundings of ancient community life are twofold: one has to conjecture what a building looked like, starting from a scratchy plan; and the functions of the building and its various sectors are often quite unclear. The relationship between form and function is difficult to establish - and this is typically a matter in which a body of theory could enlighten us[1].

Our own body of architectural theory, in which Renaissance builders like Alberti, Brunelleschi, Bramante and Palladio have been important developers, ultimately goes back to Vitruvius, who in turn was influenced by Greek architect-builders (Ten Books...VII:11-14). A key phrase in Vitruvius' work is *quod significatur et quod significat* (Ten Books... I:3), in other words, there is a meaning to be found in a building. One may safely posit that such a meaning is culturally determined. Architects will achieve their greatest success in surroundings where their 'meanings' are received most easily, i.e. within their own culture.

It would be a mistake to brush the 'reception' of a building aside as an ephemeral matter only because many of us are not actively aware of the influence we undergo from the definition of a space; but architects define space, and they do it with a culturally bound set of ground rules. These rules are set ranges of combinations of several variables: measurements and proportions, i.e. the morphology, but also the materials used (hence also acoustics), lighting, temperature. These variables act upon the sense receptors that we use for navigating through a given space: the eyes, the skin and the ears (cf. Hall 1969, chap. 4). The experience of a cultural anomaly, that is, a 'wrong' combination of these variables leads us to the realisation, however, that more factors are at play. We feel that a long echo is at home in a cathedral, but it would be anomalous in a gentleman's club in London.

And in regions where mud brick is the normal building material, the observation of a baked-brick building focuses our attention on its probable central function in the community. Hence it is clear that the cultural rules are also defined by a functional variable, or indeed by two: the utilitarian as well as the symbolic function of a building. One can think of many morphological solutions to the spatial problems set by a large congregation facing a preacher in a church. In this respect a spacious mosque is a better solution than a traditional Christian church with its narthices, naves, aisles, colonnades, curtain walls etc. However, the Christian builders of the Middle Ages preferred to build churches *in modum crucis*, the symbolic expression of Christ crucified (Wittkower 1973:30). These are cultural determinants or standards, related to symbolic functions of buildings. In whatever way these standards come to be accepted, a deviation will need a 'theoretical' rationalisation before it is accepted as a standard in its turn; it interferes with the feeling of comfort experienced before, and necessitates additional information.

Cross-culturally, as modern archaeologists concerned with the ancient Near East, we obviously require the same kind of additional information if our classification of architectural remains aims at more than a random description. Yet the relation that we posit between form and function is often based on circumstantial evidence: incomplete morphological data or a building's inventory. A case in point is the triparite ground plan in the Mesopotamian Ubaid- and Uruk periods, of which it has only relatively recently become clear that it was not exclusively reserved for religious buildings, but had a long history in domestic architecture as well (Roaf 1984)[2]. This realisation came about not through direct evidence concerning function (on the one hand, temple inventories could share household installations with domestic buildings, on the other hand some domestic buildings look rather monumental), but rather through comparative evidence of scale and frequency.

The intimate relationship between form and function may be said to concern all classes of buildings. Just as temples have to be analysed with their function in mind, since that function's symbolism could be all-pervading in the decisions defining the form, so do palaces rely heavily on functional aspects for their form. The same is true of private dwellings. Sociological determinants such as family constellation may help to define the form of private dwellings. Recently an attempt has been made to correlate the tripartite plan with a specific structure of co-residentialism (Maisels 1987). And such analyses are worthwhile, keeping in mind the role that other variables, such as climate, fashion (!), rank and individual initiative may also play (cf. Rapoport 1969). For archaeologists, such variables are to be gotten out of the equation, not put into it, lest one become enmeshed in circular argumentation.

In summary one may say that for all three main architectural classes (temples, palaces and private houses, and one could add tombs), there are four aspects involved in an analytical classification: *form, location, utilitarian function* and *symbolic function* (and note that we have no other clear names for these classes than functional ones). If a similarity is observed in the plans of two ancient buildings, these will be classified together since our control is mainly over form and location. This primacy of form over function is a legitimate, practical, but not an *a priori*, subject-intrinsic approach. Any primacy of one of these aspects or variables depends on the particular theory with which one approaches the ancient buildings.

SIMILARITY

Such a theory's powers of explanation, i.e. the validity of the conclusions based on the classification, are subject to varying considerations. In most archaeological cases, where primacy is given to the variable of form, similarity of plans will be taken to indicate either shared cultural knowledge or visual contact between the perpetrators of the plans. Note, however, that similarities have differential analytical status as they occur throughout the main classes of buildings. It would seem that similarity in the plans of temples suggests to us a shared cultural (for religious) background of the builders. We tend to infer this on the basis of our experience that temple plans reflect liturgical proceedings, and a specific liturgy is a conservative and enduring facet of a culture[3]. Wildly different cultures tend to produce temples with different plans[4]. A consequence of this is that the compared buildings need not be quite contemporaneous.

With palaces (however one defines the concept) similarity may point to another set of circumstances, reflecting a different way in which the relationship is 'set' between the variables form, utilitarian- and symbolic function. Apart from their obvious utilitarian function to accommodate the household and business traffic and industry connected with the ruler and his government, palaces tend to function as a symbol of power.

This is expressed in scale (usually larger than the utilitarian function necessitates) and building material and ornamentation. And although the utilitarian function requires a given set of spaces, such as throne room, kitchen, audience rooms, servants' quarters, chapel areas, their interconnections and placement can hardly be called dependent on liturgical custom. Consequently there is greater freedom to shape palaces according to the builder's personal tastes and requirements[5]. Thus random borrowing of foreign elements can occur, and similarity in this case does not presuppose cultural unity. On the other hand, the compared examples should overlap in time.

By virtue of their focal social function and their concomitant relative scarcity, both temples and palaces are special buildings, setting them apart form private, domestic architecture. Relatively less 'organised planning' will go into the construction of the latter, and such construction can be called more 'run of the mill'. As before, one can speak of a different set of the relationship between the variables form and function. Family size, availability and shape of a plot between adjacent plots, as well as sociological variables (even dysfunctional fashion) may all contribute to the determination of the ground plan and form of private dwellings. Presence and absence of walled courtyards, stables and storage facilities etc. depend on the ('utilitarian') functions of the household members. Hence, if two similar buildings are observed of which it is decided (!) that they are private dwellings, their similarity may derive from parallel, but not necessarily identical, variable combinations. Symbolic function in domestic architecture other than simple fashion, can only be postulated on the basis of recurrent 'improbable' plans, and very little work has been done in this field, mainly because of our lack of control over sources concerning function; Maisels' article mentioned above (Maisels 1987) is a pioneering study . Another example of symbolic functional analysis is the interpretation of certain recurring plans as reflecting the activity of a certain ethnic group (Shiloh 1970, cf. idem 1987; Margueron 1980). In such cases, shared ethnicity presupposes shared culture. This is not quite the same thing as the example of Maisels' study, for the latter's co-residentialism may, in theory, be a cross-cultural phenomenon: a case of parallel, but not identical variable combinations. Of course here also the rule applies that the more intricate a plan, the more improbable a morphological parallel, and the more detailed must the explanation be.

The main question regarding similarity is obviously the measure of the identity of buildings, that is in our case, of plans. Do we require congruence or a simple superficial likeness? Do we compare measurements? In the absence of prefabricated construction it is to be expected that no two plans are entirely congruent. It seems that the question of the

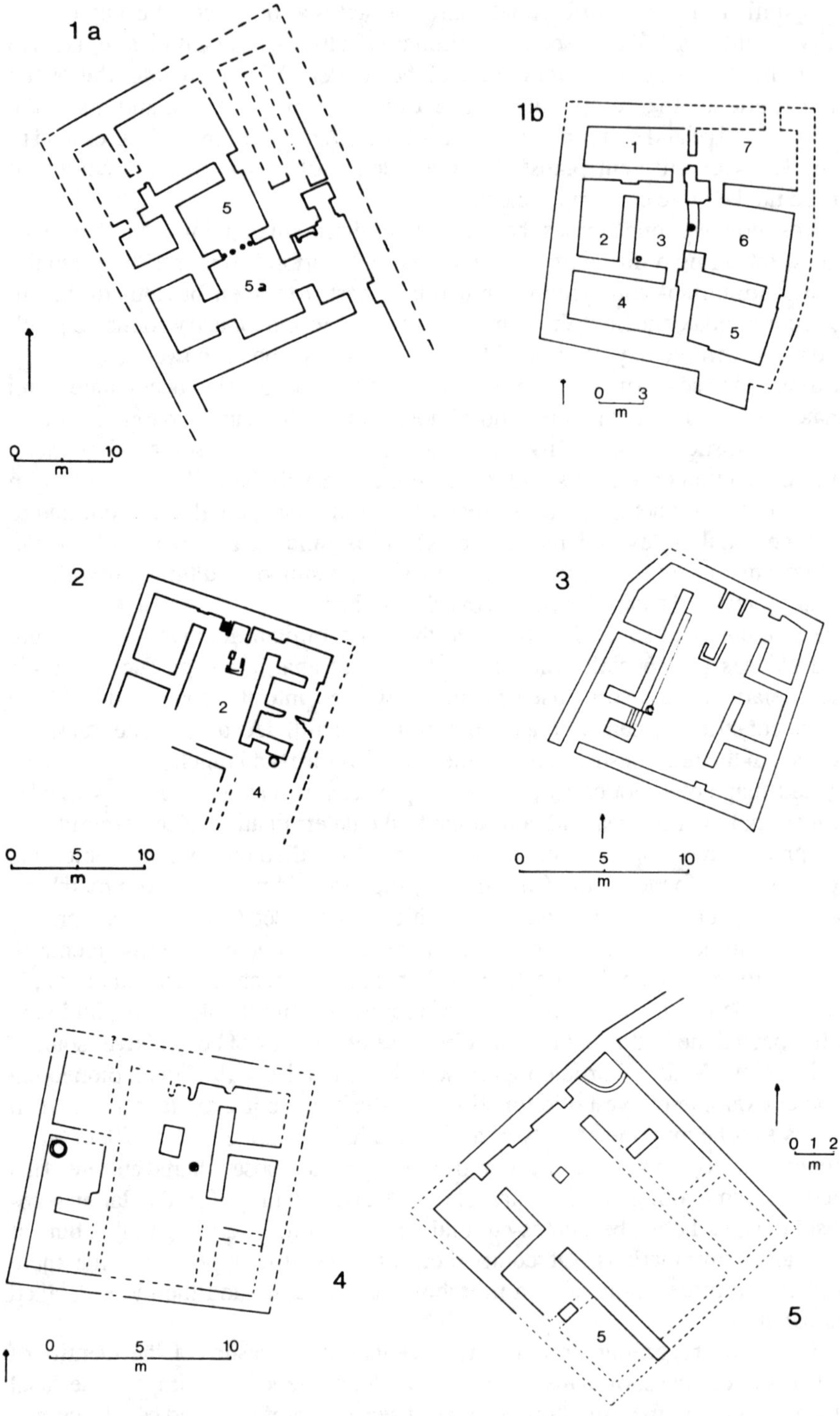

Figs 1-5. Plans of Late Bronze Age buildings along the Euphrates in Syria.

measure of similarity cannot be answered in a simple formula[6]. Layout and number of rooms, entrances, sizes per type etc. will all have to be set off against the other excavated architecture of the same site and date, in order to determine the value of observed similarities.

EXAMPLES

Bronze Age Syria provides striking examples of ground plans of buildings that are classifiable, comparable, special in their contexts - and difficult to interpret.

A. the *bīt ḫilāni* is an example of how confused we may become by the correlation between form and function. On the basis of Neo-Assyrian royal inscriptions, archaeologists have adopted the term to denote a rectangular building with a columned entrance giving access to a sequence of transverse halls (Hrouda 1975; Winter 1982). However, the precise meaning and etymology of the term *bīt ḫilāni* (b.ḫ.) are unknown (Güterbock 1975; Singer 1975; Winter 1982), and its archaeological usage tends to become functional, being equivalent to 'porticoed palace in the Northwest Syrian Iron Age'. Applying the term to ground plans seems to pivot on the presence of columns in an entrance, for in fact the number and placement of halls and side rooms can differ considerably throughout the range of examples usually subsumed under the term b.ḫ. (cf. Naumann 1971: figs. 551-553)[7]. Furthermore it is on the pillars or columns that Frankfort's idea hinges - accepted by most authorities - that a forerunner of the b.ḫ. may already be found in Yarimlim's palace at Alalakh (Frankfort 1952).

Even if ⁺*ḫilāni* were ultimately to be derived from a root ⁺*ḥln* meaning 'opening' or 'window' (cf. Winter 1982:363), the emphasis in the Assyrian texts is on columns, and suggests a loggia-like arrangement (see also Busink 1970:532). Such a loggia may have had openings to several sides that are difficult or impossible to find in plan. Presumably the Assyrian kings copied the concept to enhance the special character of their abode ('ana multa'utiịa' or 'for my relaxation' writes Tiglath Pileser III), having seen, or heard of, the well ventilated examples in the west; there the climate is milder, and the mountain winds refreshing, and such loggias would seem to be more at home there than in northern or southern Iraq, where the hot winds hardly ventilate, but blow in dust; the plan was copied for reasons of symbolic function.

In the present context the well-aeriated loggia needs further comment. If this is a good description, if not a translation, of *bīt ḫilāni* (and note that a loggia may also stand alone, as is sometimes surmised for the b.ḫ.), Frankfort's identification of the Alalakh VII palace as a ḫ. rests on the ventilation and lighting system of rooms 5 and 5a (Woolley 1955: fig. 35; see our fig. 1a). Since these rooms are surrounded by others, what ventilation there was may have come through clerestory windows or openings (⁺ḥln?).

Something quite similar can be posited, on the basis of more than purely circumstantial archaeological inference, for a building that is virtually identical in form to the Alalakh VII palace north wing. At Selenkahiye on the Syrian Euphrates Maurits van Loon excavated a building of which the ground plan corresponds remarkably to the Alalakh example (even including the placement of the earliest entrance, cf. Wooley 1955:92, note 2), but which predates it by three or four centuries (van Loon 1968:23f.; 1979:108ff.; Meijer 1984)[8]. Figure 1b shows the Mansion's layout, in which the suite with its cedar

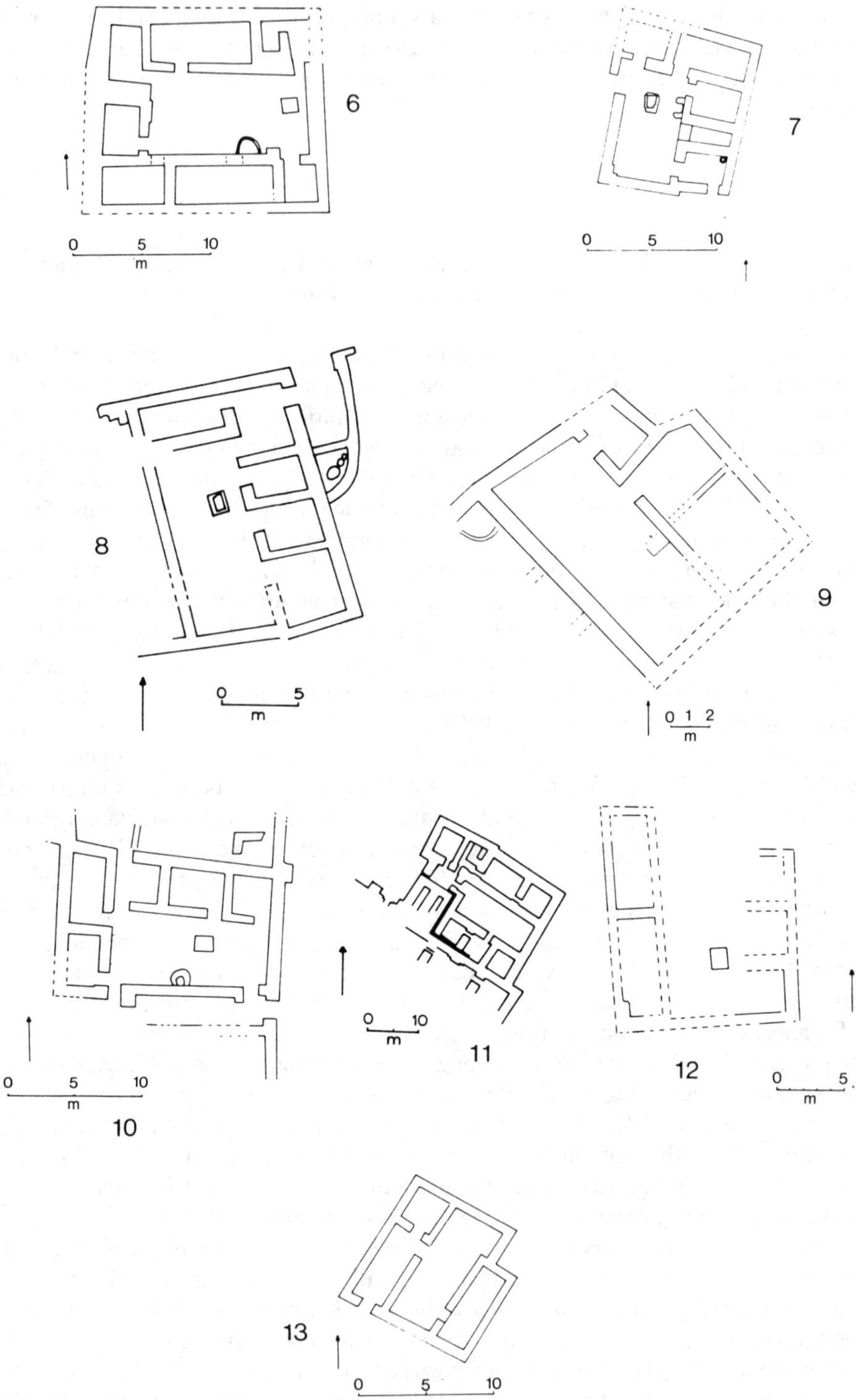

Figs 6-13. Plans of Late Bronze Age buildings along the Euphrates in Syria.

column on a round limestone base is striking (rooms 6 and 3). From the distribution of the destruction debris of the Mansion's first phase (22nd century B.C.) and that of its second phase (1st century B.C.?), it is surmised that only the western part of the building had a second storey: rooms 5, 6 and 7 contained only some 10 cm of destruction debris, whereas rooms 4, 1 and 2 and parts of 3 had some 30-50 cm. It is consistent with the disposition of finds in room 3, such as ovens, to reconstruct it as having a high roof at the same height as the roof over the rooms of the second storey above rooms 4, 2 and 1. Room 3 would then have had clerestory lighting and ventilation through its eastern upper wall, possibly above the cedar column which supported the tie beam of the roof beams pertaining to room 6. Alternatively a *partial* flooring of an upper storey above room 3 can be visualised where a veranda around a central air column could facilitate a relaxing rest near the window in the eastern upper wall.

All this is speculative, but nonetheless a feasible explanation of the archaeological findings in the Selenkahiye Mansion - by far the most palatial building excavated at the site. Whether this Mansion and the Alalakh VII palace be termed *bīt ḫilāni*'s or not, what remains is their remarkable similarity in plan (if not in scale: the palace is twice the size of the Mansion; another difference is the absence of orthostats at Selenkahiye). This similarity cannot be accidental, and leads us to postulate the existence somewhere in northern Syria of an intermediate, possibly to be found in Aleppo or in preceding strata at Alalakh itself. And whether the Selenkahiye Mansion was the local palace or not, the columned suite in any case establishes itself securely as a Syrian phenomenon[9].

B. Similarities between plans may lead to theories of ethnic movement. An example is a recent attempt to identify simple rectangular buildings with T-shaped internal partitions as a typically Hittite architectural concept (Margueron 1980). The same houses, however, occur in other places and periods than those mentioned by Margueron, such as Keisan, En Gedi or Chatal Hüyük (cf. already Braemer 1982:52-54) and early Middle Bronze I Halawa (Orthmann 1981: Tf. 77, e3, d-c2 and d3). We are dealing with a plan that is quite simple and statistically probable; it entails the easiest way to divide a rectangular plot into three easily accessible divisions: adding a fourth division or room would create incommensurate traffic problems. The sizes of most traceable examples average around 60-70 sqm, i.e. well inside Wright's class of 'plebeian housing' (Wright 1985:122). Unless the Hittite implantation visualised by Margueron is conceived of as concerning extremely modest Hittites settling abroad - but relatively massively and quite early at Halawa - it is preferable to interpret these house plans as independent, local solutions to problems of restricted means and space.

C. Even more striking examples of similarity, since they concern more intricate plans and are thus more improblable, are found in the Syrian Late Bronze Age along the Euphrates. Table 1 presents a classification of these buildings, of which sketch plans are reproduced in figures 2-10. The typology is as follows:

I The tripartite plan;

II The tripartite plan with added rooms on one short side; and

III A long room flanked by smaller rooms on one long and one short side.

TYPE I

1. Munbaqat *Haus A* (Fig. 2; Eichler et al. 1984:81-91; Machule et al. 1986:99-104). The

Table 1. Syrian Late Bronze Age buildings.

Type site		Plot size	Floor area	Rooms
I Mumbaqat	A	208(?)	?	7(?)
	B	168	96	7
	D	178	80+	7(?)
	F	170	109	7(?)
II Hadidi Tablet Bldg.		270	162	8
III Munbaqat	E	126	75	6
	SW Tor	175	113	6
	F1	140	90	?
Hadidi	Northern Bldg.	187	106	6

NB. Except in the case of Munbaqat B, D, E, F, and F1 the measurements were calculated from the published plans and cannot be called exact.

excavators tend to ascribe religious functions to at least room 2 (the main hall, Eichler et al. 1984:85f.), and substantiate this with the inventory containing fine pottery and a relief fragment, an altar-like fixture and a 'mantlepiece'. Against an exclusively religious function of the building speaks the fact that the southeastern room and perhaps the main hall's southern wall seem to be later additions (Machule et al. 1986:101), suggesting a broad rather than oblong layout, without strict parallels in religious architecture in Syria[10]. Although the published plan of *Haus A* is incomplete, it resembles that of *Haus B* to a restorable extent; however, there are important differences in the communicability between the side rooms, apart from the above- mentioned additions. Our position here is that the building was not conceived as a temple but as domestic unit, where religious functions - in a familial context - may have been fulfilled.

2. Munbaqat *Haus B* (Fig. 3; Machule et al. 1986:105-112). Smaller than *Haus A*, but apparently more regularly built, this house features comparable installations in the long room. The private character of its habitation is emphasized by the find of a business archive (Machule et al. 1986:126-131) spanning three generations. Here also, however, the suggestion implicit in Machule's note may be taken up, and a house cult may be envisaged (Machule et al. 1986:63, note 63).

3. Munbaqat *Haus D* (Fig. 4; Machule et al. 1987:112-116). About the same size as *Haus B* and with a comparable layout as well as comparable installations in the main room.

4. Munbaqat *Haus F* (Fig. 5; Machule et al. 1987:80-85). The exact reconstruction of this building is problematic (Machule et al. 1987:81, 82, note 24). Purely on the basis of formal comparison to the preceding examples, however, the longer reconstruction of F (together with F1) seems improbable. It would entail a greater number, or extreme sizes, of the southern side rooms; moreover the sizes of F, B and D would correspond nicely in the shorter reconstruction (see Table 1), where one could posit the southern wall of the building at the southern end of the southernmost stone of room 5's east wall (Machule et al. 1987, Beilage 1). The installations are limited to one 'socle'.

TYPE II
A probably related, but distinct type is the tripartite building with added rooms on one short side. The only example so far seems to be the Hadidi Tablet Building (Fig. 6; Dornemann 1979:143, fig. 31; 1981:29-41; 1985). It is striking that it has fixtures comparable to those of the type I buildings, albeit slightly differently placed (cf. Machule et al. 1987:84, note 30, where the comparison is with Munbaqat *Haus F 1*, which is here ascribed to a different type). The plan should probably be reconstructed not with a courtyard but with the central room covered (cf. already Machule et al. 1987:116, note 100).

Another example is suggested to exist at the same site, i.e. the 'Northern Building', and for instance Machule reconstructs it as similar to the Tablet Building (Machule et al. 1987:116, note 100). However, the legenda clearly indicate different phasing of the two units (Dornemann 1981, fig. 2). Although the southeastern wall of the Northern Building does continue south, there is no mention of a south wall, in its original building phase, other than the wall defining the long room with the hearth. It is unclear whether during a later phase, the more southerly wall was in double use, i.e. whether the two buildings were ever contemporaneous. The Northern Building was not originally planned on the same lines as the Tablet Building.

TYPE III
The third type consists of a long room, flanked by side rooms on one short and one long side.
1. Munbaqat *Haus E* (Fig. 7; Machule et al. 1987:116-118). The double wall stubs in the southwest (Machule et al. 1987, Beilage 2) suggest that houses E and E 1 were planned as separate units, as were D and E. Here again, the main room features installations comparable to those of the houses of the preceding types.

2. Munbaqat *Südwest Tor* (Fig. 8; Orthmann 1976:35-39). This building lacks one of the installations the others have, and only features the square socle in the long room. It also has an 'adyton' of irregular layout, which to judge from the drawing (Orthmann 1976:36), however, is clearly secondary.

3. Munbaqat *Haus* F1 (Fig. 9; Machule et al. 1987:80-85). Although its published plan is incomplete, it may be reconstructed to the same form as the two preceding plans. It would seem that here also the shorter of the two reconstruction possibilities (see above, under I,4) applies. As in F and the *SW Tor*, only one socle is in evidence. Since F and F 1 share one long wall, they were probably planned and executed together. This association recalls that of houses D and E, which were built separately, however.

4. Hadidi Northern Building (Fig. 10; Dornemann 1979; 1981; 1985). The long room of this building (which was built later than the Tablet Building) features the two installations known from the Tablet Building as well as from the Munbaqat type I buildings. Inclusion of this structure in this type depends on the completion of the plan to the north, where additional rooms could suggest a different layout.

If one looks for parallels for these types outside Mumbaqat and Hadidi themselves, the harvest is meagre. On a strictly formal basis, the northern extension of Tilmen Hüyük IIb

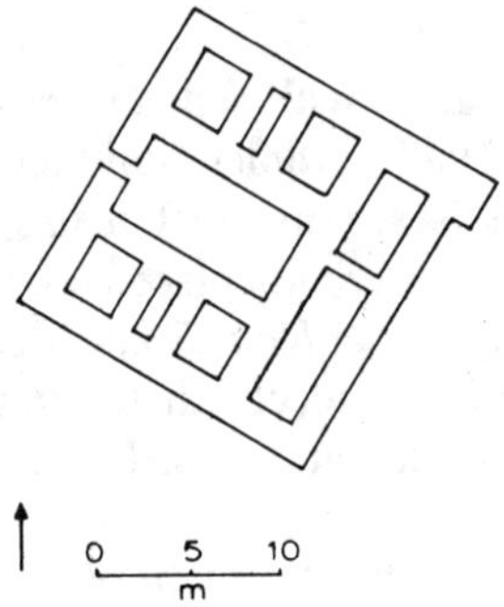

14

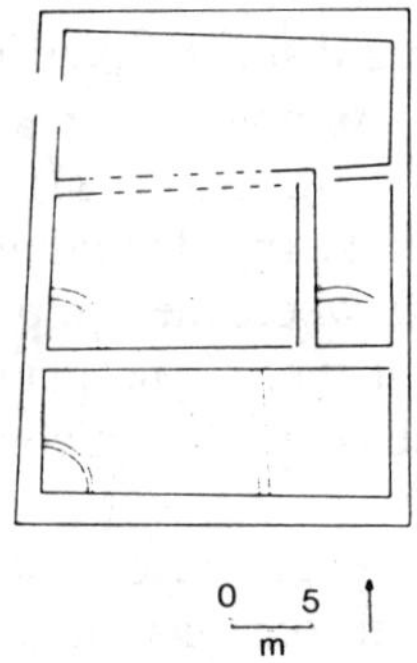

15

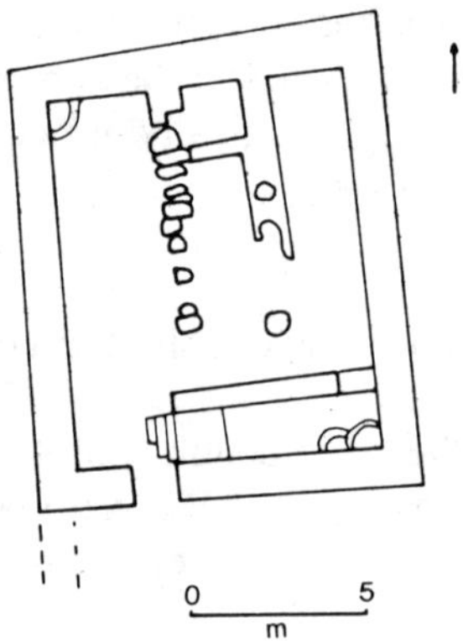

16

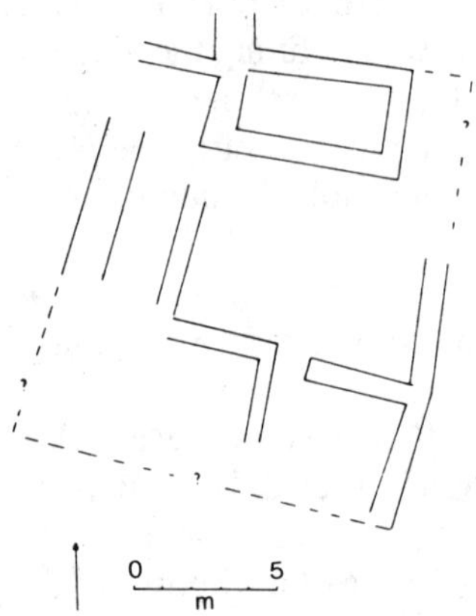

17

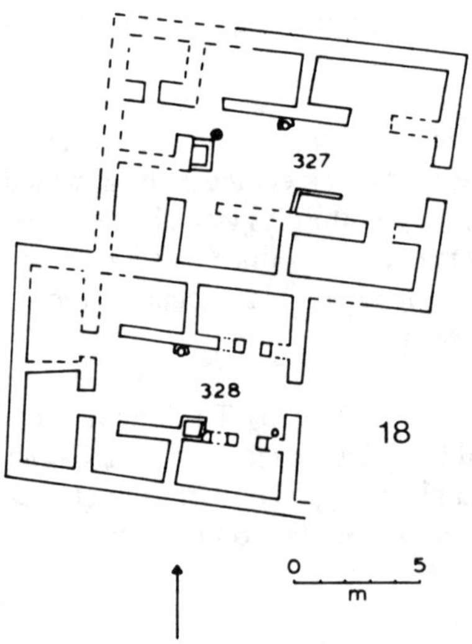

18

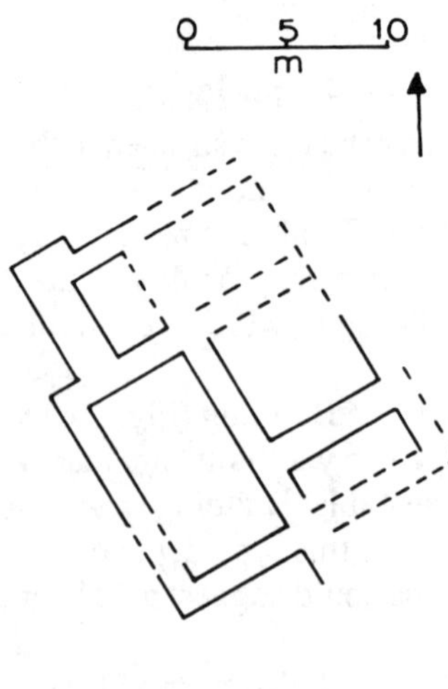

19

Figs 14-19. Plans of Late Bronze Age buildings along the Euphrates in Syria.

(Fig. 11) could be adduced as an example of our type I (Orthmann 1966:165). On the other hand, its size of ca. 21 x 19 m, and the northern lobby-*cum*-stairwell, which forms the only communication with the northern side rooms, set it apart from the Munbaqat examples. The preliminary fashion of its publication furthermore hinders a detailed assessment of the northern lobby, which, to judge from the plan, might have been appended secondarily; also a ground floor entrance is necessary if the plan is to be included in our type I. The building is dated to the 16th century B.C.

Another uncertain parallel of this type of building would be the building in square N 16 in Hama (Fig. 12; Fugmann 1958:141 and fig. 166(!)). The plan here is not quite clear; thus it is doubtful whether there was one short wall along the entire northwest width of the structure. Some 'rigoles' and drains were found, but the only installation comparable to the above examples would be the rectangular stone feature visible near the number 10 in Fugmann's plan; this same object would seem to make Braemer's reconstruction of the building (Braemer 1982:221) slightly incongruous, and that of our figure 12 is preferred. The date of this large structure (ca. 230 sqm plot size) is Late Bronze/Early Iron (Hama F2).

At the other end of the scale lies a plan like that of an Old Hittite house on Büyükkale (Fig. 13; Bittel and Naumann 1938:41, Abb. 18-2). Its plot size of ca. 100 sqm and its distance from the area under consideration make it difficult to include this building in our list, although its date fits nicely (15th century B.C.).

Our type I also has a formal correspondence with Braemer's II A 3, who notes that its usage in Palestine is strikingly restricted to large buildings (Braemer 1982:58-60).

As with type I, the parallels for type II are vague. The first is from Late Bronze Beth Shan (Fig. 14; Oren 1985); it measures 22 x 23 m. Furthermore, the eastern building at Rujm al Ḥenu in Jordan, measuring ca. 31 x 24 m (Fig. 15; McGovern 1983:109 and 126f. with fig. 2) provides a comparison in form; it dates to the Late Bronze or Early Iron Age. Both these examples, however, are wider than they are long (taking the central room as point of reference). Another parallel in Palestine is Tel Batash VII area B (Fig. 16; Kelm and Mazar 1982) where, however, the walls flanking the central hall consist of open spaces alternating with pillars. The plot size was ca. 14 x 11 m, and the date is LB II. An early 12th century example, albeit uncertain, should also be mentioned: Giloh (Fig. 17; *IEJ* 31:6f.); its plan, however, is very badly preserved.

By way of hypothesis, these vague parallels may perhaps be said to suggest connections of type II with the 'Four-Roomed' house in Palestine (Braemer 1982, *passim*; Wright 1985:294ff.; Stager 1985). This type is also called 'pillared house', because the long walls flanking the main room often consist of pillars and open spaces - possibly in connection with the use of the side rooms as stables (Stager 1985:12f.). Although it was first seen as typical for the Iron Age, more recently excavated examples tend to push that date backward. The fact remains, however that the main area of distribution is Palestine. Quite apart from this, an important question is whether the difference between pillared division and solid wall is diacritical as to a building's function (if the function is stabling, it would seem to be). A tentalizing example in this context is house 327 (Fig. 18) at Far'ah (Iron Age II; Braemer 1982:217). It has a plan quite close to the Hadidi Tablet Building but for the placement of the 'bassin', and it has solid walls. It was built adjacent to house 328, sharing a long wall. House 328 had partly solid, partly pillared divisions, as well as comparable installations. Such examples emphasize the need for careful analysis of all finds from all rooms, certainly including bone and vegetable remains, in order to determine function more precisely; Kent (1984, chap. 5) shows, however, that even then answers can be equivocal. It should be noted that the side rooms of the Hadidi Tablet Building produced much pottery, making stabling an unfeasible option; the small room in the southeast had been used for grain storage (Dornemann 1981:32). The time lag of at least 400 years between Hadidi and Far'ah, as well as the distance between them would make it extremely hazardous to posit any other than the loosest of cultural ties between the inhabitants of the two buildings.

In looking for parallels of our type III, the Northern Gate Building at Alalakh comes to mind (Fig. 19; Woolley 1955:158 and fig. 57, VUT 6-7). This building is older than Niqmepa's palace and younger than the level VII gateway, and thus seems to constitute an early Late Bronze building of a locally exceptional plan. At ca. 245 sqm, its plot size is much greater than those of the examples discussed above, making the comparison difficult to evaluate. Although its position at the gate in our eyes might suggest otherwise, there is no argument against an at least partially domestic function. The type corresponds to Braemer's type II B 2 for Iron Age Palestine (1982:65f.), but the similarity may best be seen as fortuitous until more tangible proof of contacts between the two areas - and periods - becomes available.

The three types isolated here must be called local to Mumbaqat and Hadidi. They do not occur in any of the major LB towns such as Ras Shamra, Byblos or even Nuzu. They seem to spring up just after the Middle Bronze Age as new concepts in new town extensions on virgin soil or leveled earlier remains. The conjunction of types I and III in Mumbaqat is striking (houses D and E, F and F 1); if indeed the Tablet Building and the Northern Building at Hadidi did overlap in time (Dornemann 1981:32), they provide a conjunction of types II and III. One cannot escape the conclusion that their proximity, contemporaneity, local distinctness and the above mentioned conjunctions must reflect direct acquaintance of the builders with their examples, perhaps even with their colleagues. In the case of the Mumbaqat and Hadidi examples, such acquaintance is easy to posit; in the case of Late Bronze Alalakh it is more difficult but not impossible in view of the pottery parallels (Dornemann 1981) and those in the glyptic finds (Machule et al. 1987:124ff.). Contacts are also easier to visualise if the class of houseowners is taken into account. Sizes of the covered areas as well as the inventories of many of the examples - among which tablets and sealings are important - suggest relatively rich inhabitants with business dealings. These presumably widened their horizons beyond their own city proper. The relative paucity of the three types seems to preclude their interpretation a a reflection of a special, recurring mode of co-residence, although a secure pronouncement or hypothesis on the social position of the inhabitants is hampered by the very same paucity of examples. It seems clear, however, that all three types represent mainly domestic buildings, be they 'high class' ones. Such variables as upper storeys and possible number of inhabitants are extremely difficult to establish, although at Mumbaqat a few staircases were found; these may, however, only have led to the roof, not necessarily to a second floor. The average covered living floor space of type I is some 95 sqm, of type III 96 sqm (counting all rooms). If indeed an average family size of eight is too large for the Ancient Near East (Stager 1985:18, note 10), second storeys would not be necessary to house a nuclear family. Perhaps a careful analysis of the archival finds at Mumbaqat and Hadidi will later throw light on this problem.

The houses apparently were self-contained units without special arrangements such as pens for livestock or open yards. If present, animals were probably held in the side rooms, but these buildings are easy to visualise as town houses of the well-to-do, whose herds were kept for them by others in the countryside[16].

If types I, II and III are foremost a local phenomenon, their similarities due to direct acquaintance, radiation of the types to the west and south is at least a theoretical possibility; since type III is simpler than the other two, however, independent invention is relatively less improbable in its case.

CONCLUSION

In the above examples, form, location and function have been considered. On the basis of form alone, ties might be postulated between the Syrian Euphrates area and regions more to the west and south. However, uncertainties in both utilitarian and symbolic function (the latter mainly in the sense of reflecting a certain pattern of residentialism) make the postulate difficult to substantiate. The tripartite houses discussed above cannot be compared with the tripartite houses of the Ubaid and Uruk periods, except in form alone: both visual and cultural acquaintance are improbable in view of the 'void' of some 1500 years. The superficiality of form as a tool for analytical classification is obvious, and knowledge of function is indispensible. The latter must be obtained through careful description and analysis of all contents of all rooms.

NOTES

1. For one class of buildings we have at our disposal a number of text loci that bear on the spiritual or philosophical background of their appearance: temples figure prominently in building- and votive inscriptions, hymns etc. Although these references may touch on details of construction or on parts of the building in question, they most often concern the symbolic function of the institution as a whole rather than its real physical idiosyncrasies.
2. Roaf's remark (1984:88) about the easy distinction between Ubaid houses and temples may need to be qualified, certainly for the Uruk period. We have very few direct clues for religious functions of architecture (Tunca 1984:199). The house at CC X in Habuba South (Strommenger 1980, map inside back cover), for instance, is larger and has thicker walls than most similar buildings at the site, as well as a few niches; if the Qannas temples had not been found, it would certainly have been called a temple solely on the basis of its plan and relative size.
3. Our axiom is obviously that form must reflect function in a standardised way. That this may not always be the case is shown by the change during the Renaissance from the Latin cross church plan to the Greek cross and round church plans (Wittkower 1973:30-32).
4. Unfortunately, this cannot be automatically turned about. In Mesopotamia, temple plans tend to change, toward the end of the third millennium B.C., from long rooms with bent-axis approach (the long room being incorporated in an otherwise non- uniform complex) to an axial type where the broad cella is reached axially through a court. From the textual evidence, no correponding change in religion or liturgy can be documented. Thus different temple plans may reflect the 'same culture'.
5. Cf. Margueron 1982:575: 'Les palais sont tous différents les uns des autres, mais ils sont l'expression d'un mème programme général'.
6. Note, however, that the locational choices for each room added to an existing compound can be approached in terms of statistical probability (Dickens 1977).
7. In this respect an etymology related with biblical Aramaic *'ilan* 'tree' might be feasible, if the gap between the western semitic transscription of an early word *?ln* and the eastern semitic ḫ could be bridged by phonetic precedents.

8. Now called 'Mansion' in the final report (to be published shortly), this building was first presented as 'portico-building' and as 'south west fortress' (Van Loon 1968; 1979).
9. This would make a Hittite derivation of the term *hilāni* less probable (given the above suppositions), although it certainly does not disprove it. The Hittite root and the Syrian architectural concept might well have existed independently before they would have merged.
10. In this context, the excavator's grounds for an 'original' square plan of 16.25 x 16.25 m are unclear to me. If room 4 were later, what traces indicate a forerunner extending south in the first phase? (cf. Machule et al. 1986:101 and 104, note 60). It should be noted also that temple buildings in the Syrian Late Bronze Age tend to have excessively thick walls.
11. Beside the ones mentioned above, other constructional details of the four-roomed house have already been treated by Braemer (1982) and Stager (1985). In his discussion of the various types, Braemer convincingly suggests that earlier interpretations of the four-roomed house may have been wrong in several regards. He disputes Shiloh's (1970) ethnic association as well as the open character of the central largest space (1982:84-88; cf. Stager 1985:15; and Shiloh's unconvincing reply (1987)). Especially functional correspondences depending on formal ones, hinge on a careful consideration of the directions the traffic inside a building may take. Main entrance doorways play an important part in this, and when these are not located with certainty, comparisons lose some of their vigour. That many of the Iron Age four-roomed houses have the main entrance via the main long room may prove to be a meaningful difference with the Syrian examples from the Late Bronze Age.

REFERENCES

BITTEL K. AND NAUMANN R.
1938 *Boghazköy-Hattuscha II*. Berlin.

BRAEMER F.
1982 *L'architecture domestique du Levant à l'Âge du Fer*. Paris.

BUSINK TH.A.
1970 *Der Tempel von Jerusalem, I*. Leiden.

DICKENS P.
1977 'An Analysis of Historical House Plans: a Study at the Structural Level', in: D. Clarke (Ed.), *Spatial Archaeology*. London. Pp. 33-45.

DORNEMANN R.
1979 'Tell Hadidi: A Millennium of Bronze Age City Occupation', *Annual of the American Schools of Oriental Research* 44: 113-151.
1981 'The Late Bronze Age Pottery Tradition at Tell Hadidi, Syria', *Bulletin of the American Schools of Oriental Research* 241: 29-47.
1985 'Salvage Excavations at Tell Hadidi in the Euphrates River Valley', *Biblical Archaeologist* 48/1: 49-59.

EICHLER S. ET AL.
1984 'Ausgrabungen in Tall Munbāqa 1983', *Mitteilungen der deutschen Orientgesellschaft* 116: 65-93.

FRANKFORT H.
1952 'The Origin of the Bīt Ḫilāni', *Iraq* 14: 120-131.

FUGMANN E.
1958 *Hama. L'architecture des périodes préhellénistiques*. Copenhagen.

GÜTERBOCK H.
1975 'Hilammar', *Reallexikon der Assyriologie IV, 6/7*: 404f.

HALL E.
1969 *The Hidden Dimension*. Garden City: Anchor Books.

HROUDA B.
1975 'Bīt Ḫilāni'. *Reallexikon der Assyriologie IV,6/7*: 406ff.

KELM G. AND MAZAR A.
1982 'Three Seasons of Excavations at Tel Batash', *Bulletin of the American Schools of Oriental Research* 248: 1-36.

KENT S.
1984 *Analyzing Activity Areas. An Ethnoarchaeological Study of the Use of Space.* Albuquerque.

MACHULE D. ET AL.
1986 'Ausgrabungen in Tall Munbāqa 1984', *Mitteilungen der deutschen Orientgesellschaft* 118: 67-145.
1987 'Ausgrabungen in Tall Munbāqa 1985', *Mitteilungen der deutschen Orientgesellschaft* 119: 73-134.

MAISELS C.
1987 'Models of Social Evolution: Trajectories from the Neolithic to the State'. *Man N.S.* 22: 331-359.

MARGUERON J.
1980 'Emar: Un example d'implantation hittite en terre Syrienne', in: J. Margueron (Ed.) *Le Moyen Euphrate. Zone de contacts et d'échanges.* Actes du colloque de Strasbourg, 10-12 mars 1977. Leiden. Pp. 285-312.
1982 *Recherches sur les palais mésopotamiens de l'âge du bronze.* Paris.

MEIJER D.J.W.
1984 'Tell Selenkahiye, Syria. Architecture and Stratigraphy of the Excavations by the University of Amsterdam at Tell Selenkahiye on the Euphrates', Amsterdam. (part of dissertation, to be published in M. Van Loon (Ed.) *Excavations at Tell Selenkahiye*).

MCGOVERN P.
1983 'Test Sounding of Archaeological and Resistivity Survey. Results at Rujm al Henu', *Annual of the Department of Antiquities of Jordan* 27: 105-141.

NAUMANN R.
1971 *Architektur Kleinasiens*^{2}. Tübingen.

OREN E.
1985 'The Architecture of 'Egyptian Governor's Residencies' in Late Bronze Age Palestine', *Eretz Israel* 18: 183-199.

ORTHMANN W.
1966 'Tilmen Hüyük', *Arciv für Orientforschung 21: 165-166.*
1976 'Mumbaqat 1974 - ...', *Mitteilungen der deutschen Orientgesellschaft* 108: 25-44.
1981 *Halawa 1977-1979.* Bonn.

RAPOPORT A.
1984 *House Form and Culture.* Englewood Cliffs.

ROAF M.
1984 'Ubaid Houses and Temples', *Sumer* 43: 80-90.

SHILOH Y.
1970 'The Four-Room House', *Israel Exploration Journal* 20: 180-190.
1987 'The Casemate Wall...', *Bulletin of the American Schools of Oriental Research* 268: 3-15.

SINGER I.
1975 'Hittite Hilammar and Hieroglyphic Luwian *hilana', *Zeitschrift für Assyriologie* 65: 69-103.

STAGER L.
1985 'The Archaeology of the Family...', *Bulletin of the American Schools of Oriental Research* 260: 1-35.

STROMENGER E.
1980 *Habuba Kebira. Ein Stadt vor 5000 Jahren.* Mainz.

TUNCA Ö.
1984 *L'architecture religieuse protodynastique en Mésopotamie.* Leuven.

VAN LOON M.N.
1968 'First Results of the 1967 Excavations at Tell Selenkahiye', *Annales Archéologiques Arabes Syriennes* 18: 21-32.

1979 '1974 and 1975 Preliminary Results of the Excavations at Selenkahiye near Meskene, Syria', *Annual of the American Schools of Oriental Research* 44: 97-112.

WINTER I.

1982 'Art as Evidence for Interaction', in: H.J. Nissen and J. Renger (Eds.), *Mesopotamien und seine Nachbarn*. Berlin. Pp. 355-382.

WITTKOWER R.

1973 *Architectural Principles in the Age of Humanism*. London: Academy Editions.

WOOLLEY L.

1955 *Alalakh*. Oxford.

WRIGHT G.

1985 *Ancient Building in South Syria and Palestine*. Leiden.

The *Aussenstadt* settlement of Munbaqa, Syria

T. DE FEYTER

INTRODUCTION

Tell Munbaqa is a site on the left bank of the Euphrates in the Lake Assad area, and lies about 100 km north of Meskene/Emar and just south of Tell Hadidi/Azu. The site seems to have first been occupied during the EB IV period. The presence of a citadel-like structure, temples and a defense wall, indicates the importance of the town in the third millennium B.C. The occupation, however, was limited (in extension) to the stretch of land between the citadel and the Euphrates. During the transition from the MB to the LB period the expansion of the settlement at Munbaqa resulted in three areas: the *Innenstadt* and *Ibrahim's Garten*, both strengthened with a defense wall, and slightly later in time the *Aussenstadt*. Excavations at Munbaqa were carried out under the direction of Heinrich, Orthmann and Machule ((Heinrich 1969; 1970; 1971; 1974; Orthmann 1974; 1976; Machule 1978; 1986; 1987). Further reports have been published by Frank (1982) and Eichler (1984).

Whereas the EB IV settlement shows a very complicated stratification, the sequence of layers in the late MB and LB extensions is in general less complicated (3 to 4 phases). In the *Aussenstadt*, only one main building phase is present which can be subdivided in ASI-AS III. The two lower phases (AS I and AS II) are associated with architectural features whereas the third phase consists of trodden surfaces covering the larger part of the ruins of the ASI-II houses; a small section of these houses was reused. The ceramics found in the *Aussenstadt* are highly homogeneous. These are all indications of a short-lived settlement extension. During the last stage of settlement in the *Aussenstadt* (AS III), occupation was probably of a dispersed nature. Evidence in other excavated areas also points to a desintegration of the settlement in this period (Machule 1987: 101-2).

In the following the pottery of the *Aussenstadt* will be presented with an emphasis on form parallels. In the conclusions a short excursion will be made into the historical backgrounds to try to explain under what conditions and in which period Munbaqa could have reached such a sudden and short-lived flowering.

STRATIGRAPHY AND ARCHITECTURE

A south-north test trench was made from the supposed defense wall of the *Aussenstadt*, visible in the tell surface as a long dike, to investigate the stratigraphic connection of this

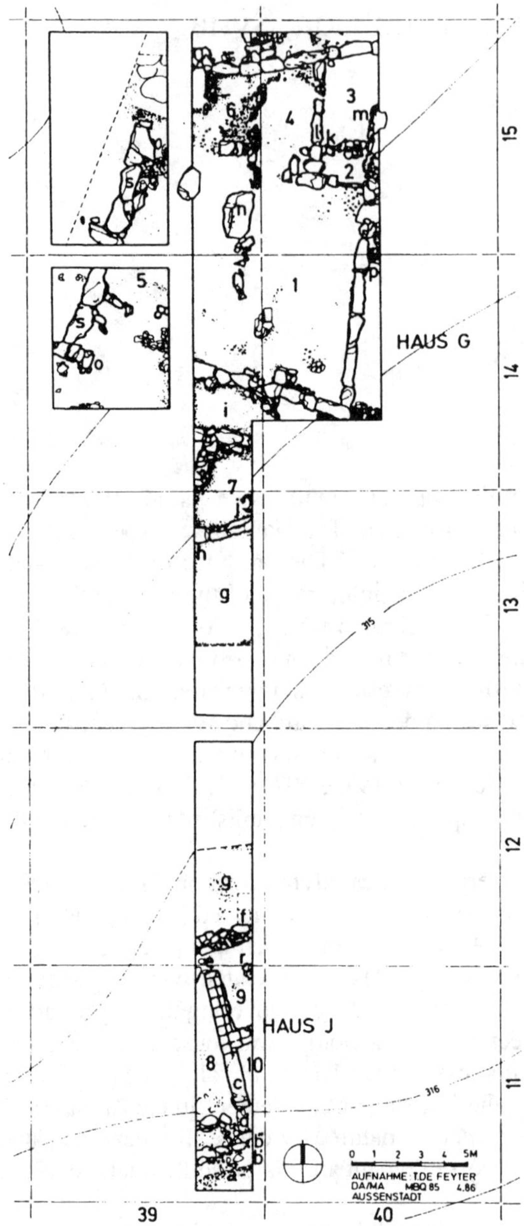

Fig.1 Plan of *Aussenstadt* Houses J and G
From: Machule 1987:93

supposed city wall with the settlement in the *Aussenstadt*. This trench extended from 39/11 to 39/13; later the trench was extended in 39/14, 39/15, 40/14 and 39/15 to a horizontal excavation to encompass the stone structures visible on the surface in those squares (see fig.1).

In 39/11, 39/12 and 39/15 virgin soil was reached at resp. 3.50 m., 1.50 m. and 0.50 m. below modern surface (Machule et al. 1987: 92 ff.). In 39/11 and 39/12 a few rooms of a structure (House J) have been exposed which can be connected to what might be the city wall: the southern doorpost of entrance d was built into the city wall. North of it were situated a courtyard (g), a room (7), an alley (i) and complex G. This complex may represent either one or two houses. The destruction of wall n (possibly the partition between the two houses), and of the south-west corner of the complex, where there may have been an entrance, makes it impossible to decide clearly on the existence of one or two houses. Walls were built with unworked or roughly worked stones neatly laid to form a base, sometimes as high as 1.00 m., for the mudbrick walls. In the sides of the dooropenings oblong stones were erected to serve as doorposts. Walls and floors were mud plastered, the latter on a base of stone rubble. Outer courts and alley were covered with rubble.

In the excavation of the *Aussenstadt* two phases (AS I and II) could be determined. AS I was founded on virgin soil. The AS I walls were used in AS II as well. A third phase (AS III) consisted of surfaces which covered the AS II ruins in 39/11 to 39/13. In this third phase a part of complex G was reused, another part may have been used as a stone quarry (for more details Machule et al. 1987:92 ff.).

In comparison to *Ibrahim's Garten* the architectural features of the *Aussenstadt* are rather simple. For instance, there is no representational room with benches and podium in complex G such as found in the houses of *Ibrahim's Garten* (Machule et al. 1986:99 ff.; 1987:101 ff.). A possible exception seems to be the more elaborate House F. This house had been constructed on top of the enclosure wall of the *Innenstadt*. This house was accessible through an entrance from the *Aussenstadt*, and it therefore belongs topographically to the latter area (Machule et al. 1987:80 ff.). On the basis of the 1985 architectural features the *Aussenstadt* seems to have been a quarter with simple households. This hypothesis is corroborated by the pottery and finds.

POTTERY

1. Procedures

The 1985 excavation in the *Aussenstadt* yielded 1300 diagnostic sherds. The sherds from topsoil layers and other disturbed contexts have been excluded from the analysis. The remaining 970 diagnostic sherds, retrieved from reliable contexts, have been subdivided into rim (638), base (214), handle (95) and lid fragments (23). In a next stage, the 638 rim sherds have been classified as jars, pots, bowls and large storage vessels. This classification is based on the ratio between the rim diameter and the estimated vessel height, according to the following scheme: jars h = 2d, pots h = d, bowls h = d. Goblets (h = d < 12 cm) have been neglected as a category, because only four rim sherds were found that fit into this category. Large storage vessels were separated on the basis of their obvious function. They have average rim diameters of 40 cm. The so called 'bath tubs' (storage jars with oval mouths) are also classified into this category.

The presentation of the *Aussenstadt* pottery is based on the rim sherd catalogue with eight objects added. First a general description of surface treatment, temper and decoration will be given. Subsequently, information about percentage, surface treatment and temper for each individual type group within these categories will be listed. Parallels in other contemporary assemblages complete the discussion of the defined rim types.

2. Surface treatment

Surface treatment is very uniform: 86.8 % of the rim sherds have a green-yellowish or yellow irregularly applied wash. 12.4 % of the pottery has no surface treatment (mainly cooking pots). A few rim sherds have a red or black slip, sometimes polished in irregular rings or strokes. The occurrence of a red polish is also rare at neighbouring sites in the Euphrates area like at Hadidi and Qitar (Dornemann 1979:141; Culican 1984:55; McClellan 1985:49). In assemblages from sites outside the Euphrates area, a polish on a red wash or slip, especially in use on bowls, seems to be more common. At Hammam et-Turkman on the Balikh and at Alalakh the polish on a red wash is described as a common feature (Smit, in press; Woolley 1955:319); in the Amuq plain this technique is described as 'an innovation for phase M' (Swift 1958:22).

3. Temper

Four types of temper can be discerned in the *Aussenstadt* ceramics: (1) A fine lime-sand temper with particles up to 0.5 mm. in diameter, (2) a medium temper with particles between 0.5 and 1.0 mm. in diameter, (3) a coarse temper with particles over 1.0 mm. in diameter, and (4) a mixture of lime, sand and calcite, mainly used for the cooking pot ware.

The fine and medium lime-sand temper are both more or less evenly represented among the rim sherds, resp. 44.0 % and 42.1 %. The coarse lime-sand temper and the cooking pot ware have percentages of resp. 5.8 % and 7.3 %. A shift can be noticed in texture, from predominantly fine texture for jars (average diameter 14 cm) through fine to medium texture for pots (average diameter 28 cm) to medium and relatively coarse texture for large storage vessels (average diameter 40 cm).

The lime-sand temper with calcite addition was used mainly for cooking pots; however, one bowl and one jar show this type of temper as well.

Plant temper is very rare, and was noticed only in four handmade rim fragments of the so called 'bath tubs'.

4. Decoration

Decoration mainly consists of comb-incised wavy and/or horizontal bands. Relief bands with incised oblique incisions (rope-patterned ridge), sometimes in combination with a comb-incised wavy band, represent another type of decoration. The base of one cooking pot was provided with the above described rope pattern in concentric circles (cf. Type II.2).

5. Rim types and pottery shapes

Table 1 presents the frequencies of the defined rim types per phase.

In order to detect a chronological order in the assemblage a seriation matrix was constructed on the basis of presence/absence of the defined rim types (Marquardt 1978:274). Several types appear to be earlier or later. The early types are illustrated in fig. 2:11, 12; fig. 5:4, 8; fig. 6:7, 14; and fig. 7:8, 10. The later types are illustrated in fig. 2:18; fig. 3:8; fig. 4:9, 10; fig. 5:3; fig. 6:8; and fig. 7:5.

There are some correlations with the stratified ceramics of Hammam at-Turkman. For example some of the early bowl rims (fig.7: 8, 10) are present in strata dated to MB II at that site (Curvers, in press, Pl. 126: 37-46, especially 44 and 45 of phase VII:5). The early horizontal rims of pots (fig.5:4) are present in MB II strata at Hammam et-Turkman (Curvers in press, Pl. 134: 123-124; Pl. 135: 125-134, especially no. 127 of phases VII:4 and 5). Ridges in the zone below the rim of large storage vessels (cf. fig.8:4) are a common feature in the MB assemblage at Hammam et-Turkman (Curvers, in press).

From the seriation matrix can be deduced that 48% of the rim types are present in two or more phases, 14% in phase II only and 38% of the types either in the first or te last phase. This seems to point to a chronological differentiation within the ceramic material. One can doubt, however, the archaeological implications of these statistical observations, as they are based on a presence/absence matrix, and not adduced from quantitative analysis. In the pottery sample used, early and late types consisted of one or a few sherds only. Based on sherd counts, therefore, the *Aussenstadt* material is much more homogeneous: 80 % of the sherds appear in two phases or more, 4 % in phase II and 16 % in the first and the last phase.

LIST OF RIM TYPES AND POTTERY FORMS

I. Jars

I.1 *Jars with straight or flaring necks and simple plain rims* (23.0 % of jars). The rim diameter vary from 15 to 20 cm. The surface treatment is a green-yellowish wash. Temper: predominantly fine lime-sand texture with only a few medium and coarse sherds (fig. 2:1, 3-5, 7, 8).

I.2 *Jars with straight or flaring necks and rolled or folded rims* (53.1 % of jars). Rim diameters vary from 10 to 15 cm. Temper: predominantly fine lime-sand temper. At about one quarter of the rims shows a medium lime-sand temper Surface treatment consists almost exclusively of a green-yellowish wash (fig. 2:2, 6, 9).

I.3 *Jars with straight necks and horizontally flattened or rounded rims* (8.0 % of jars). Rim diameters vary from 10 to 15 cm. Temper: fine and medium lime-sand temper. Surface treatment: green-yellowish wash (fig. 2:13, 14, 16, 17).

Types I.1-3 include 84.1 % of all jar rims. Rim type I.1 with the straight neck (fig. 2:1) could have belonged to complete shapes such as found in the so called *Topflager* in House F at Munbaqa (Heinrich 1970:78, fig.34:b). Parallels are to be found in Baghouz (de Mesnil du Buisson 1948, Pl. LXXIX:forme R, z 220), at Hadidi (Dornemann 1981, fig. 16:2-6) and at Alalakh (Woolley 1955, type 106a). In Palestine this rim-neck type occurs in the LB period as well, although with handles, e.g. at Lachish (Tufnell 1940, Pl. XLIX:253); see also Amiran (1970, Pl. 41:3).

Rim type I.1 with the flaring neck (fig. 2:5) is a typical form for the Khabur ware pottery, for instance at Chagar Bazar (Mallowan 1936, fig.17:4, 5, 6, 9) and at Baghouz (de Mesnil du Buisson 1948, Pl. LXXIX:forme R); see also Hamlin (1971:71 and fig.6:jartype 1). Also at Hammam et-Turkman (Curvers, in press), at Hadidi (Dornemann 1979, fig.23:4, 46) and at Alalakh (Woolley 1955, type 97a) this type occurs in the

Fig.2 Jars. Scale 1:4.7, or as indicated. Nrs.1, 3-5, 7, 8 (I.1); nrs.2, 6, 9 (I.2); nrs.13, 14, 16, 17 (I.3); nrs.15, 18 (I.4); nrs.10-12 (I.5).

assemblages dated to the MB period. This shape, however, also occurs in the LB assemblages of Hadidi (Dornemann 1979, fig. 20:31) and in the Hamrin area at Tell Zubeidi (Boehmer and Dämmer 1985, Pl. 117:216). A similar occurrence in MB and LB assemblages is true for the small jars (fig. 2:3). They have parallels at Hadidi (Dornemann 1981, fig.4:6), at Qitar (McClellan 1986, fig.10:5). McClellan considers the goblets of 'pottery groups B-D', dated to the MB IIB period, as 'an earlier form of the small jars found in Building 15' (McClellan 1986:14, fig. 8:1-3). Other parallels can be found in the Amuq plain (Swift 1958, fig. 14), at Alalakh (Woolley 1955, type 91b) and at Tell Rifa'at

Table 1. Frequencies of categories per *Aussenstadt* phase

count % row % column % tot. %	AS I	AS II	AS III	total
bowls	62	73	72	207
	29.9	35.3	34.8	
	31.1	31.3	34.9	
	9.7	11.4	11.3	32.4
pots	77	93	75	245
	31.4	37.9	30.6	
	38.7	39.9	36.4	
	12.1	14.6	11.7	38.4
jars	32	43	38	113
	28.3	38.0	33.6	
	16.1	18.4	18.4	
	5.0	6.7	5.9	17.6
storage jars	28	24	21	73
	38.3	32.9	28.8	
	14.1	10.3	10.2	
	4.4	3.8	3.3	11.5
total	119	233	206	638
total %	31.2	36.5	32.2	

(Matthers 1981, fig. 220:5). This type of small jar occurs in other areas of Munbaqa as well, for instance in the *Innenstadt* (House F) and in *Ibrahim's Garten* (resp. Machule 1987:88, fig.9:2; 1986, fig. 27:3, 4).

Jars with rim types of group I.2 (the rolled out type) and I.3 (fig. 2:2, 6, 9, 13, 14, 16, 17) are very common in the LB period. They already occur, however, at the end of MB (Dornemann 1979:139, fig. 23:44; 1981, fig.15:13, 15, 18, 19). J.-W. Meyer (*apud* Orthman 1981:27) considers this jar to be a characteristic of the MB-LB transition at Halawa. Other parallels are to be found at Qitar (McClellan 1986, fig.10:7), at Alalakh (Woolley 1955, type 103a), and at Ugarit (Schaeffer 1949, fig. 6:6). In this context, I would like to stress that the actual evidence is opposed to Dornemann's observation that the occurrence of these forms in the Lake Assad area during the LB I period is a sign of influence from the coastal region and Palestine. Important parallels are already to be found in the Khabur ware pottery at Chagar Bazar (Mallowan 1936, fig.16:5; 1937, fig.23:10); see also Hamlin (1971:71, fig.6:jartype 11 and rim-neck shape E).

In general the jars with straight or flaring necks are considered by Gates to be the result of influence from inland North Syria, especially the Khabur region, during Alalakh VI and V (Gates 1976:189).

I.4 *Jars with outward-bent rims* (6.1 % of jars). Rim diameters circle around 12 cm. Fine and medium lime-sand temper and green-yellowish wash (fig. 2:15, 18).

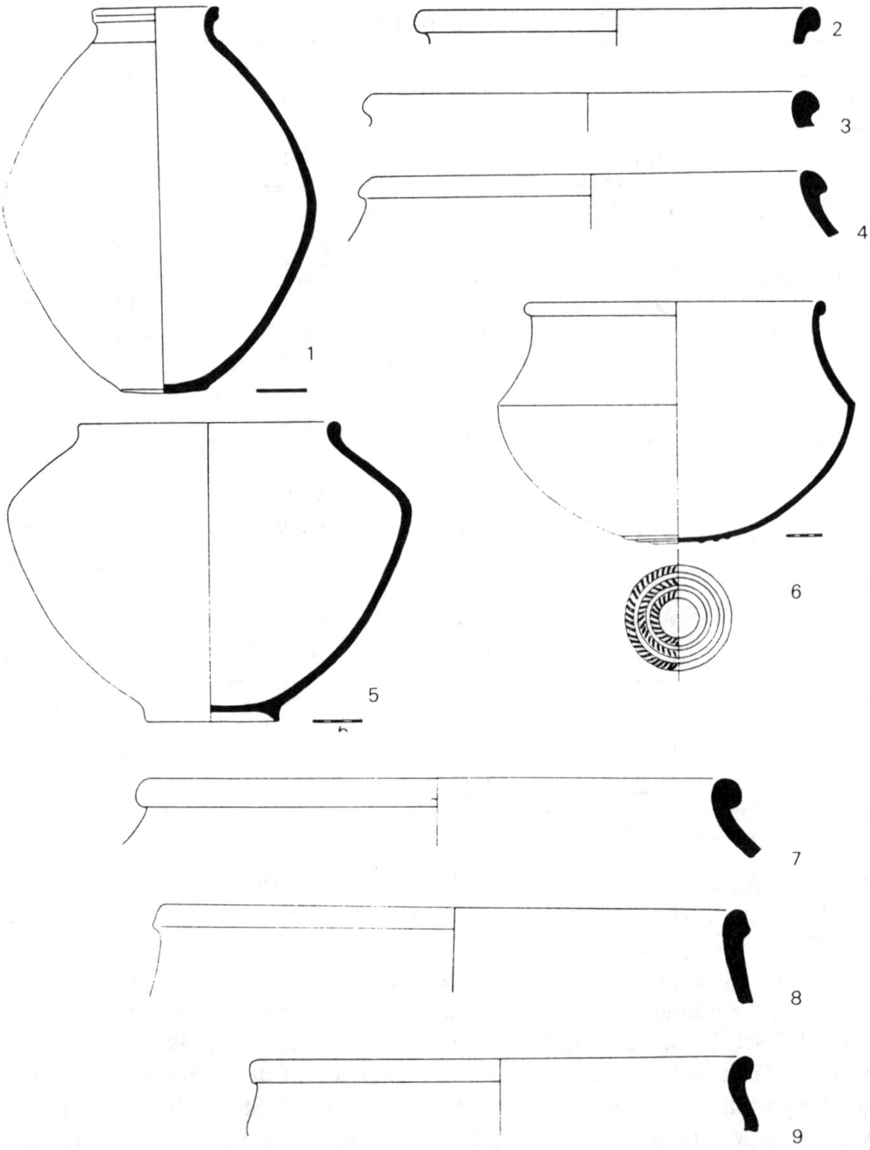

Fig.3 Jar (nr.1:1.5) and pots. Scale 1:4.7, or as indicated. Nrs.2-4, 7 (II.2); nr.5 (II.1); nrs.8, 9 (II.4).

I.5 *Jars with folded, grooved rims* (6.1 % of jars). Rim diameters vary from 13 to 18 cm. Fine lime-sand temper and green-yellowish wash (fig. 2:10)12).

Jars with folded and sometimes grooved rims (i.e. without the straight or flaring necks) were found at Munbaqa in other squares as well (Frank 1982:11, 36, figs. 1-4, figs. 20-28). This assemblage is related to the older phases of Hama G (Frank 1982:47).

A parallel for the small storage jar with folded, grooved rim (fig.3:1) can be found at Hammam et-Turkman MB II (Curvers, in press).

II. Pots

II.1 *Pots with simple plain rims* (1.2 % of pots; not illustrated). This group consists of a few sherds only. Rim diameters average around 26 cm.; green-yellowish wash and medium lime-sand temper are typical. Cooking pot ware is represented by one sherd. A pot with angled body profile, ring base, no surface treatment and coarse lime-sand temper has a rim of this type (fig. 3:5). For general parallels of this pottery see below (II.5).

II.2 *Pots with outward rolled rims* (36.8 % of pots). The majority of rims belongs to cooking pots (calcite temper, no surface treatment, black or grey-brown surface colour). Rim diameters vary from 30 to 40 cm. The other rims in this group have predominantly fine lime-sand temper with a green-yellowish wash and rim diameters from 20 to 30 cm. (fig. 3:2-4, 7)

One specimen of a cooking pot, which is usually very brittle, could be reconstructed (fig. 3:6). It had a carinated body, the upper part having a concave, the lower part a convex profile. The base is round which is a common feature for cooking pots, and it is decorated with three applied relief bands with oblique incised lines. Cooking pots of this form appear in MB contexts, and continue into LB. General parallels are to be found at neighbouring sites such as Hadidi, and they also occur at Alalakh and in Palestine. At Alalakh this type was found from phase XVI till phase I, but predominantly in phase IV (Woolley 1955:133, 339, type 153; Gates 1976:73). Parallels in Palestine are listed by Amiran (1970:135). She considers the outward-rolled rim as an LB innovation. These rims, however, already occur at Hadidi in MB II contexts (Dornemann 1979, fig.21:19, 23; fig.22:35, 39; fig.23:30, 31; 1981, fig.5:4; fig.10:3). Similar cooking pots were found at Kamid el-Loz in MB IIB to Iron I contexts, especially in the LB layers of IF 13 (Hachman 1970, Pl.6:3; 1980:105 and Pl.35:2, 5, 6, 8, 9). The custom to decorate the bases of these pots with applied bands in concentric circles and/or cross patterns is attested at Kamid el-Loz LB I and II (Hachman 1980, Pl.6:3), at Lachish LB I (Tufnell 1940, Pl.LV:357,358), and at Megiddo X (Loud 1948, Pl.46). Amiran suggests that the applied band pattern on bases could have been intended more for strengthening of the base than for decoration (Amiran 1970:135).

II.3 *Pots with inward rolled rims* (1.2 % of pots). A very small group of three sherds. The average rim diameter is 36 cm. Fine lime-sand temper and green-yellowish wash. (fig.4:1, 2)

II.4 *Pots with folded rims* (31.0 % of pots). Rim diameters circle around 30 cm. Fine and medium lime-sand temper and green-yellowish are typical. (fig.3:8, 9 and fig.4:3-5, 7, 9)

The pottery shapes to which this type of rim is attached are hard to determine. One reason to ascribe them to the category of pots is the large average diameter of 30 cm. A possible candidate would be the pot with angled body profile and ring base (fig.3:5; Type II.1 and fig.4:8; Type II.5). Some of the rim types in this group, however, occur on jars as well, i.e with smaller diameters (cf. fig.4:5, 7, 9).

II.5 *Pots with folded, grooved rims* (4.0 % of pots). A small group with the same characteristics as described under II.4. (fig.4:6, 10)

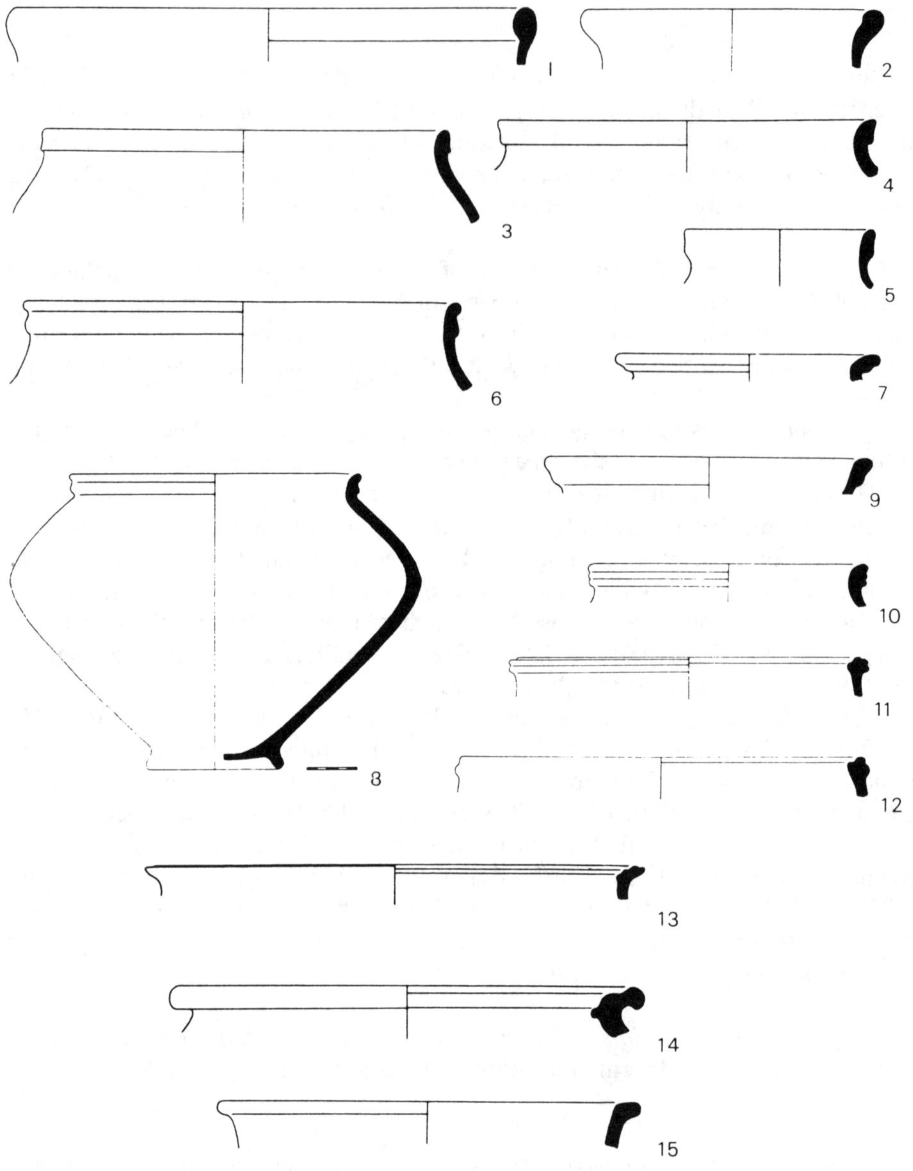

Fig.4 Pots. Scale 1:4.7, or as indicated. Nrs.1, 2 (II.3); nrs.3-5, 7, 9 (II.4); nrs.6, 8, 10 (II.5); nrs.11-14 (II.7); nr.15 (II.6).

A pot with angled body profile and ring base has a rim which belongs to this group (fig.4:8). This shape is very similar to the pot illustrated in fig.3:5 (Type II.1). Both pots were found in room 1 of House G. They are also attested in House F (Heinrich 1970, fig.34:a; Machule 1987:87 and fig.7:4). Parallels for this type of pot can be found at Hadidi in MB II context (Dornemann 1979, fig.23:3) and in LB IB context (ibid.1981, fig.7:2), at Baghouz in the MB IIA or B period (de Mesnil du Buisson 1948,

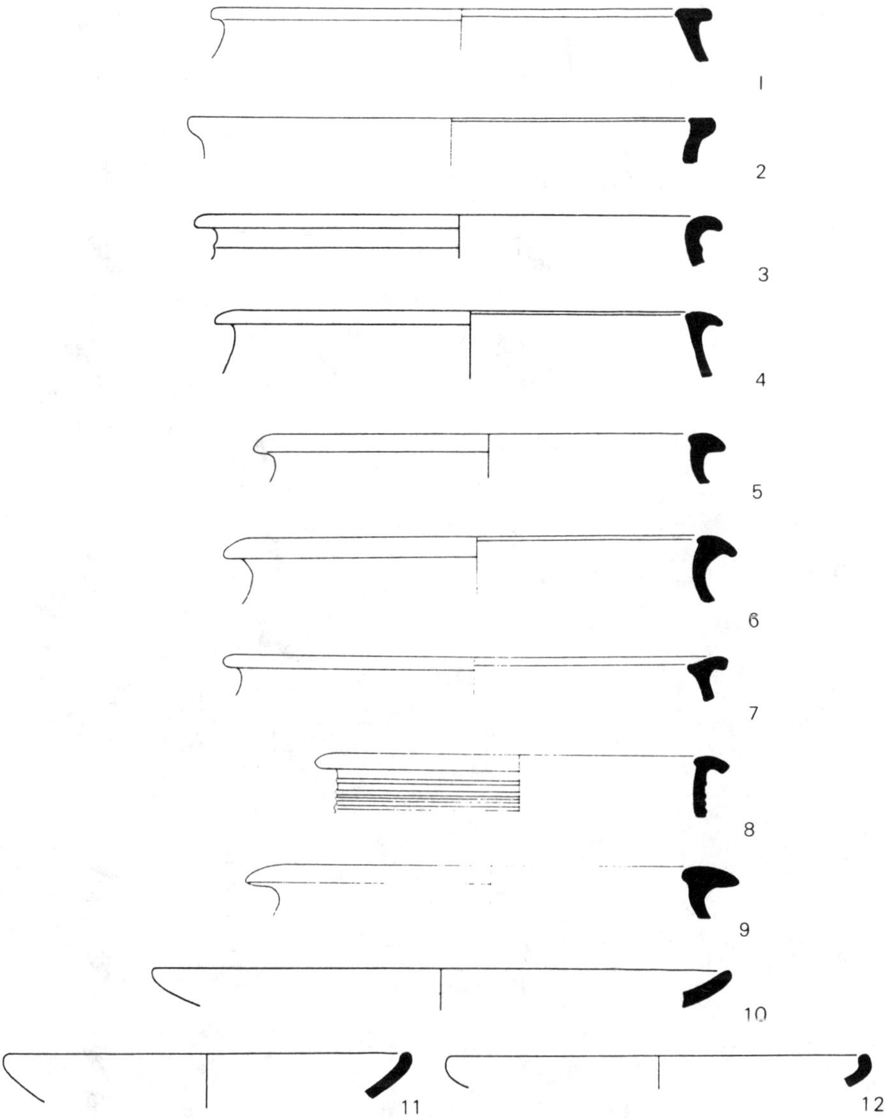

Fig.5 Pots and bowls. Scale 1:4.7. Nrs.1, 2 (II.6); nrs.3-9 (II.8); nrs.10-12 (III.1).

Pl.LXXVI:forme G, z 129), at Alalakh in the levels VIII, V and IV (Woolley 1955, type 26a), and at Lachish in LB IIb context (Tufnell 1940, Pl.XLII:149). At Chagar Bazar level 1, dated ca. 1450 BC., a miniature of this pot-type was found in grave 204 (Mallowan 1947, Pl.LXV).

II.6 *Pots with flat rims* (4.0 % of pots). The average diameter is 30 cm. Fine and medium lime-sand temper and green-yellowish wash. A few sherds have a red slip. (fig.4:15 and fig.5:1, 2)

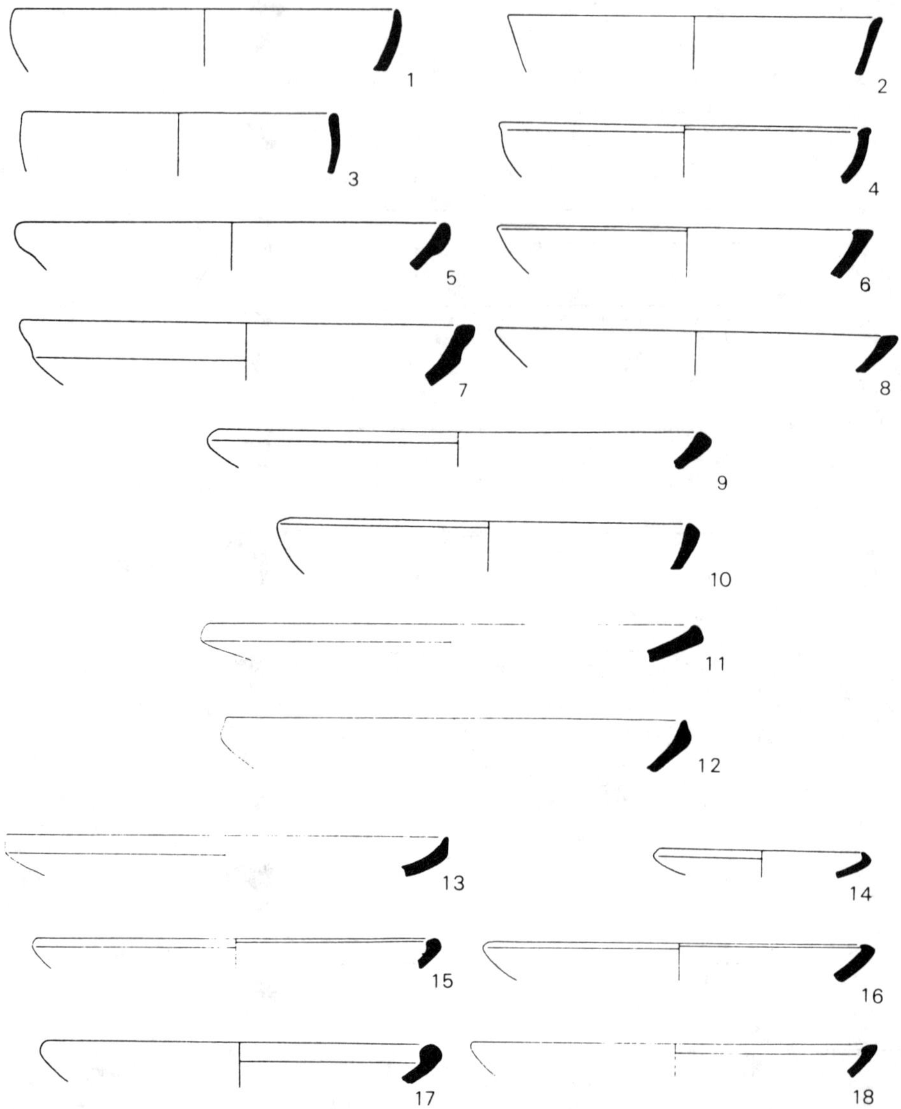

Fig.6 Bowls. Scale 1:4.7. Nrs.1-3 (III.2); nrs.4, 5 (III.3); nrs.6-8 (III.4); nrs.9-13 (III.5); nrs.14, 16 (III.7); nrs.15, 17, 18 (III.8).

II.7 *Pots with ribbed rims* (2.4 % of pots). Characteristics as described under II.6, but no red slip. (fig.4:11-14)

II.8 *Pots with horizontally rounded rims* (19.2 % of pots). Rim diameters vary from 26 to 35 cm. Green-yellowish wash and fine and medium lime-sand temper are typical. (fig.5:3-9). Some of these types could possibly be reconstructed as jars, if one looks at the angle of the rim towards the body fragment.

III. Bowls

III.1 *Shallow bowls with simple, plain rims* (9.7 % of bowls). Rim diameters vary from 29 to 33 cm. Surface treatment is the green-yellowish wash. Fine lime-sand temper dominates. About one quarter of the sherds have medium texture. (fig.5:10-12)

III.2 *Deep bowls with simple, plain rims* (13.5 % of bowls). Rim diameters vary from 21 to 30 cm. Green-yellowish wash and fine and medium lime-sand temper are typical. (fig.6:1-3)

III.3 *Bowls with thickened rims* (9.7 % of bowls). Characteristics as described under III.2. (fig.6:4, 5).

III.4 *Bowls with flat rims*. (5.3 % of bowls). Characteristics as described under III.2. The average diameter is 30 cm. (fig.6:6-8)

III.5 *Bowls with beveled rims*. (18.4 % of bowls). Characteristics as described under III.4. (fig.6:9)13)

III.6 *Bowls with outward-bent rims*. A very small group of only two sherds. Characteristics as described under III.4 (not ill.)

III.7 *Bowls with inward-bent rims* (5.3 % of bowls). Characteristics as described under III.2. Rim diameters vary from 14 to 36 cm. (fig.6:14, 16)

III.8 *Bowls with inward-rolled rims* (27.5 % of bowls). Characteristics as described under III.4. (fig.6:15, 17, 18 and fig.7:1, 2) The rims illustrated in fig.6:15, 18 and fig.7:2 are most characteristic; they make out 70 % of this rim type.

A black polished tripod bowl has a rim type of this group (fig.7:4).

Bowls with inward-rolled rims commonly occur in the ceramic assemblages of the LB period. Their first appearance has already been attested in assemblages dated to the end of the MB period. This common shape has a wide distribution in Anatolia, the Lebanon, Palestine as well as east of the Euphrates. The majority of the parallels can be found in the south and the west, for instance at Kamid el-Loz in the beginning of LB I (Frisch 1985, Pl.23:5), at Gezer MB IIB-C (Dever 1970, Pl. 31:19; 1974, Pl.18, 19), at Ugarit Moyen 2-3 (Schaeffer 1949, fig.105:17, 19, 27), Moyen 2 (Courtois 1978, fig.3:6, 7) and during the transition Moyen to Récent 1 (ibid.1978, fig.7:9; fig.22:6), at Tell Rifa'at during the transition of MB to LB (Matthers 1981, fig.220:13; fig.223:6; fig.225:6), at Hama H1 (Fugman 1958, fig.127:4B178, 2C908), and at Alalakh IX till V (Woolley 1955, types 2a, 3a and 5). In other areas this rim type is less common in the late MB and LB assemblages, for instance at Korucutepe during LB I (Van Loon 1980, Pl.15:J) and during the transition MB to LB (ibid. 1980, Pl.16:A), at Hammam et-Turkman phase VIIC (end of MB) (Curvers, in press) and phase VIIIA (LB I) (Smit, in press), at Mari from an uncertain context, but probably MB II (Parrot 1958-59:129 and fig.89), at Baghouz MB IIA-B (de Mesnil du Buisson 1948, Pl.LXXVI:forme T) and in the Hamrin region at Tell Zubeidi in layers dated from the end of the 14th cent. till 1225 BC. (Boehmer and Dämmer 1985, Pl.105:23). Important parallels occur at sites such as Hadidi in MB (Dornemann 1979,

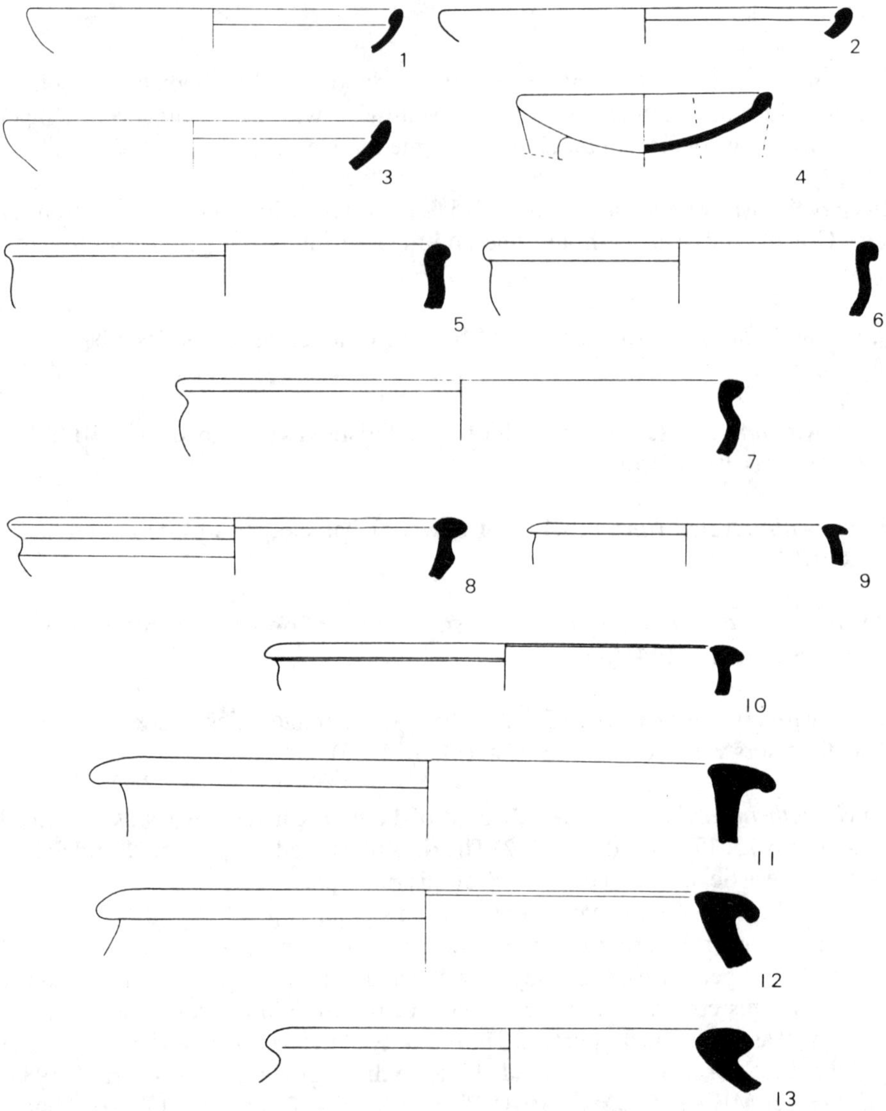

Fig.7 Bowls and big storage jars. Scale 1:4.7. Nrs.1, 2, 4 (III.8); nr.3 (III.9); nrs.5-7 (III.10); nrs.8-10 (III.11); nrs.11-13 (IV).

fig.22:13) and in LB IA context (ibid.1979, fig.19:13; fig.20:16-18; 1981, fig.13:30, 32) and at Qitar LB IA (McClellan 1985, fig.5:1-8).

III.9 *Bowls with inward-folded rims* (2.4 % of bowls). Characteristics as described under III.2 with rim diameters around 27 cm. (fig.7:3)

III.10 *Deep bowls with S-profiles* (3.9 % of bowls). Characteristics as described under III.2. Rim diameters vary from 25 to 32 cm. (fig.7:5-7)

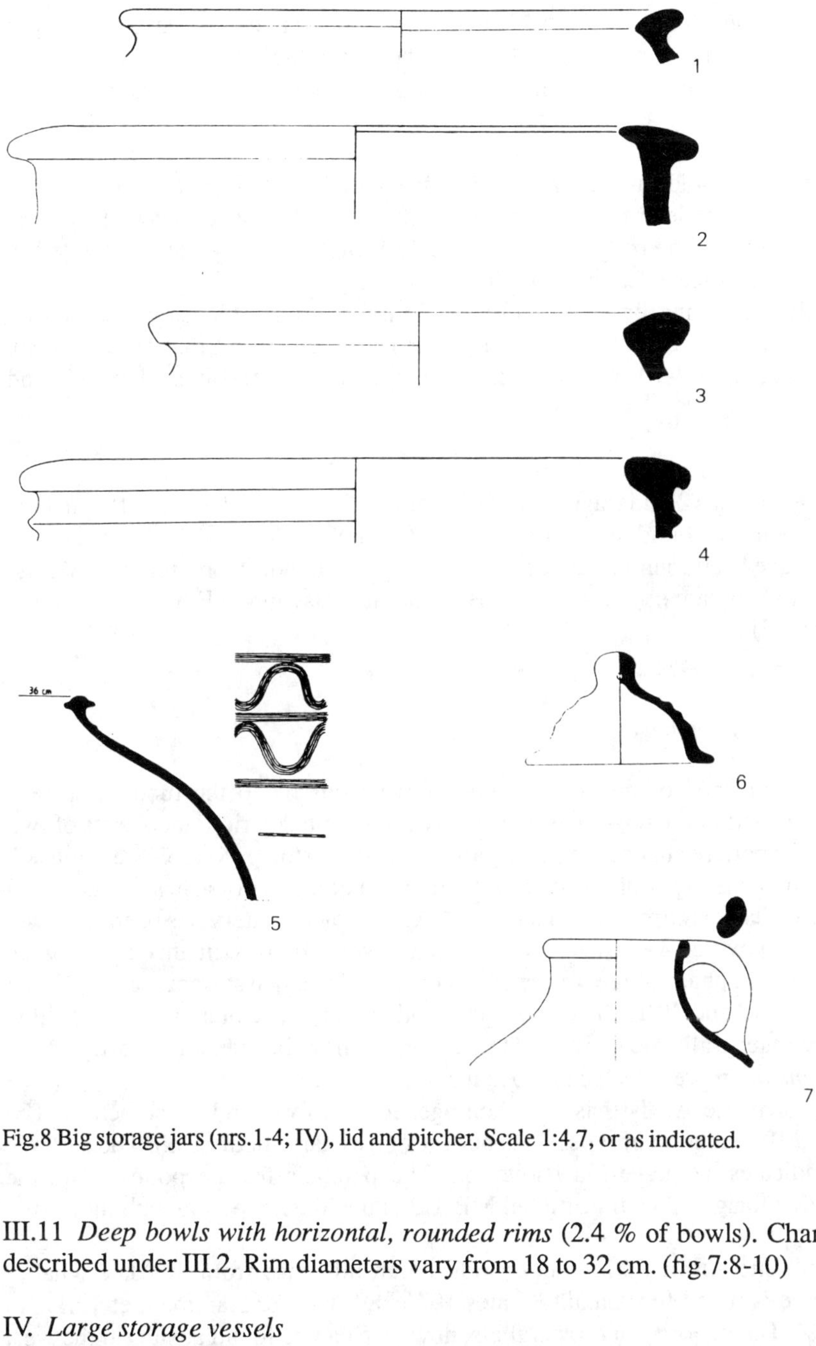

Fig.8 Big storage jars (nrs.1-4; IV), lid and pitcher. Scale 1:4.7, or as indicated.

III.11 *Deep bowls with horizontal, rounded rims* (2.4 % of bowls). Characteristics as described under III.2. Rim diameters vary from 18 to 32 cm. (fig.7:8-10)

IV. *Large storage vessels*
The average rim diameter is 40 cm. The surface treatment consists of a green-yellowish wash. About half of the sherds has the medium lime-sand temper. The other half is evenly divided between fine and coarse texture. (fig.7:11-13 and fig.8:1-4). The large storage vessels have sagging bases. The comb-incised wavy and horizontal band decoration commonly occurs on these vessels (fig.8:5).

The great variation of rim types does not allow a reasonable typology: many type groups would consist of one or two sherds only. Apparently the shape of the rim did not become standardized in the course of manufacturing of the larger storage vessels.

V. *Handles*

The total number of handles in the *Aussenstadt* assemblage is 95. They probably belong to pitchers, as could be determined from those fragments that were attached to rims. Many of the jar rims with straight necks (Type I.2) might actually have belonged to pitchers. A few incomplete specimen were found (fig.8:7).

Pitchers with flat or ring bases are common in the LB assemblages of the coastal region, in the Lebanon and in Palestine. They occur in Hadidi and Emar as well (resp. Dornemann 1981, fig.4:7-10, fig.9:9, 12; 1979, fig.24:2, 3, 5; Beyer 1982:80 and fig.20).

VI. *Lids*

The assemblage includes 23 lid fragments. The average diameter of the lids is 10 cm. One complete specimen was found in room 3 of House G (fig.8:3).

Similar lids have been found in the LB I assemblage at Hadidi (Dornemann 1981, fig. 9:10, fig.6:4), at Hammam et-Turkman VIIIB (Smit, in press) and in Emar LB II (Beyer 1982:83 and fig.35).

CONCLUSIONS

The very short sequence of the *Aussenstadt* from virgin soil to the modern surface indicates that this part of the town was in use only for a limited period. It consists of two phases without important changes in plan; some of the large storage vessels were reused. After phase II there are signs of destruction in the form of a layer of ashes and charcoal that covers the walls of House J and fills room 9. All complete pottery has been retrieved from phase AS II which seems to indicate that several houses in this phase were abandoned. The third phase of the *Aussenstadt* consists of trodden surfaces, covering the ruins in 39/11, 39/12 and 39/13. These surfaces without any trace of architecture follow the slope of the outer wall embankment. House G might have been in reuse partly. Phase III in the *Aussenstadt* represents 'scatter occupation'.

The pottery of phase AS I-III is very homogeneous in form and manufacture. The chronological differentiation of types does not warrant a sharp subdivision. The ceramic assemblage indicates a one-period habitation. The parallels for the pottery date the *Aussenstadt* assemblage in the transitional MB-LB and in the LB period without further specification.

The jars with straight or flaring necks indicate an influence from the 'late Khabur ware'. The same is noted for Alalakh (Gates 1976:189) and for Hammam et-Turkman (Smit, in press). The majority of the parallels, however, can be found at sites in the west and south. The characteristic shallow bowls with inward-rolled rims commonly occur at sites in Palestine, the Lebanon and West Syria. The closest parallels, however, occur in the MB II and LB assemblages of Hadidi and Qitar. The parallels for the pitchers and cooking pots show a similar geographical and chronological distribution; especially the bases with rope pattern.

The other finds from Munbaqa corroborate the orientation to the west and south, e.g. the scripture of the tablets found in House B second phase of *Ibrahim's Garten* (Mayer 1986:128-29) and the cylinder seals and bullae, among others in the 'Mitannian elaborate style', but with actual parallels in the coastal region (Karstens 1987:128-30). Furthermore the molded figurines have parallels in West Syria and the Lebanon (Machule 1987:99-100 and fig.14).

Munbaqa has strong similarities with neighbouring sites like Hadidi and Qitar, not only in material culture, but in settlement development and even in administration as well. The occupational history of Hadidi goes back to the EB IV period, after which a drastic contraction in size occurred. During the MB-LB transition and in the LB period Hadidi, like Munbaqa, experienced an explosive growth (Dornemann 1984:11). At both sites the textual evidence seems to point to an administration held by a group of men called the LÚ.MEŠ. aḫ-ḫe, 'the Brothers' (Dornemann 1979:146; Mayer 1986:129)

All evidence points to a date for the Munbaqa town extensions within the second half of the 16th and the first half of the 15th cent. BC. *Ibrahim's Garten* second phase is dated by Mayer from 1531 till 1458. In his hypothesis IG II was destroyed by Thoetmosis III in 1458. The proposed ancient name for Munbaqa, Uru, occurs in the list of destroyed towns in the account of the 8th campaign of this pharao. Counting back three generations, mentioned in the Munbaqa tablets (approximately 60-80 years), Mayer concludes that IG II might have been destroyed during the raid of Mursilis I in 1531, and thereby adheres to the low Mesopotamian chronology (Mayer 1986:126-31). I cannot go into the question of chronology here. A review of the matter, relating especially to the Kingdom of Yamhad and Alalakh, is given by Gates (1976:268 ff.). She herself defends the low chronology. Arguments against the low chronology (and in favor of the middle chronology: Mursilis' raid in 1595), based on the Alalakh stratigraphy and architecture are provided by Kempinski (1973:212 ff.) and by Na'aman (1974:267 and n.10; 1976:140 ff. ; 1979:103-13). The last article contains a discussion of Collon's view of the chronology of Alalakh VII (Collon 1975; 1977).

Most scholars assume that Aleppo resumed an independant status after the destruction by Mussilis, which put an end to the Kingdom of Yamhad of which Aleppo was the capital (a.o. Klengel 1965:175, 219; Astour 1969:381; Na'aman 1974:267). In the so-called *Idrimi* text there are allusions to a continuing dynasty at Aleppo from the 17th century on. This dynasty probably held the same territory, which extended to the Euphrates in the east, until the days of Idrimi's father, Ilimilimma. A new edition of the text has been published by Dietrich and Loretz (1981:201 ff., see esp. p.250). See also Klengel (1981:277 and n.54) and Na'aman (1974:266).

Yet the *Idrimi* text not only alludes to a continuation of the Aleppo kingdom, but to important changes as well. The Hurrian king Barattarna put an end to the independant kingdom, and subsequently Idrimi fled, first to the related dynasty of Emar, to land ultimately in Alalakh. In founding his kingdom there he recognized the suzerainty of the Hurrian king. Aleppo does not appear in the list of towns belonging to his kingdom, whereas the names of the other towns and regions are all interpreted as situated to the west of Aleppo. So by this time Aleppo and the region east of it were probably under Hurrian control. (On this problem see Na'aman (1974:268 and n.18) and Klengel (1981:276 ff.). Klengel stresses the unclear relation between Barattarna, the adversary of Idrimi, and the Mitanni dynasty.)

The destruction of IG II and relating levels in *Innenstadt* and *Aussenstadt* could be interpreted as caused by the struggle of a Hurrian king to gain control over the Euphrates region, which had strong ties with the west, and not necessarily as being caused by Mursilis, as Mayer says (cf. Mayer 1986:130-31). The *Idrimi* text provides evidence for such struggles in the region to which Munbaqa belongs. If one interpretes this first destruction as a sign of the expanding power of the Hurrians, then one can at the same time explain the sudden economic flowering of Munbaqa and nearby towns, which preceded the destruction, as a result of the location of these sites exactly between these emerging new powers in northeastern Syria on the one hand, and the old kingdom of Aleppo and the coastal region in the west and south on the other. Before the confrontations mentioned in the *Idrimi* text the Euphrates may have functioned as the border line between these spheres of political influences; it would not be the first time and surely not the last that the Middle Euphrates functioned as such.

Archaeologically one could date the beginning of the Munbaqa town extensions very well in this period, the second half of the 16th century B.C., in which the site became of new interest as a crossing place and harbour on the Euphrates.

BIBLIOGRAPHY

AMIRAN R.
1969 *Ancient Pottery of the Holy Land*. Jerusalem.

ASTOUR M.C.
1969 'The partition of the confederacy of Mukish-Nukhashshe-Nii by Shuppiluliuma. A study in political geography of the Amarna Age' *Orientalia (n.s.)* 38:381-414

BADRE L.
1980 *Les figurines anthropomorphes en terre cuite à l'Age du Bronze en Syrie*. Institut français d'archéologie du Proche Orient, Beyrouth, Damas, Amman. Bibliothèque archéologique et historique, 103. Paris.

BEYER D. (ED.)
1982 *Meskene-Emar, dix ans de travaux (1972-1982)*. Paris.

BOEHMER R.M. AND DÄMMER H.-W.
1985 *Tell Imlihiye, Tell Zubeidi, Tell Abbas*. Hamrin report 13. Mainz am Rhein.

COLLON D.
1975 *The Seal Impressions from Tell Atchana/Alalakh*. Alter Orient und Altes Testament 27. Kevelaer.
1977 'A new look at the chronology of Alalakh level VII: a rejoinder', *Anat.St.* 27: 127-31.

COURTOIS J.-C.
1978 'Corpus Céramique de Ras Shamra-Ugarit: niveaux historiques d'Ugarit', *Ugaritica* 7: 191-370. Paris.

CULICAN W. AND MCCLELLAN T.L.
1983-84 'El-Qitar: First season of excavations, 1982-83', *Abr-Nahrain* XXII: 29-63.

CURVERS H.H.
in press 'The period VII pottery', in: M. N. van Loon (ed.) *Hammam et-Turkman I, Interim report on the University of Amsterdam's 1981-84 excavations in Syria*. Istanbul.

DE MESNIL DU BUISSON
1948 *Baghouz: l'ancienne Corsote*. Leiden.

DEVER W.G. (ED.)
1974 *Gezer II*. Jerusalem.

DIETRICH M. AND LORETZ O.
1981 'Die Inschrift der Statue des Königs Idrimi von Alalakh', *UF* 13: 201-68.

DORNEMANN R.H.
1979 'Tell Hadidi: A Millennium of Bronze Age Occupation', *AASOR* 44: 113-151.

1981 'The Late Bronze Age Pottery Tradition', *BASOR* 241: 29-47.

1984 *Tell Hadidi: One Bronze Age Site Among Many in the Tabqa Dam Salvage Area*. Unpublished paper.

EICHLER S. ET AL.

1984 'Ausgrabungen in Tall Munbaqa 1983', *MDOG* 116: 65-93.

FRANK D.R. ET AL.

1982 'Tall Munbaqa 1979', *MDOG* 114: 7-71.

FRISCH B. ET AL.

1985 *Kamid el-Loz 6. Die Werkstätten der Spätbronzezeitlichen Paläste*. Bonn.

FUGMANN E.

1958 *Hama, Fouilles et Recherches 1931-1938*, II,1. Copenhagen.

GATES M.H.C.

1976 *Tell Atchana, levels VI and V: A reexamination of a mid-second millennium B.C. Syrian site*. University Microfilms. Ann Arbor.

HACHMANN R.

1970 *Kamid el-Loz 1968-70*. Bonn.

1980 *Bericht über die Ergebnisse der Ausgrabungen in Kamid el-Loz in den Jahren 1968 bis 1970*. Bonn.

HAMLIN C.

1971 *The Khabur ware ceramic assemblage of Northern Mesopotamia. Analysis of its distribution*. Ann Arbor.

HEINRICH E. ET AL.

1969 'Bericht über die von der Deutschen Orient- Gesellschaft mit Mitteln der Stiftung Volkswagenwerk in Euphrattal bei Aleppo begonnenen archäologischen Untersuchungen erstattet von Mitgliedern der Expedition', *MDOG* 101: 27-67.

1970 'Zweiter vorläufiger Bericht über die von der Deutschen Orient-Gesellschaft mit Mittel der Stiftung Volkswagenwerk in Habuba Kebira und in Mumbaqat unternommenen archäologischen Untersuchungen (Herbstkampagne 1969), erstattet von Mitgliedern der Mission', *MDOG* 102: 27-85.

1971 'Dritter vorläufiger Bericht über den von der Deutschen Orient-Gesellschaft mit Mitteln der Stiftung Volkswagenwerk in Habuba Kebira und in Mumbaqat unternommenen archäologischen Untersuchungen (Herbstkampagne 1970), erstattet von Mitgliedern der Mission', *MDOG* 103: 5-58.

1974 'Vierter Vorläufiger Bericht über die von der Deutschen Orient-Gesellschaft mit Mitteln der Stiftung Volkswagenwerk in Habuba Kebira (Habuba Kabira, Herbstkampagnen 1971 und 1972 sowie Testgrabung 1973) und in Mumbaqat (Tall Munbaqa, Herbstkampagne 1971) unternommenen archäologischen Untersuchungen, ertstattet von Mitgliedern der Mission', *MDOG* 105: 5-68.

KEMPINSKI A.

1983 *Syrien und Palästina (Kanaan) in der letzten Phase der mittelbronze IIB Zeit (1650-1570 v. Chr.)*. Wiesbaden.

KLENGEL H.

1965 *Geschichte Syriens im 2. Jahrtausend v.u.Z. I, (Nord Syrien)*. Berlin.

1981 'Historischer Kommentar zur Inschrift des Idrimi von Alalakh', *UF* 13: 269-78.

KÜHNE H. AND STEUERWALD H.

1980 'Das Nordost-Tor von Tell Munbaqat', in: J. Margueron (Ed.) 1980: 203-15.

VAN LOON M.N. (ED.)

1980 *Korucutepe*, Vol. 3. Amsterdam.

in press *Hammam et-Turkman I*, Istanbul.

LOUD G.

1948 *Megiddo*, II. OIP 62. Chicago.

MACHULE D. ET AL.

1978 'Tall Munbaqa 1978', *MDOG* 110: 73- 6.

1986 'Ausgrabungen in Tall Munbaqa 1984', *MDOG* 118: 67-145.
1987 'Ausgrabungen in Tall Munbaqa 1985', *MDOG* 119: 73-134.

MALLOWAN M.E.L.
1936 'The Excavations at Tell Chagar Bazar and an Archaeological Survey of the Habur Region 1934-5', *Iraq* 3: 1- 87.
1937 'The Excavations at Tell Chagar Bazar and an Archaeological Survey of the Habur Region', *Iraq* 4: 91-178.
1947 'Excavations at Brak and Chagar Bazar', *Iraq* 9: 1-259.

MARGUERON J.-CL (ED.)
1980 *Le Moyen Euphrate*. Zone de contacts et d'échanges. Acte du Colloque du Strasbourg 10-12 mars 1977. Leiden.

MARGUERON J.-CL.
1980 'Emar: un exemple d'implantation Hittite en terre Syrienne', in: J-C. Margueron (Ed.) 1980.

MARQUARDT W.H
1978 'Advances in Archaeological Seriation', in: M.B. Schiffer (Ed.) 1978: 266-314.

MATTHERS J. (ED.)
1981 *The River Qoueiq, Northern Syria, and its Catchment*. Studies arising from the Tell Rifa'at Survey 1977-79. BAR International Series 98. Oxford.

MAYER W.
1986 'Die Tontafelfunde von Tall Munbaqa 1984', *MDOG* 118: 126-31.

McCLELLAN T.L.
1984-85 'El-Qitar: Second Season of Excavations, 1983-84', *Abr Nahrain* 23: 39-72.
in press 'El-Qitar: Third Season of Excavations, 1984-85'. *Abr Nahrain* 24.

NA'AMAN N.
1974 'Syria at transition from the Old Babylonian Period to the Middle Babylonian Period', *UF* 6: 265-74.
1976 'A new look at the chronology of Alalakh level VII', *Anat.St.* 26: 129-43.
1979 'The chronology of Alalakh level VII once again', *Anat.St.* 29: 103-13.

ORTHMANN W.
1976 'Munbaqat 1974. Vorläufiger Bericht über die von der Deutschen Orientgesellschaft mit Mitteln der Stiftung Volkswagenwerk unternommenen Ausgrabungen', *MDOG* 108: 25-44.

ORTHMANN W. AND KÜHNE H.
1974 'Mumbaqat 1973, Vorläufiger Bericht über die von der Deutschen Orient-Gesellschaft mit Mitteln der Stiftung Volkswagenwerk unternommenen Ausgrabungen', *MDOG* 106: 53-97.

PARROT A.
1938 'Nouvelles Archéologique. Mari et Chagar Bazar', *Syria* 1938: 308-10.
1958 *Mission Archéologique de Mari II, Le Palais*. Paris.

SCHIFFER M.B. (ED.)
1978 *Advances in Archaeological Method and Theory, Vol. 2*. New York et. al.

SCHAEFFER C.F.A.
1949 'Corpus céramique de Ras Shamra', *Ugaritica* 2: 131-300. Paris.

SMIT F.
in press 'The period VIII pottery', in: M. N. van Loon (ed.) *Hammam et-Turkman I, Interim report on the University of Amsterdam's 1981-84 excavations in Syria*. Istanbul.

SWIFT JR. G.F.
1958 *The Pottery of the 'Amuq Phases K to O, and its historical relationships*. Unpublished Ph.D. thesis, University of Chicago. Chicago.

TUBB J.N.
1980 'A Reconsideration of the Date of the Second Millennium Pottery from the Recent Excavations at Terqa', *Levant* 12: 61- 69.

TUFNELL O. ET AL.
1940 *Lachish II. (Tell ed Duweir). The Fosse Temple*. London.

WOOLLEY C.L.
1955 *Alalakh, an account of the excavations at Tell 'Atchana in the Hatay, 1937-1949*. Oxford.

Zur Datierung der *altmonochromen* Ware von Tell Halaf

KARIN BARTL

1 EINLEITUNG

Die stratigraphische Abfolge des Tell Halaf gliedert sich nach Aussagen der Publikationen (von Oppenheim 1931; Schmidt 1943; Langenegger et al. 1950; Moortgat 1955; Hrouda 1962) in zwei deutlich voneinander getrennte Komplexe:

1. die prähistorischen Schichten des 6. bis 4. Jahrtausends
2. die historischen Schichten aus dem 1. Jahrtausend

Daneben weisen einige Kleinfunde auf die Existenz von Schichten des dritten und zweiten Jahrtausends hin (Hrouda 1962). Das fast vollständige Fehlen von Hinweisen auf Schichten des dritten Jahrtausends erscheint besonders erstaunlich, da Oberflächenuntersuchungen im Habur-Dreieck und am Habur in diesem Zeitraum eine relativ dichte Besiedlung zeigen (vergl. Meijer 1986, Fig. 33; Kühne 1974-77:32, 1978-79:186). Zum zweiten Jahrtausend gehören neben kappadokischen auch einige Mitanni-Siegel, die eine Besiedlung im 15. bis 13. Jahrhundert nahelegen. Mitannizeitliche Schichten sind auch aus dem benachbarten Tell Fakhariyah bekannt (vergl. Kantor *apud* McEwans 1958:21).

Hinsichtlich der Fundumstände, die die Grundlage für die Datierung der Keramik und Kleinfunde bilden, bestehen jedoch Unterschiede. Die prähistorischen Schichten sind fast ausschliesslich durch die Keramik und Kleinfunde bezeugt, die zumeist aus unklaren Fundzusammenhängen stammen. Hier sind eine ältere Schicht mit altmonochromer Ware und eine jüngere mit Buntkeramik (Schmidt 1943:25 ff) zu trennen. Grössere Baubefunde sind für beide Perioden nicht belegt. Jüngere Nachuntersuchungen am Tell Halaf haben gezeigt, dass die Buntkeramik hier zusammen mit Lehmziegelbauten auftritt. Ebenso fanden sich Baureste in der darunterliegenden Schicht mit ausschliesslich altmonochromer Ware (vergl. Davidson 1977:160 ff). Die als 'buntkeramisch' angesprochenen Grossbauten (Langenegger et al. 1950:251 ff) stammen sicher aus jüngeren Perioden.

Für die historischen Schichten bilden die Repräsentations- und Befestigungsbauten aus aramäischer Zeit die umfangreichsten freigelegten Befunde (Langenegger et al. 1950:17 ff; von Oppenheim 1931:260 ff; Hrouda 1962:76 ff; Moortgat 1955). Einige Bauten und Gräber, sowie Kleinfunde und Keramik datieren in die neuassyrische Zeit (Langenegger et al. 1950:203 ff, 349 ff; von Oppenheim 1931:262 ff; Hrouda 1962:81 ff). Für das späte erste Jahrtausend sind einige hellenistische Gräber, Keramik und Kleinfunde dokumentiert (Langenegger et al. 1950:361 ff; von Oppenheim 1931:263 ff;

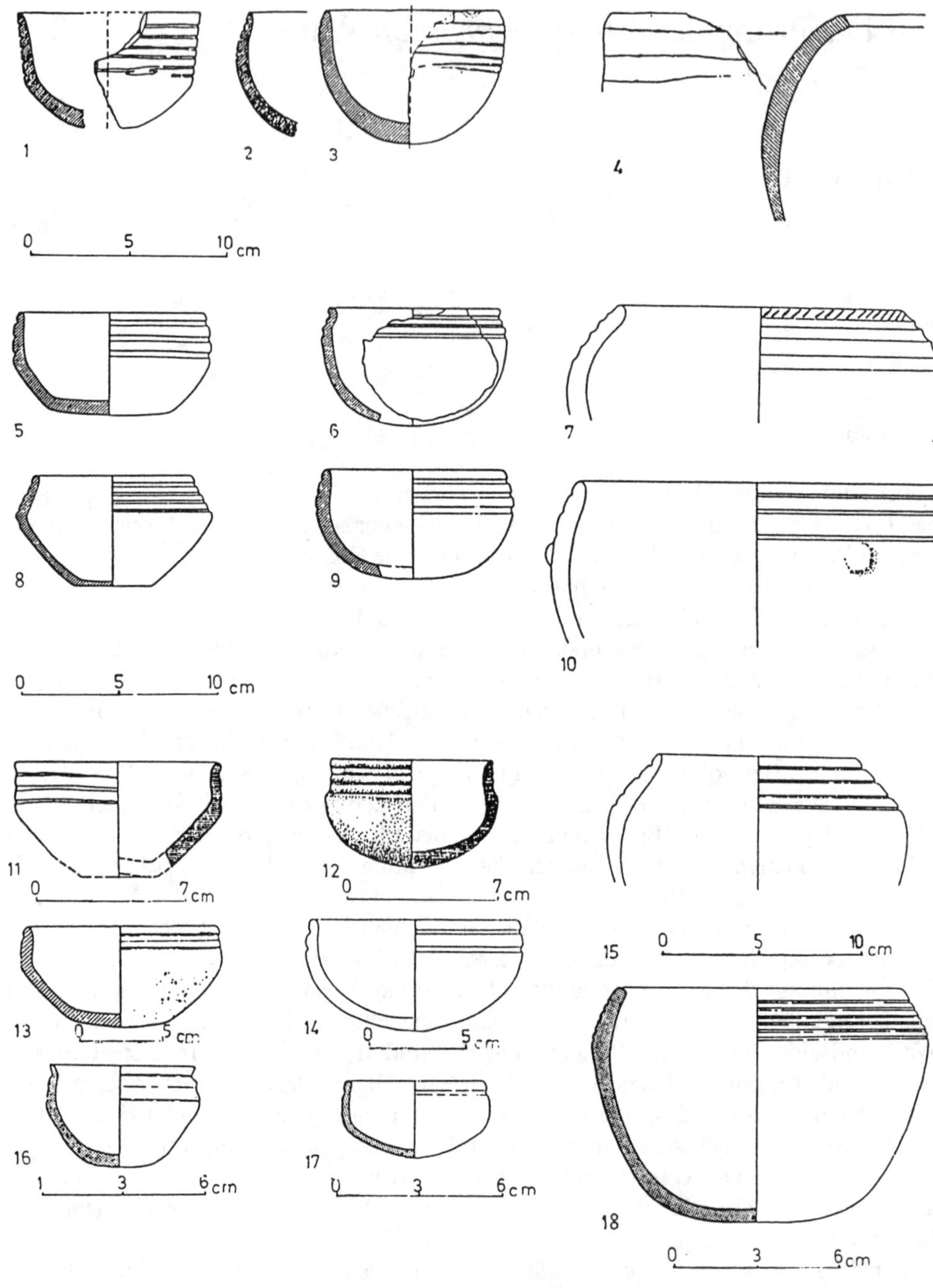

Figur 1. 1. Schmidt 1943, Tf. IV. 13; 2. Schmidt 1943, Tf. IV. 14; 3. Schmidt 1943, Tf. IV. 10; 4. Schmidt 1943, Tf. III. 7; 5. Hauptmann 1969-70, Abb. 18. 4; 6. Hauptmann 1969-70, Abb. 18. 6; 7. Bartl 1988, Fig. 37. 24a; 8. Hauptmann 1969-70, Abb. 16. 10; 9. Hauptmann 1969-70, Abb. 18. 5; 10. Bartl 1988, Fig. 39. 26; 11. Duru 1979, Pl. 63. 2; 12. Duru a.O. Pl. 61. 29; 13. Hauptmann 1972, Tf. 54. 11; 14. Bartl 1988, Fig. 33. 16a; 15. Bartl 1988, Fig. 30. 9a; 16. Winn 1982, Pl. 57. 6; 17. Winn 1982, Pl. 54. 3; 18. Winn 1982, Pl. 57. 9.

Hrouda 1962:13 ff, 85 ff), während aus parthischer Zeit nur Kleinfunde und Keramik belegt sind (von Oppenheim 1931:265; Hrouda 1962:14, 90 ff).

Hinsichtlich der stratigraphischen Zusammenhänge bestehen sowohl in den prähistorischen als auch in den historischen Schichten zahlreiche Unklarheiten, die immer wieder zu Diskussionen geführt haben (siehe unten).

Neben dem fast vollständigen Fehlen von Hinweisen auf Schichten der Mitanni-Zeit sind auch keine Funde aus jüngeren Straten, die in den Zeitraum zwischen 1200 und 900 v. Chr. zu datieren wären, belegt. Neuere Ausgrabungen in Südostanatolien haben jedoch in früheisenzeitlichem Kontext (ca. 1200-900 v.Chr.) eine Keramik hervorgebracht, die auffallende Parallelen zu einigen Formen der *altmonochromen* Waren aufweist. Es soll daher im folgenden untersucht werden, ob und inwieweit ein Datierungsansatz für Teile der *altmonochromen* Waren in das 12. bis 10. Jahrhundert möglich ist.

2 WAREN UND FORMEN DER *ALTMONOCHROMEN* KERAMIK

Die *altmonochrome* Keramik am Tell Halaf wird in zwei Gruppen unterteilt: Aa und Ab, wobei Aa den grösseren Anteil ausmacht (Schmidt 1943:25).

Gruppe Aa Magerung: sand- und kalkgemagert, grob geschlämmt. Oberflächenfarben: grau bis schwarz, gelblich-braun, rot, aussen und oft auch innen Flecken in schwarz, bräunlich oder rötlich. Oberflächenbehandlung: Tonüberzug in verschiedener Ausführung, oft poliert. Brand: unregelmässig, wobei die rot monochrome Ware am besten gebrannt ist.

Gruppe Ab Gruppe Ab wird als Wirtschaftskeramik bezeichnet, deren hauptsächliches Merkmal die primitive Tonbereitung ist und die keine Politur aufweist. Eine detailliertere Beschreibung findet sich nur bei der ersten Formengruppe, den kugelbauchigen Töpfen, so dass unklar bleibt, ob diese Klassifizierung für alle Gefässe der Gruppe Ab gilt (Schmidt 1943:28). Magerung: grob, stark mit Sand versetzt. Oberflächenfarben: grauschwarz bis grau, gerötet, bräunlich oder heller (gelblich oder graugelb). Oberflächenbehandlung: meist keine Politur, manchmal jedoch Politurstriche. Brand: primitiv, schwarze Flecken auf den Aussenflächen.

Innerhalb der beiden Waren werden die Gefässe nach Formgruppen unterteilt, wobei in Ware Aa ausschliesslich Schalen, Näpfe und Töpfe vorkommen, teilweise mit Henkeln, während Giessgefässe vollständig fehlen. In dieser Warengruppe scheinen dickwandige rundbauchige Töpfe mit einziehender Gefässmündung, die manchmal mit Griffknubben im oberen Gefässdrittel versehen sind, vorrangig zu sein (Schmidt 1943, Tf. I, nos. 1-13). Weitere Schalenformen sind dickwandige, flache Becken (Schmidt 1943, Tf. II, nos. 10-11, 13-14), weit geöffnete Schalenformen (Schmidt 1943, Tf. II, no. 8), Schalen mit scharfem Schulterknick (Schmidt 1943, Tf. II, nos. 18-22), sowie als Vorratsgefässe bezeichnete Töpfe mit weiter Randöffnung, die zapfenartige Griffe oder Handhaben aufweisen. In Warengruppe Ab tauchen ebenfalls flache Becken (Schmidt 1943, Tf. IV, nos. 16-18, 20-21) und kurzhalsige Topfformen mit Henkel auf (Schmidt 1943, Tf. IV, no. 23), sowie Gefässe mit stark einziehender Mündung und dicker Wandung auf (Schmidt 1943, Tf. IV, no. 24). Zwei weitere Formen, die in Gruppe Aa nicht auftreten, gehören ebenfalls in diese Reihe; ein sehr grober Kochtopf mit umlaufen-

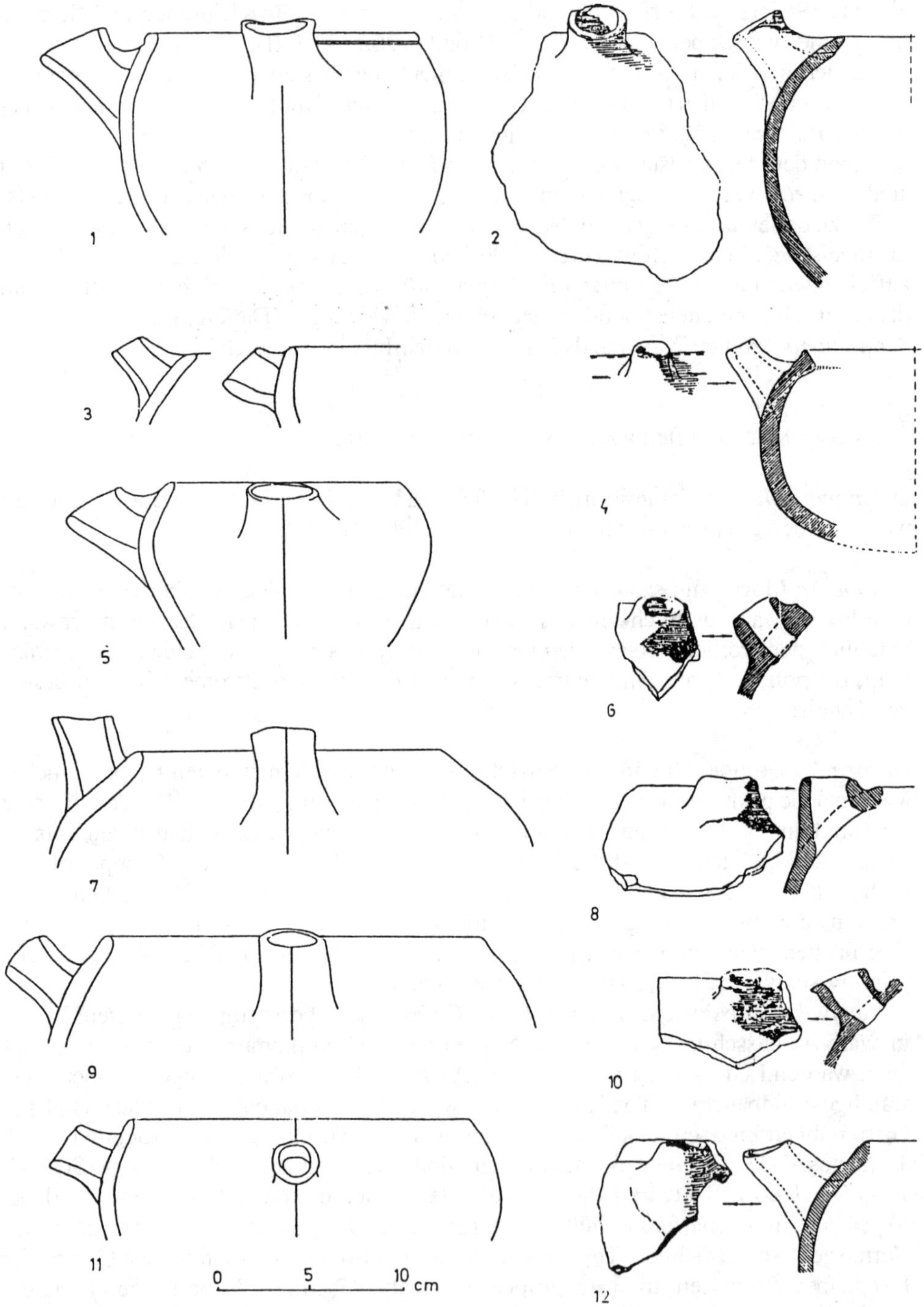

Figur 2. 1. Bartl 1988, Fig. 58. 209a; 2. Schmidt 1943, Tf. III. 12; 3. Bartl 1988, Fig. 60. 216; 4. Schmidt 1943, Tf. III. 19; 5. Bartl 1988, Fig. 57. 203; 6. Schmidt 1943, Tf. III. 3; 7. Bartl 1988, Fig. 58. 208; 8. Schmidt 1943, Tf. III. 1; 9. Bartl 1988, Fig. 60. 215; 10. Schmidt 1943, Tf. III. 2; 11. Bartl 1988, Fig. 59. 210; 12. Schmidt 1943, Tf. III. 4.

dem Wulstrand (Schmidt 1943, Tf. XLI, no. 4) und eine flache Beckenform mit Rippenwülsten auf der Gefässinnenseite, die den 'husking trays' ähnlich ist (Schmidt 1943, Textabb. 2; vergl. Lloyd and Safar 1945, Fig. 3, no. 10).

Neben diesen Formen, die sicher in die älteste Siedlungsperiode des Hügels datieren, erscheint eine zweite Gruppe, deren zeitliche Einordnung nicht so eindeutig ist: Schalen mit scharfem Wandungsknick (Schmidt 1943, Tf. IV, nos. 1-3), manchmal mit ausgeprägter Randlippe (Schmidt 1943, Tf. IV, nos. 4-8), Kochtöpfe mit Fingertupfenhenkeln (Schmidt 1943, Tf.IV, nos. 9, 11-12) und ein singuläres Gefäss mit anthropomorpher Randfigur (Schmidt 1943, Textabb. 1).

Eine dritte Formengruppe stellen die Giessgefässe dar, die als kugelbauchige Töpfe mit einziehender Mündung und hochangesetzten Ausgusstüllen ausgebildet sind Schmidt 1943, Tf. III, nos. 4, 6, 12, 18-19). Als Verzierung zeigen diese Gefässe häufig mehrere horizontal umlaufende Rillen, die in unregelmässiger Anordnung den Topfrand akzentuieren (Schmidt 1943, Tf. III, nos. 3, 5, 11; Tf. XLI, no. 2). Dieses Muster erscheint auch bei einigen Henkelfragmenten, die offenbar ebenfalls zur Gruppe der kugelbauchigen Gefässe gehören. Unklar ist hier jedoch, ob diese Stücke ursprünglich zu Tüllengefässen gehörten oder ob sie eine eigene Gruppe von Henkeltöpfen bildeten (Schmidt 1943, Tf. III, nos. 13-14). Die Rekonstruktion zweier Fragmente zu einem Tüllengefäss mit Henkel erscheint nicht ganz schlüssig, da das Fragment mit dem Tüllenansatz eine deutlich dickere Wandung als der gegenüberliegende Henkelteil aufweist (vergl. Schmidt 1943, Tf.III, 19-20; Tf. XLI, 2-3). Sonderformen innerhalb der Henkelfragmente sind eine Handhabe mit Fingertupfenleiste (Schmidt 1943, Tf. III, no. 17), sowie Henkel ohne Durchlochung (Schmidt 1943, Tf. IV, nos. 9, 11-12, Tf. III, no. 16). In die Gruppe der rillenverzierten Gefässe sind auch drei Näpfe einzuordnen, die entweder in Kalottenform vorkommen oder einen Wandknick aufweisen (Schmidt 1943, Tf. IV, nos. 10, 13-14). Bei allen Gefässen finden sich nur selten nähere Angaben hinsichtlich der Ware. Bei den wenigen beschriebenen Stücken handelt es sich entweder um Ware mit rotem überzug (Schmidt 1943, Tf. III, nos. 4, 17) oder um rötlichen Ton (Schmidt 1943, Tf. III, nos. 19-20).

3 FUNDSTELLEN DER *ALTMONOCHROMEN* WARE

Innerhalb der Publikation finden sich einige allgemeine Angaben zu den Fundstellen. Danach soll in den Tiefgrabungen 'am tiefsten immer die altmonochrome Töpferware, darüber die Buntkeramik mit altmonochromen gemeinsam' gelegen haben, 'bis schliesslich die altmonochrome gänzlich aufhörte' (Schmidt 1943:25). Dazu ist zu bemerken, dass während der Ausgrabungen sehr wahrscheinlich mehr Tiefgrabungen angelegt wurden, als später dokumentiert wurden. Dabei wurden häufig rezente Getreidespeicher-Brunnen, die das Gelände an allen Stellen teilweise mehr als 5 m tief durchlöcherten, weiter eingetieft (vergl. Langenegger et al. 1950:3). Nur selten wurden diese Ergebnisse in die Publikation aufgenommen (Langenegger et al. 1950, Abb. 4 und Abb. 35). Im allgemeinen sind diese Schnitte nicht systematisch dokumentiert und finden sich auch nicht auf den Plana.

Innerhalb des Keramikkataloges werden nicht bei allen Stücken Angaben zur Herkunft gemacht. Folgende Fundstellen werden genannt:

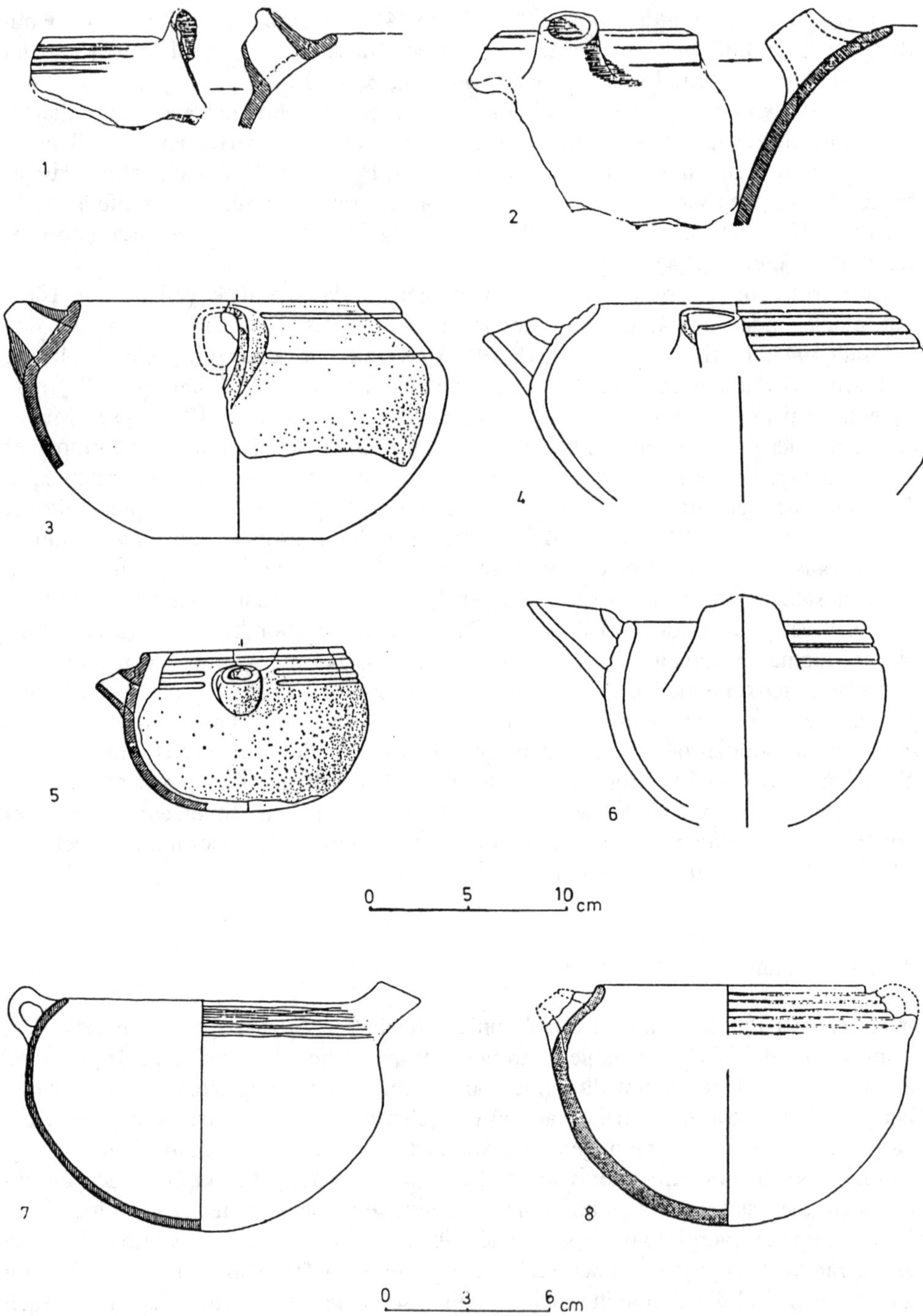

Figur 3. 1. Schmidt 1943, Tf. III. 5; 2. Schmidt 1943, Tf. III. 11; 3. Hauptmann 1969-70, Abb. 17. 3; 4. Bartl 1988, Fig. 58. 206; 5. Hauptmann 1969-70, Abb. 26. 12; 6. Bartl 1988, Fig. 57. 204; 7. Winn 1982, Pl. 54. 6; 8. Winn 1982, Pl. 52. 4.

1. Ravine auf dem Burghügel in C 1/VI 2 beim assyrischen Haus, unbekannte Tiefe (Schmidt 1943, Tf. IV, 2).

2. Nordabhang des Hügels von Suchgraben 3, auf D 1 oder D 2/IV 3, bzw. Nordabhang des Hügels bei Suchgraben 3 (Schmidt 1943, Tf. IV, nos. 7, 10; Textabb. 4; eventuell auch Tf. IV, nos. 3-6). Die Fundstelle geht nicht ganz klar aus dem Text hervor. Es wird jedoch hervorgehoben, dass diese Formen 1929 in den Buntkeramikschichten vorkamen (Schmidt 1943:29, Fn. 1; auch Tf. IV, nos. 17-18, 21 wurden offenbar zusammen mit Buntkeramik gefunden. Vergl. Schmidt 1943:30).

3. Tiefgrabung von 1929, nordwestlich des Tempelpalastes (Schmidt 1943, Tf. I, no. 14; Tf. II, nos. 8, 10, 20-22 (Gruppe Aa); Tf. IV, nos. 15-16, 24; Tf. XLI, no. 4 (Gruppe Ab); Tf. XLI, no. 5; Textabb. 2).

4. Brunnen 1 und 2 in der Nähe des Tempelpalastes (Schmidt 1943, Tf.II, nos. 5-7 (Gruppe Aa); Tf. IV, 20 (Gruppe Ab)). Hier werden jedoch nur allgemein 'tiefste Schichten der Brunnen' angegeben.

5. Stadtgebiet westlich des Expeditionshauses, aus etwa 2,50 m Tiefe (Schmidt 1943, Tf. IV, no. 11).

Fundstelle 1. Zwar lässt sich der Hügelbereich aufgrund der Planquadratangabe lokalisieren, da jedoch genauere Eingrenzungen und Tiefenangaben fehlen, sind die Angaben für weitere Schlussfolgerungen zu ungenau. Unklar ist zudem, aus welchem Zusammenhang das Stück stammt. In der Publikation findet sich eine Tiefgrabung nur im Inneren des assyrischen Hauses (Langenegger et al. 1950:201). Es wird allerdings eine Brunnentiefgrabung nördlich des langen Anbaus am assyrischen Haus beschrieben (keine Abb.) mit der Abfolge: Kaparazeitliche Keramik, Buntkeramik, *altmonochrome* Ware, sterile Schicht, gewachsener Fels.

Fundstelle 2. Es handelt sich um den nördlichen Teil des grossen Suchgrabens, der in Nordsüd-Richtung über den gesamten Burghügel verläuft und zum Fluss hin abfällt (Langenegger et al. 1950:10 und Pl. 2). Hier wurde im Bereich D 1/VI 3, D 2/ VI 3, D 3/IV 3 viel Buntkeramik gefunden. Hinsichtlich der Planquadrate kann keine genauere Eingrenzung vorgenommen werden.

Fundstelle 3. Diese Tiefgrabung von 1929[1] ist für die *altmonochromen* Waren die wichtigste dokumentierte Fundstelle (Langenegger et al. 1950:10; Schmidt 1943, Beilage 4). Ihre Masze betrugen 17x9,50 m. Die Profilzeichnung zeigt eine sehr starke Buntkeramikschicht mit nur geringen Mauerresten. Die darunterliegende dünne Schicht wird der 'Urbewohnung' zugesprochen. Da die Buntkeramikschicht keine weiteren Unterteilungen in Form von Mauerresten oder Fussböden aufweist, liegt die Vermutung nahe, dass es sich hier um eine mit Buntkeramik vermischte Schuttschicht handelt. Die Lage des Tiefschnittes im Bereich der Terrassenaufschüttungen lässt starke Störungen der ursprünglichen Schichtenlage vermuten. Hinweise für diese Annahme finden sich bereits im Text. So wird berichtet, dass es sich die Kapara-Terrasse deutlich von der Altbau-Aufschüttung mit der Buntkeramikvermischung durch unterschiedliche Erdfärbung unterschied (Langenegger et al. 1950:97 ff) und dass die Altbauaufschüttung 'aus Schutt, Scherben, altem Bauabraum und teilweise starken Scherbenmengen der voraufgehenden Buntkeramikzeit' bestand (Langenegger et al. 1950:29). Es ist fraglich, in wieweit hier Aufschüttungsschutt und Buntkeramikschutt zu trennen waren, da ja in der Buntkeramikschicht keine Mauerreste erreicht wurden, sieht man von den als 'buntkeramisch' angesprochenen Mauerzügen in der nördlichen Tiefgrabung ab. Diese Mauerzüge könnten zu einem unterhalb der Altbauzeit liegenden Mauerrest gehören, der vor

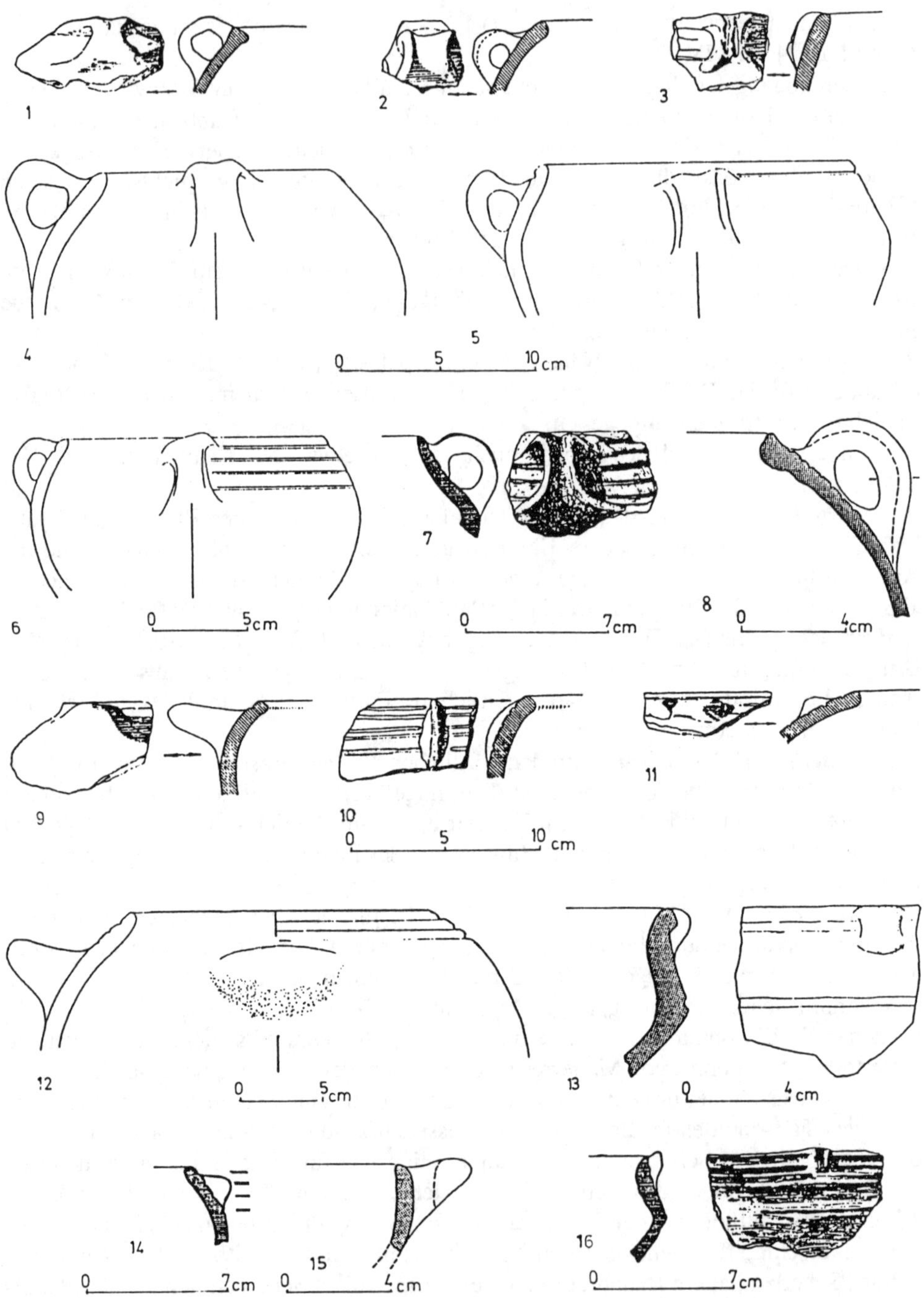

Figur 4. 1. Schmidt 1943, Tf. III. 14; 2. Schmidt 1943, Tf. III. 13; 3. Schmidt 1943, Tf. III. 15; 4. Bartl 1988, Fig. 53. 109; 5. Bartl 1988, Fig. 53. 110b; 6. Bartl 1988, Fig. 54. 110a; 7. Duru 1979, Pl. 61. 44; 8. Winn 1982, Pl. 59. 63; 9. Schmidt 1943, Tf. III. 10; 10. Schmidt 1943, Tf. III. 9; 11. Schmidt 1943, Tf. III. 8; 12. Bartl 1988, Fig. 55. 114b; 13. Winn 1982, Pl. 59. 68; 14. Duru 1979, Pl. 61. 25; 15. Winn 1982, Pl. 59. 65; 16. Duru 1979, Pl. 61. 44.

dem Südost-Eckturm liegt und 0,90 m tiefer als der Altbau selbst gründet. Möglicherweise sind diese Baureste einer direkt vor der Altbauzeit liegenden Periode zuzuordnen (Langenegger et al. 1950:33, Pl. 11, Schnitt 7-8). Innerhalb dieses Baurestes liegt, nach Auskunft der Profilzeichnung ebenfalls eine Tiefgrabung mit der Abfolge: Buntkeramik, grobwandige Ware (altmonochrom ?), gewachsener Boden.

Die Schwierigkeiten einer eindeutigen zeitlichen Zuordnung der unterhalb des Tempelpalastes liegenden Schichten werden auch durch den Hinweis deutlich, dass die in der Tiefgrabung, unterhalb der der Nordfront liegenden Baureste 'völlig zusammengequetscht' waren. Trotzdem meinten die Ausgräber, dass sie verschiedenen Zeitabschnitten der Buntkeramikepoche zu zuordnen seien (Langenegger et al. 1950:33 ff, Abb. 7). Die Mauerunterkanten liegen mindestens 2 m tiefer als das Altbaufundament, das sich bereits 1 m tief in (den Schutt) dieser Schicht gründet (Langenegger et al. 1950:35). Die Problematik der mit den Terrassierungen und Aufschüttungen, die ja auch im südlichen Burghügelgebiet nachgewiesen wurden (Langenegger et al. 1950:144 ff), verbundenen Schuttmengen und die schwierige Trennung von Bau- und Schuttschichten lassen es zweifelhaft erscheinen, ob die Abfolge: aramäisch/assyrisch, Buntkeramik, *altmonochrome* Ware in dieser Form wirklich überall auf dem gesamten Burghügel und im Stadtgebiet vorliegt oder ob nicht möglicherweise die an einer Stelle festgestellte Stratigraphie auf andere Fundstellen schematisch übertragen wurde. Für das Gebiet des Tempelpalastes stellt sich die Frage, ob hier nicht die unterhalb des Altbaus liegenden Schichten anhand der Schuttkeramik datiert wurden - diese Schichten jedoch tatsächlich viel junger sind (vergl. Naumann apud Langenegger et al. 1950:371, der bezweifelt, ob die unterhalb des Südwest-Eckturms gelegene Mauer wirklich in die Buntkeramikzeit zu datieren ist).

Eine Datierung in die Buntkeramikzeit durch Funde aus Schuttschichten wurde bei der alten Befestigung unter dem Wohnpalast vorgenommen (Langenegger et al. 1950:251). Die ursprünglich vorgeschlagene Datierung der östlichen Burgumwallung in das Chalkolithikum wurde bereits durch Naumann schlüssig widerlegt (Naumann apud Langenegger et al. 1950:371ff).

Fundstelle 4. Es wird nicht ganz deutlich, wo diese Brunnen liegen. Möglicherweise handelt es sich um die beiden Getreidebrunnen auf der Mittelterrasse vor der Nordfront des Tempelpalastes (Langenegger et al. 1950:67, Abb. 34). Ebenso ist unklar, wo sich der erwähnte Brunnen 3 befindet.

Fundstelle 5. Im Bereich westlich des Expeditionshauses liegen mehrere grosse Suchschnitte (Langenegger et al. 1950:343ff, Pl. 1) in den Bereichen C V und C VI, deren grosse Ausdehnung keine genauere Eingrenzung erlaubt.

Die beschriebenen Angaben machen folgendes deutlich:

1. Die genaue Lokalisierung der im Text angebenen Fundstellen ist oft nicht möglich.

2. Die Ausgräber scheinen teilweise auch die um bestimmte Gebäude bei Bau- und Aufschüttungsmassnahmen angehäuften Erdmengen mit ihrem Scherbeninhalt zur Datierung von Bauten herangezogen zu haben.

3. Die durch die grossen Erdbewegungen angefallenen Schuttmengen und Störungen der Schichten lassen vermuten, dass auch jüngere Scherben in ältere Straten gekommen sind, bzw. umgekehrt. Es erscheint daher möglich, dass der *altmonochromen* Ware, die in den Tiefgrabungen vermutlich teilweise aus schichtbestimmtem Kontext stammt, auch jüngere Keramik archaischen Aussehens typologisch zugeordnet wurde.

4 DATIERUNG UND VERBREITUNG DER *ALTMONOCHROMEN* WAREN

Wie bereits angemerkt, soll die *altmonochrome* Ware sowohl allein als älteste Keramikgruppe, als auch zusammen mit Buntkeramik vorkommen. Bereits Schmidt (1943:25, 70) nahm also ein zeitweiliges Nebeneinander beider Keramiktypen an und darüber hinaus sogar eine zeitweilige Beeinflussung der *altmonochromen* Waren durch die Buntkeramik. Für die Heranziehung von Parallelbeispielen sind zwei Punkte von Bedeutung:

1. Handelt es sich bei dieser Keramik um eine östliche Variante der 'Dark-Faced Burnished Ware' (DFBW) des Amuq-Gebietes (Amuq A-B) oder stellt diese Keramik eine eigenständige Gruppe dar ? Während ältere Zusammenfassungen häufig von einer Gleichsetzung der *altmonochromen* Ware mit der DFBW der Periode Amuq A ausgehen (vergl. Watson 1965:65 ff; Mellaart 1975, Fig. 81), wird sie in neueren Bearbeitungen für eine eine eigenständige Keramikentwicklung des ostsyrischen Raums gehalten (Davidson 1977:146; Copeland and Hours 1987:403). Offensichtlich ist es jedoch sowohl schwierig, die beiden Keramikgruppen voneinander zu unterscheiden als auch die *altmonochromen* Waren von jüngeren monochromen Waren zu trennen (Davidson 1977:88).

2. Welche Formen der *altmonochromen* Ware finden Parallelen in anderen Siedlungen? Die als Vergleichsbeispiele herangezogenen Gefässtypen der *altmonochromen* Waren beschränken sich offenbar auf die einfachen Formen, d.h. insbesondere die o.g. erste Formengruppe der dickwandigen, rundbauchigen Töpfe mit einziehender Mündung, sowie die Formen mit gerader und ausladender Wandung (Davidson 1977:146; Davidson and Watkins 1981:7; Akkermans 1987:39).

Diese o.g. erste Gefässgruppe der Ware Aa ist aus spätneolithischem und frühhalafzeitlichem Kontext neuerer Grabungen bekannt. In Habech, nahe Tell Aqab, kommen einfache Waren in der untersten Schicht allein vor, in der darüber liegenden Schicht zusammen mit Frühhalaf-Keramik (Davidson 1977:88). In Tell Aqab finden sich einfache gradwandige oder ausladende Schalen in strohgemagerter geglätteter Ware zusammen mit Frühhalafscherben (Davidson and Watkins 1981:7). Ein einzelnes Stück mit 'pattern burnish' (Amuq B) wird als Import angesprochen. Eine unterhalb der Frühhalafschicht liegende ältere Periode ist hier nicht vorhanden. Chagar Bazar zeigt unbemalte, geglättete Waren zusammen mit Halafkeramik in den ältesten Schichten (Mallowan 1936:10 und Pl. III, no. 8). Tell Sabi Abyad im Balikh-Tal ist ein weiterer Fundort, an dem eine grobe, unbemalte Ware am übergang zwischen Spätneolithikum und Frühhalafzeit zusammen mit Frühhalafscherben vorkommt (Akkermans 1987:39).

Die Datierung dieser Keramik wird entsprechend der Fundsituation in den o.g. Orten an den Übergang zwischen Spätneolithikum und Frühhalafzeit gesetzt. Gestützt auf ein C-14 Datum vom Tell Halaf (Vogel and Waterbolk 1964:349 ff) bedeutet das den Zeitpunkt um 5620 v.Chr. Copeland und Hours (1987:402 ff) rechnen die *altmonochromen* Waren an das Ende ihrer Periode 6, Anfang 5, d.h. Ende Amuq A /Anfang Amuq B (5620 - 5000 v.Chr.).

Zusammenfassend lässt sich sagen, dass die Stratigraphie anderer Fundorte der Jezirah und des Balikh-Tals die am Tell Halaf in verschiedenen Tiefschnitten gemachten Beobachtungen zu bestätigen scheinen. Unklar bleibt vorerst allerdings Art und Umfang älterer neolithischer Schichten (Watkins and Campbell 1987:434), sowie die Trennung von älteren neolithischen Keramikformen und Gefässtypen der *altmonochromen* Ware, die am Übergang zwischen Neolithikum und Halafzeit stehen.

Im Gegensatz zu den o.g. Gefässformen finden sich für die ebenfalls zu den *altmonochromen* Waren gerechneten Tüllengefässe und rillenverzierten Fragmente in keiner neolithischen Siedlung der näheren und weiteren Umgebung Parallelen. Eine sehr ähnliche Keramik kennzeichnet jedoch die früheisenzeitlichen Schichten zahlreicher Fundorte Südost- und Ostanatoliens.

5 FORMEN UND WAREN DER FRÜHEISENZEITLICHEN KERAMIK

Ebenso wie am Tell Halaf[2] handelt es sich hier um handgemachte Waren, die verschiedene Unterteilungen aufweisen und unter denen die Waren mit mehr oder weniger dichtem Überzug den grössten Raum einnehmen (Hauptmann 1969:70-58; Winn 1982:156; Bartl 1988:160 ff). Innerhalb dieser Gruppe sind Stücke mit grobem Überzug in verschiedenen Beigetönen am häufigsten. Sie weisen oft durch die Brenntechnik erzeugte graue Flecken auf den Aussenseiten auf. Von dieser Ware unterscheidet sich eine Variante mit dunkelrotem bis dunkelbraunem, sehr dichten Überzug, die eine sorgfältige, teilweise wie Politur wirkende Glättung, aufweist. Die daneben vorkommende Ware ohne Überzug ist von geringerer Bedeutung. Alle Waren sind mit Sand, Steingrus und Häcksel relativ grob gemagert. Das Formenspektrum ist eng begrenzt und vor allem durch Napf- und Schalenformen in unterschiedlicher Grösse und Ausprägung gekennzeichnet. Die unverzierten und rillenverzierten Tüllen- und Henkelgefässe bilden eine prägnante Formengruppe, die jedoch weniger häufig vorkommt.

Die kleinen Näpfe und Schalen haben Parallelen in Norşun-Tepe (Hauptmann 1969-70, Abb. 16:10-11, Abb. 18:4-7), Değirmentepe (Duru 1979, Pl. 61:4, 7), Korucutepe (Winn 1982, Pl.57:G, Pl. 58:12), Tepecik (Esin 1970, Pl.6:6), Imikuşaği (Sevin 1982:399, Res. 7)), Tille (French 1986:211, Res. 4:1,2,6) und Van-Dilkaya (Cilingiroglu 1986:94, Res. 15:1, Reihe rechts, Res. 15:2, Reihe rechts und links).(Fig. 5)

Diese Gefässtypen, die auch in sehr viel grösserer Ausprägung vorkommen (z.B. Hauptmann 1969-70, Abb. 18:10)), können als Leitformen der frühen Eisenzeit gelten. Für die Tüllengefässe ohne Rillenverzierung (Fig. 6) zeigen Beispiele aus Norşun-Tepe die besten Parallelen (Bartl 1988, Fig. 57:203), während die rillenverzierte Variante (Fig. 7) ausser in Norşun-Tepe (Bartl 1988, Fig. 57:202) auch in anderen Fundorten bekannt ist, z. B. in Değirmentepe, Korucutepe, Tille (Duru 1979, Pl. 61. 63 - French 1986, Res. 4. 4 - Winn 1982, Pl. 52. 4.60; Pl. 54. 6).. Auch die rillenverzierten Henkeltöpfe ohne Ausgusstüllen (Fig. 8) kommen in den Fundorten des oberen Euphrat-Gebietes vor, wie Beispiele aus Norşun-Tepe, Korucutepe und Değirmentepe belegen.

6 VERBREITUNG DER FRÜHEISENZEITLICHEN KERAMIK

Früheisenzeitliche Schichten mit der typischen rillenverzierten Keramik konnten vor allem im Gebiet des oberen Euphrats in zahlreichen Ausgrabungen nachgewiesen werden:

1. Keban-Staudammgebiet.

Korucutepe (van Loon (Ed.), 1978, 1982), Norşun-Tepe (Hauptmann 1969-70), Tepecik (Esin 1970), Değirmentepe (Esin 1983), Aşvan Kale (Mitchell 1980).

2. Karakaya-Staudammgebiet.

Habibuşaği (P 52/2; Ögün 1983), Değirmentepe (P 50/7; Esin 1983), Imikuşaği (O 50/1; Sevin 1983), Köşkerbaba (P 51/17;Bilgi 1985), Imamoğlu (P 50/3; Uzunoğlu 1985), Pirot Höyük(P 51/15; Karaca 1984).

3. Karababa-Staudammgebiet.

Tille Höyük (T 52/11; French 1986).

Die früheisenzeitlichen Leitformen wurden während der verschiedenen Oberflächenuntersuchungen, die in Vorbereitung der Rettungsgrabungen stattfanden, auch auf zahlreichen anderen Siedlungshügeln gefunden. Erstmals publiziert wurde diese Keramik von Burney. Er datierte sie jedoch, aufgrund damals fehlender Vergleichsbeispiele, in die frühe Bronzezeit (vergl. Burney 1958:201, Abb. 216-218; Russell 1980). Zu weiteren Surveys s. Whallon and Kantman 1979; Serdaroğlu 1977; Özdoğan 1977).

Die Siedlungsverteilung zeigt das relativ häufige Vorkommen im Bereich des oberen Euphrat, wobei Malatya offenbar bereits nicht mehr zum engeren Verbreitungsbereich gehört. In Arslantepe wurde in einer durch C. Schaeffer angelegten Sondage ein rot-auf-creme bemaltes Stück gefunden, das den bemalten Stücken am Norsun-Tepe entspricht. Ein rillenverziertes Fragment wurde dicht unter der Oberfläche gefunden (nicht publiziert). Die neueren Ausgrabungen haben jedoch fur die späthethitischen Schichten nur Keramik in zentralanatolischer Tradition hervorgebracht (vergl. Pecorella 1975, Fig. 7. 8-9). Als südlichster Fundort kann derzeit Tille Höyük gelten, östlichster Fundort Anatoliens ist nach augenblicklichem Kenntnis stand Van-Dilkaya. Dieser Fundort zeigt deutlich, dass die Absenz von Siedlungen im ostanatolischen Raum zwischen Elaziğ und Van zwar den heutigen Forschungsstand repräsentiert, aber wahrscheinlich nicht der tatsächlichen Siedlungsverteilung entspricht. In den Gebieten Südost) und Ostanatoliens weist die früheisenzeitliche Keramik eine bemerkenswerte Homogenität auf, die sich sowohl in den Waren als auch in den Formen beobachten lässt. Daneben kommen jedoch auch Gefässformen, wie Vasen und Krüge vor, die deutlich in spätbronzezeitlicher Tradition stehen (Winn 1982, Pl. 51.5; Pl. 55:11; Pl. 56:1; Esin 1970, Pl. 6:2-3; Hauptmann 1969-70, Abb. 17:5).

Im nordwestiranischen Raum findet sich in früheisenzeitlichem Kontext (Ende Eisenzeit I/Beginn Eisenzeit II) in einigen Orten Keramik, die den ostanatolischen Formen ähnlich ist, so z.B. in Geoy Tepe A (Burton-Brown 1951, Fig. 36:643; Fig. 39:915) und in Tappeh Gijlar (Pecorella und Salvini 1984, fig.62:15) sowie in Kordlar Tepe (Lippert 1979, Abb. 14:4, 7), und in Schichten des 8. Jahrhunderts am Zendan-i Suleiman (Boehmer 1973:4, 504 f, Abb. 31 rechts). Ein ähnliches Stück stammt aus Mudjesir (Boehmer 1973, Abb. 31 links. Die am Zendan i-Suleiman gefundenen Fragmente werden als Importstücke angesprochen). Sehr ähnliche Schalenformen, allerdings aus einer schwarzen, glänzend polierten Ware, sind in den 'prä-urartäischen' Schichten Karmir Blurs (Sorokin 1958:2, Fig. 2:1-3; Fig. 7: 1-2) belegt, sowie, neben einigen anderen Fundorten, in Metsamor (Chanzadian et al. 1973, Pl. 39:5).

7 ABSOLUTE DATIERUNG DER FRÜHEISENZEITLICHEN WAREN

Der Beginn der frühen Eisenzeit Ostanatoliens liegt etwa bei 1200/1150 v.Chr. Da innerhalb der Stratigraphie von Korucutepe und Norsun-Tepe kein auf die Spätbronzezeit folgender Hiatus festgestellt wurde (van Loon 1978:40 ff; Hauptmann 1970:111), ist mit einer kontinuierlichen Weiterbesiedlung auch nach dem Ende der Spätbronzezeit zu rechnen. Eine Besiedlungsdauer kann aufgrund stratigraphischer Beobachtungen bis etwa 950/900 v.Chr. angenommen werden. Der darauffolgende Zeitraum bis zum Beginn der mitteleisenzeitlicher Besiedlung um 850/800 ist jedoch nur schwer zu fassen. Weder am Norsun-Tepe, der eine mitteleisenzeitlichen Besiedlung aus dem 8.und 7. Jahrhundert aufweist, noch in den Karakaya-Grabungen Değirmentepe und Köşkerbaba, wo ebenfalls beide Perioden vorhanden sind, lässt sich stratigraphisch ein Übergang zwischen beiden Perioden beobachten. Auch in der Keramiktypologie zeigen sich neben den völlig verschiedenen Keramikgruppen beider Perioden keine Übergangsentwicklungen (vergl. Hauptmann 1972:83).

8 DATIERUNG DER JÜNGEREN SCHICHTEN AM TELL HALAF

In Tell Halaf finden sich für den vor 900 v.Chr. liegenden Zeitraum keine Datierungsgrundlagen (zur Datierungsdiskussion und Einzelaspekten s. von Oppenheim 1931:266; Herzfeld 1934:111 f.; Bowman 1941:359 ff; Akurgal 1949:139 ff; Moortgat 1955:1 ff; Albright 1956:75 ff; Kantor 1956:171 ff; Mellink 1958:438; Canby 1964:71 f; Moortgat 1970:211 ff; Porada 1970:97 ff; Orthmann 1971:119 ff; Winter 1973:158 ff; Dankwarth und Müller 1987, im Druck). Die Orthostatenreliefs werden aufgrund stilistischer Analysen allgemein in das 9. Jahrhundert datiert (Moortgat 1955:31; Orthmann 1971:128 f). Ebenso erlauben auch die Kleinfunde aus aramäischem Kontext keinen wesentlich anderen zeitlichen Ansatz, wenn auch verschiedene Autoren auf ältere Traditionen bestimmter stilistischer Merkmale hingewiesen haben. (Canby (1964:72) weist auf das Überleben stilistischer Eigenarten aus dem 2. Jahrtausend hin, besonders in Bezug auf die Kultraumstatuetten. Winter (1973:163 f) betont das Weiterleben von Motiven der Mitanni-Glyptik auf den Tell Halaf-Reliefs (z.B. geflügelte Sonnenscheibe). Die Datierung in die Mitte des 9. Jahrhunderts legt auch die aramäische Ältar-Inschrift nahe (Dankwarth und Müller 1987). In assyrischen Quellen findet sich die Nennung Guzanas ebenfalls erstmals Anfang des 9. Jahrhunderts im 2. Feldzugsbericht Adad-Niraris II (zusammenfassende Darstellung der historischen Ereignisse s. von Oppenheim in Schmidt 1943:13 ff).

Für den fraglichen Zeitraum zwischen 1200 und 900 finden sich in Tell Halaf weder inschriftliche noch archäologische Hinweise. Verschiedene stilistische Merkmale von Fundgruppen des ersten Jahrtausends belegen jedoch Einflüsse aus dem nordsyrischen Raum des 2. Jahrtausends.

9 SCHRIFTLICHE QUELLEN

Während um die Mitte des 2. Jahrtausends die Beziehungen zwischen dem anatolischen und nordsyrischen Raum geprägt sind durch die Auseinandersetzungen zwischen Hatti und Mitanni, über die eine Reihe schriftlicher Quellen vorliegen (Wilhelm 1982:28ff; Klengel 1968:28 ff), gibt es für den Zeitraum zwischen 1200-900 v. Chr. nur spärliche Nachrichten.

In zahlreichen Siedlungen der Levanteküste und des zentralantolischen Raumes haben die Ereignisse um 1200 v. Chr., die häufig mit dem 'Seevölkersturm' in Verbindung gebracht werden, zu einem deutlichen Hiatus geführt.

Inschriftlich belegt und lokalisiert sind am Ende des 2. Jahrtausends nur zwei Fundorte, die zu diesem Zeitpunkt bereits eine lange Besiedlungsdauer aufweisen: Karkemisch und Malatya. Beide Orte werden in den Feldzugsberichten Tiglat-Pilesars I erwähnt. Hier findet sich die Nennung eines Assurtributpflichtigen Herrschers Ini-Tešub, der die Titulatur 'König des Landes Hatte' trägt und wahrscheinlich als König von Karkemisch zu verstehen ist (vergl. Hawkins 1976-80:434; 1974:69f). In der daran anschliessenden Genealogie ist Uru-Tarhundas wohl um 970 v.Chr. anzusetzen. Malatya/ Melidu wird im Rahmen der Feldzugsberichte Tiglat-Pilesars I. mehrmals genannt. Der melitenische Herrscher Allumari wird als Tributpflichtiger Assurs beschrieben (vergl. Hawkins 1974:76 ff). Nach einer längeren Lücke in der schriftlichen Überlieferung ist die nächste Nennung die Tributzahlung des Lalli von Milidia und des Sangara von Karkemisch an Salmanassar III in der Mitte des 9.Jahrhunderts (vergl. Goetze 1928-29:59). Die gleichzeitige Nennung dieser beiden Herrscher hat dazu geführt, für den Zeitraum zwischen 1100 und 850 eine zeitweilige Oberherrschaft von Karkemisch über Malatya anzunehmen (vergl. Wäfler 1980-81:87).

Für den Zeitraum um 1200 v.Chr. sind Beziehungen von Karkemisch zum weiter nördlich gelegenen Lidar Höyük durch den Fund eines Königssiegels bezeugt. Der hier genannte Kuzi-Tešub ist ein Sohn von Talmi-Tešub und ungefähr in das beginnende 12. Jahrhundert zu datieren (Sürenhagen 1986:183 ff).

Die wenigen Nachrichten aus assyrischen Quellen bringen also wenig Klärung hinsichtlich der Verbindungen Nordsyriens und Südost-Anatoliens. In den Feldzugsberichten Tiglat-Pilesars I. findet sich die mehrmalige Nennung der Landschaft Alzi am oberen Euphrat, die wohl mit dem Išuwa der hethitischen Quellen identisch ist (vergl. Luckenbill 1926:361). Die Erwähnung der Schlacht gegen die Mushki, die häufig mit den späteren Phrygern gleichgesetzt werden, hat dazu geführt, die Ansiedlung dieses Volkes ca. 300 Jahre vor ihrem Auftauchen in Zentralanatolien im Gebiet des oberen Euphrats anzunehmen (vergl. Luckenbill 1926:221). In einem Kommentar zu den Ausgrabungen in der Altinova meinte Burney, die groben handgemachten Waren der frühen Eisenzeit mit den Mushki in Verbindung bringen zu können (Burney 1980:157 ff).

10 ZUSAMMENFASSUNG

Die Untersuchung der Fundumstände der altmonochromen Waren hat gezeigt, dass die unklare Stratigraphie der ältesten Schichten am Tell Halaf die Einordnung des gesamten Keramikkomplexes in die frühesten Besiedlungsperioden zweifelhaft macht. Die Einordnung eines Teils der *altmonochromen* Waren in den Zeitraum am Übergang von Spätneolithikum und Frühhalafzeit ist durch neuere Ausgrabungen in der Jezirah und im Balikh-Tal bestätigt worden. Die Vergleichsstücke umfassen nur einfache, dickwandige Gefässformen, jedoch keine Tüllengefässe und ritzverzierten Fragmente. Für diese Gefässtypen zeigen früheisenzeitliche Siedlungen in Südost- und Ostanatolien die besten Parallelen. Obwohl im Zeitraum zwischen 1200-900 v. Chr. sowohl die archäologische als auch die schriftliche Quellenlage bisher nur sehr wenig Hinweise auf Verbindungen zwischen dem nordsyrischen und südostanatolischen Raum bietet, erscheinen engere Beziehungen zwischen den beiden Regionen, die im 2. Jahrtausend zum Einflussgebiet von Mitanni und Hatti gehörten und zu dieser Zeit zahlreiche Kontakte hatten, sehr wahrscheinlich. Die Verbindungen am Ende des 2. Jahrtausends lassen sich jedoch augenblicklich nur durch einige archäologische Hinweise am Tell Halaf vermuten. Einen eindeutigen Beweis für den o.g. Datierungsansatz einiger altmonochromer Gefässtypen können jedoch nur Nachgrabungen am Tell Halaf erbringen, deren Ziel die Untersuchung der vollständigen stratigraphischen Abfolge ist.

NOTEN

1. In der Baubeschreibung finden sich folgende Hinweise auf weitere Tiefgrabungen: (1) Nördlich des Tempelpalastes, wo starke Lehmziegelmauern gefunden wurden. Langenegger (1950:10) nahm an, dass es sich um Monumentalbauten der Buntkeramikzeit handele.
(2) Innerhalb des Nordsüd-Suchgrabens 3 liegen 2 Tiefgrabungen (Langenegger et al. 1950:10, Pl.16 unten). Auf Planquadrat C2 liegt ein Suchbrunnen mit der bekannten Abfolge, in C3 eine weitere Tiefgrabung.
(3) Viele Suchschächte von 2 m Durchmesser wurden durch einheimische Brunnenbauer bis auf die Felsbank eingetieft (Vergl. Langenegger et al. 1950:10).
(4) In der Mittelterrasse vor der Nordfront des Tempelpalastes waren zwei Getreidebrunnen eingetieft, in denen sich die Kopfsäulchen der Karyatiden befanden. Möglicherweise handelt es sich um die bei Schmidt (1943:25) beschriebenen Brunnen 1 und 2. (Vergl. Langenegger et al. 1950:65, Abb. 34).
(5) Tiefgrabung ausserhalb der Burgumwallung (unterste Schicht Buntkeramik; vergl. Langenegger et al. 1950:200, Pl. 2, Schnitt 3-4; Abb. 97).
(6) Brunnentiefgrabung nördlich des langen Anbaus an das assyrische Haus (westliche Halle). Die Abfolge ist hier: Kaparazeitliche Keramik, Buntkeramik, altmonochrome Ware, sterile Schicht, gewachsener Fels (vergl. Langenegger et al. 1950:201, Abb. 102).
(7) Über dem Innenhof des Palastes befand sich eine starke Schuttschicht, in die 5 m tiefe Getreidebrunnen eingeteuft waren. Zu unterst soll sich Buntkeramik befunden haben (Langenegger et al. 1950:294; Abb.134-135).
(8) Im Innenhof des Palastes wurden zwei Versuchsschächte eingetieft: A und B. In Schächt B fand sich Buntkeramik (Langenegger et al. 1950:297ff.; Abb. 138).
(9) Tiefschnitt vor der Südfront des Tempelpalastes (Langenegger et al. 1950, Pl. 11).
2. In diesem Zusammenhang danke ich Frau Dr. Moortgat-Correns, die mir freundlicherweise die in Berlin verbliebene unbemalte Keramik des Tell Halaf zeigte. Unter den noch vorhandenen Stücken fand sich jedoch kein rillenverziertes Stück mehr. Auch die noch vorhandenen Gefässtüllen gehören wahrscheinlich in jüngere Perioden. Einige sehr grobe Stücke sind sicher der altmonochromen Ware zuzuordnen.

BIBLIOGRAPHIE

AKKERMANS P.M.M.G.
1987 'A Late Neolithic and Early Halaf Village at Tell Sabi Abyad, Northern Syria', *Paléorient* 13: 23-40.
ALBRIGHT W.
1956 'The Date of the Kapara Period at Gozan (Tell Halaf)', *Anatolian Studies* 6: 75-85.
BARTL K.
1988 *Norşun-Tepe, Die frühe Eisenzeit*, Berlin: unpublished Dissertation.
BILGI O.
1985 '1984 Dönemi Malatya-Köşkerbaba Höyük Kurtama Kazilari', *Kazi Sonuçlari Toplantasi* 7: 143-149.
BOEHMER R.M.
1973 'Forschungen in und um Mudjesir (Irakisch-Kurdistan)', *Archäologischer Anzeiger* 4: 479-521.

BOWMAN R.A.
1941 'The Old Aramaic Alphabet at Tell Halaf. The Date of the 'Altar' Inscription', *American Journal of Semitic Languages* 58: 359-367.
BURNEY C.
1958 'Eastern Anatolia in the Chalcolithic and Early Bronze Age', *Anatolian Studies* 8: 157-209.
1980 'Aspects of the Excavations in the Altinova, Elazig', *Anatolian Studies* 30: 157-167.
BURTON-BROWN T.
1951 *Excavations in Azerbaidjan 1948*, London.
CANBY J.V.
1964 Review of 'Tell Halaf IV, Die Kleinfunde aus historischer Zeit', *American Journal of Archaeology* 68: 71-73.
CHANZADIAN C. ET AL.
1973 *Metsamor*, Erevan.
CILINGIROĞLU A.
1986 'Van-Dilkaya Höyüğü Kazilari, 1985', *Kazi Sonuçlari Toplantasi* 8.I: 81-93.
COPELAND L. AND HOURS F.
1987 'The Halafians, their Predecessors and their Contemporaries in Northern Syria and the Levant. Relative and Absolute Chronologies', in: O. Aurenche et al. (Eds.), *Chronologies in the Near East*, BAR International Series 379. Pp. 401-425.
DANKWARTH G. AND MÜLLER C.
1987 'Zur altaramäischen ''Altar''-Inschrift vom Tell Halaf', *Archiv für Orientforschung* 34, in press.
DAVIDSON T.
1977 *Regional Variation within the Halaf Ceramic Tradition*, Univ. Edinburgh, PH.D. thesis.
DAVIDSON T. AND WATKINS T.
1981 'Two Seasons of Excavations at Tell Aqab in the Jezirah, N.E. Syria', *Iraq* 43: 1-18.
DURU R.
1979 *Keban Project Değirmentepe Excavations 1973*, Keban Project Publications Series III, no. 2, Ankara.
ESIN U.
1970 'Tepecik Excavation, 1968 Preliminary Report', *1968 Summer Work*, Keban Project Publications 1. Pp. 159-172.
ESIN U. AND ARSEBÜK G.
1983 ' 1982 yili Değirmentepe (Malatya) kurtama kazisi', *Kazi Sunuçlari Toplantasi* 5 (1983): 71-79.
FRENCH D.H.
1986 'Tille', *Kazi Sonuçlari Toplantasi* 8.I: 205-212.
GOETZE A.
1928 'Die historische Einleitung des Aleppo-Vertrages (KBOI/6), *Mitteilungen der Altorientalischen Gesellschaft* 4: 59-66.
HAUPTMANN H.
1969 'Die Grabungen auf dem Norşun-Tepe', *Türk Arkeoloji Dergisi* 18/2: 111-121.
1969-70 'Norşun Tepe. Historische Geographie und Ergebnisse der Grabungen 1968/69', *Istanbuler Mitteilungen* 19-20: 21-78.
1971 'Die Grabungen auf dem Norşun-Tepe 1969', *Keban Project 1969 Activities*, Keban Project Publications 2, Ankara. Pp. 81-89.
1972 'Die Grabungen auf dem Norşun-Tepe 1970', *Keban Project 1970 Activities*, Keban Project Publications 3, Ankara. Pp. 103-117.
1974 'Die Grabungen auf dem Norşun-Tepe 1972', *Keban Project 1972 Activities*, Keban Project Publications 4, Ankara. Pp. 71-90.
1974 'Die Grabungen auf dem Norşun-Tepe. Bericht 'uber die 5. Kampagne', *Türk Arkeoloji Dergisi* 21/1: 59-73.
HAWKINS J.D.
1974 'Assyrians and Hittites', *Iraq* 36: 67-83.

1980 'Karkamiš', *Reallexikon der Assyriologie* 5: 426-446.

HERZFELD E.

1934 'Tell Halaf und das Problem der hethitischen Kunst', *Archäologische Mitteilungen aus Iran* 6: 111-223.

HROUDA B.

1962 *Tell Half IV. Die Kleinfund aus historischer Zeit*. Berlin.

KANTOR H.J.

1956 'Syro-Palestinian Ivories', *Journal of Near Eastern Studies* 15: 153-174.

KLENGEL H.

1968 'Die Hethiter und Išuwa', *Oriens Antiquus* 7: 63-76.

KARACA Ö.

1984 'Pirot Höyük 1983 Kazilari', *Kazi Sonuçlari Toplantasi* 6: 37-48.

KÜHNE H.

1977 'Zur historischen Geographie am unteren Habur', *Archiv für Orientforschung* 25: 249-255.

1978-79 'Zur historischen Geographie am unteren Habur, 2. vorläufiger Bericht über eine archäologische Geländebegehung', *Archiv für Orientforschung* 26: 181-195.

LANGENEGGER F. ET AL.

1950 *Tell Halaf II, Die Bauwerke*, Berlin.

LLOYD S. AND SAFAR F.

1945 'Tell Hassuna', *Journal of Near Eastern Studies* 4: 255-289.

LIPPERT A.

1979 'Die Östereichischcen ausgrabungen am Kordlar Tepe in Persisch-Westaserbeidjan (1971-1978)', *Archäologische Mitteilungen aus Iran* 12: 103-137.

VAN LOON M. (ED.)

1978 *Korucutepe*, II, Amsterdam.

LUCKENBILL D.D.

1926 *Ancient Records of Assyria and Babylonia*, SAOC I, Chicago.

MALLOWAN M.E.L.

1936 'The Excavations at Tell Chagar Bazar and an Archaeological Survey on the Habur Region, 1934-5', *Iraq* 3: 1-59.

McEWANS C.W. ET AL.

1958 *Soundings at Tell Fakhariyah*, OIP 79, Chicago.

MEIJER D.J.W.

1986 *A Survey in Northeastern Syria*, Istanbul.

MELLAART J.

1975 *The Neolithic of the Near East*, London.

MELLINK M.

1958 Review of A. Moortgat and D. Opitz, Tell Halaf III, Die Bildwerke (1955), *American Journal of Archaeology* 62: 438.

MITCHELL S.

1980 *Aşvan Kale*, Oxford: BAR International Series 80.

MOORTGAT A.

1955 *Tell Halaf III, Die Bildwerke*, Berlin.

1970 'Noch einmal zur Datierung des Kapara', in: A. Kuschke and E. Kutsch (Eds.), *Archäologie und Altes Testament*, Festschrift K. Galling, Tübingen. Pp. 211-217.

ÖĞUN B.

1983 'Habibusagi Kazisi 1982', *Kazi Sonuçlari Toplantasi* 5: 237-239.

ÖZDOĞAN M.

1979 *Lower Euphrates Basin 1977 Survey*, METU Lower Euphrates Project Publication Series I, No. 2, Istanbul.

OPPENHEIM M. VON

1931 *Der Tell Halaf. Eine neue Kultur im ältesten Mesopotamien*. Berlin.

ORTHMANN W.
1971 *Untersuchungen zur späthethitischen Kunst*, Saarbrücker Beiträge zur Altertumskunde Bd. 8, Bonn.
PECORELLA P.E.
1975 'Malatya III', *Oriens Antiqui Collectio* 12, Roma.
PECORELLA P.E. UND SALVINI M.
1975 Tra lo Zagros e l'Urmia. Ricerche Storiche ed Archeologiche Nell' Azerbaidian Irianno. Rom.
PORADA E.
1970 Review of: B. Hrouda, Tell Halaf IV, Die Kleinfunde aus historischer Zeit, *Artibus Asiae* 32: 97-99.
RUSSELL H.F.
1985 *Preclassical Pottery in Eastern Anatolia*, Oxford: BAR International Series 85.
SCHMIDT H.
1943 *Tell Halaf I, Die prähistorischen Funde*, Berlin.
SERDAROĞLU Ü.
1977 *Aşaği Firat Havzasinda Araştirmalari 1975*, Ankara.
SEVIN V.
1982 'Imikusagi Kazilari 1981', *Kazi Sonuçlari Toplantasi* 4: 121-129.
1983 'Imikusagi Kazilari 1982', *Kazi Sonuçlari Toplantasi* 5: 137-142.
SOROKIN W.
1958 'Remains of the Oldest Village in Karmir Blur', *Sowjetskaya Archaeologia* 2/1: 149-163.
SÜRENHAGEN D.
1986 'Ein Königssiegel aus Kargamiš', *Mitteilungen der Deutschen Orientgesellschaft* 118: 183-190.
UZUNOĞLU E.
1985 'Imamoğlu Kazilari 1984 Yili Calişmalari', *Kazi Sonuçlari Toplantasi* 7: 181-199.
VOGEL H. AND WATERBOLK H.T.
1964 'Groningen Radiocarbon Dates V', *Radiocarbon* 6: 349-369.
WÄFLER M.
1980-81 'Zum assyrisch-urartäischen Grenzkonflikt', *Acta Praehistorica et Archaeologica* 11-12: 86-97.
WATKINS T. AND CAMPBELL S.
1987 'The Chronology of the Halaf Culture', in: O. Aurenche et al. (Eds.), *Chronologies in the Near East*, BAR International Series 379. Pp. 427-464.
WATSON P.J.
1965 'The Chronology of North Syria and North Mesopotamia from 10.000 to 2.000 B.C.', in: R.W. Ehrich (Ed.), *Chronologies in Old World Archaeology*, Chicago. Pp. 65-100.
WHALLON R. AND KANTMAN S.
1979 *An Archaeological Survey to the Keban Dam Reservoir Area of East-Central Turkey*, Memoirs of the Museum of Anthropology, University of Michigan, Ann Arbor.
WILHELM G.
1982 *Grundzüge der Geschichte und Kultur der Hurriter*, Darmstadt.
WINN M.
1982 'The Iron Age Pottery', in: M.N. van Loon (Ed.), *Korucutepe*, 3, Amsterdam. Pp. 155-175.
WINTER E.
1973 *North Syria in the Early First Millenium with Special Reference to Ivory Carving*, New York.

The origins of the Aramaeans in Syria and northern Mesopotamia: Research problems and potential strategies

GLENN M. SCHWARTZ

The ancient Mesopotamian written sources refer to a succession of peoples from outside the urban sphere who pose a military threat to that sphere and periodically assume political power within the urban context themselves. In the later third millennium B.C., the Gutians are credited with the destruction of the Sargonic state and the establishment of a ruling dynasty in Mesopotamia, and the Amorites are viewed as similarly dangerous intruders in the succeeding Ur III period. Sutians, Ahlamu, and others are cited in the mid-to-late second millennium, and by the twelfth century the Aramaeans appear in the cuneiform record, eventually emerging as rulers of states throughout Syria and north Mesopotamia. Although these groups are conceptualized as outsiders by the urban scribes, a closer examination of the evidence often reveals a complex symbiotic relationship between the urban states and the 'alien' groups; thus, the Amorites are regarded as an external military threat, yet Amorite individuals simultaneously occupy diverse positions within Ur III society.

The Aramaeans are of particular interest in Near Eastern scholarship because their language, Aramaic, was in extensive use throughout the Near East for a period of a millennium or more. But of equal importance is the association of the early Aramaeans with the 'Dark Age' of the late second millennium B.C. Ancient Near Eastern urban civilizations appear to have experienced a cyclical pattern of centralization and fragmentation (Adams 1981:131ff.; Yoffee 1977), and the reasons for decentralization, the consequent social changes, and the precise extent of urban decline in the 'dark ages' are important issues directly tied to the problem of Aramaean origins. But while considerable attention has been given the Amorites, for example (Kupper 1957; Luke 1965; Buccellati 1966; Rowton 1976 with bibliography; Matthews 1978), relatively little has been determined of the origins of the Aramaeans and the nature of their 'takeover' of Syria and north Mesopotamia in the late second millennium B.C. Recent excavation projects in Syria have achieved signal progress in the study of the developments of the third and early second millennium at such sites as Mardikh-Ebla, Mari, Hammam-et-Turkman, Bi'a, Leilan and Brak, but little attention has been paid to the late second and early first millennium B.C. The following discussion is offered to evaluate the current state of the evidence, the benefits to be gained from further research, and potential avenues of investigation.

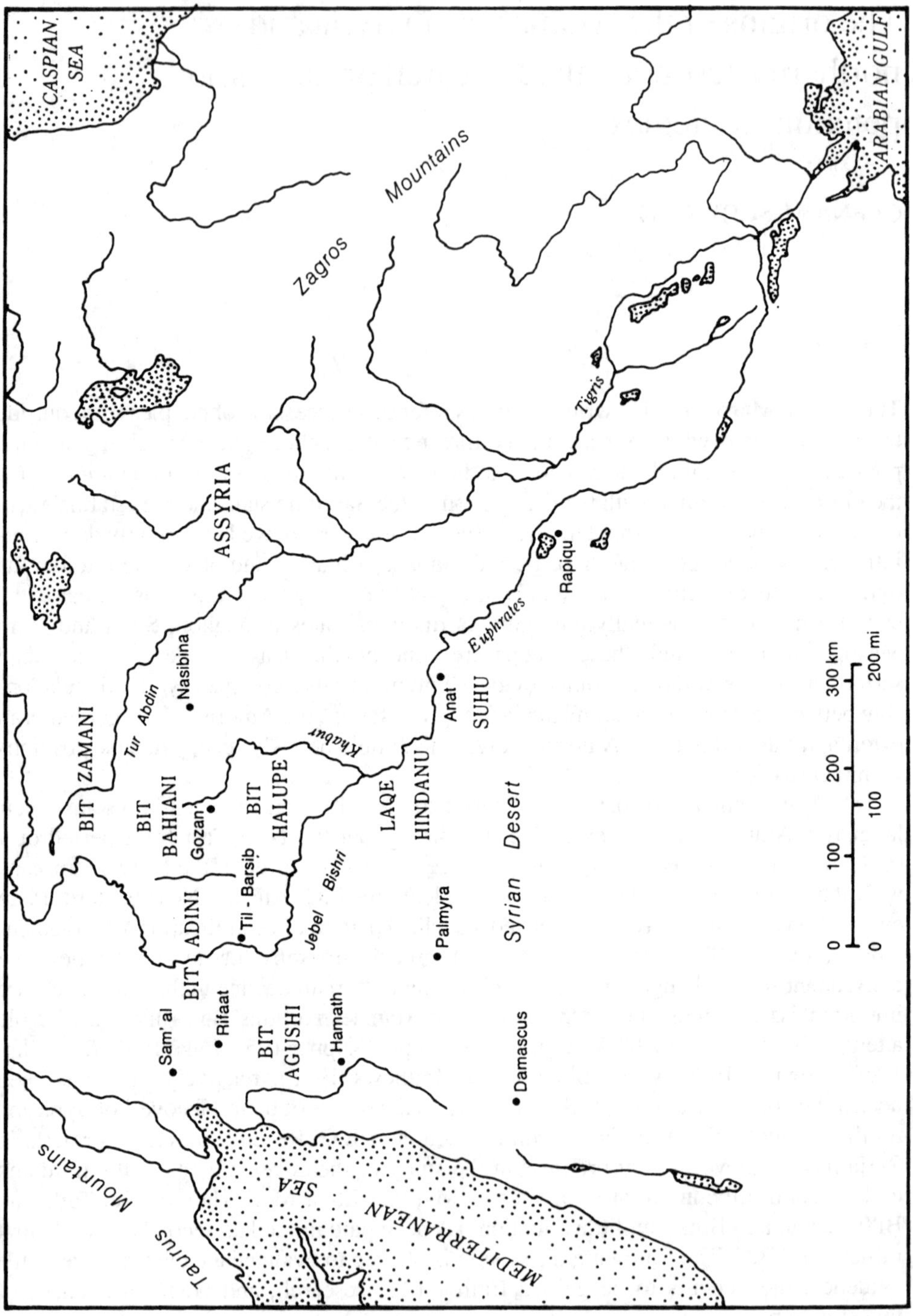

Fig. 1. The Near East in the period of the early Aramaeans.

THE NATURE OF THE EVIDENCE

The evidence for the early Aramaeans consists of historical and archaeological data, which I shall review before proceeding to an interpretive discussion.

Texts which mention names and regions bearing a resemblance to 'Aram' date back as early as Sargonic times, but these widely dispersed attestations are of dubious significance to the study of the Aramaeans of the late second millennium (Gelb 1961:28; Albright 1975:529). The first relevant references appear in the annals of Tiglath-pileser I (1114-1076). In the fourth year of his reign, the king conducted a campaign against the 'Ahlamu-Aramaeans' (*ah-la-me-i* ^{KUR}*ar-ma-a-ia*^{MEŠ}) (Grayson, *ARI* 2, §34), in which the desert from Suhu on the middle Euphrates to Carchemish was raided 'in one day', the river was crossed, and six settlements in the vicinity of Mount Bishri pillaged. In all, Tiglath-pileser crossed the Euphrates 28 times in campaigns against the Ahlamu-Aramaeans, confronting them in an area extending from Rapiqu on the middle Euphrates to Palmyra in the Syrian desert and even as far as the mountains of Lebanon (*ARI* 2, §83, 97)[1].

Towards the end of the reign of Tiglath-pileser I, a massive Aramaean onslaught on both Assyria and Babylonia occurred coincidentally with a disastrous crop failure and famine; in the next year, another famine and another invasion took place, forcing the Assyrian king to retreat to Katmuhi (*ARI* 2, §200) (Brinkman 1968:387). This same period of famine also witnessed an Aramaean incursion into Babylonia, then under the rule of Marduk-nadin-ahhe of the Isin II dynasty. (Brinkman 1968:92, 130, 126)[2].

The 'Broken Obelisk', attributable to the reign of Ashur-bel-kala (1073-1056), describes the continual conflict of the Assyrians against the Aramaeans, now without the Ahlamu designation, who are active in an area from the Tur Abdin (Mt. Kashiyari) to the Tigris as well as in the Habur valley (*ARI* 2, §239-247). Sources are few for the remainder of the eleventh century: Shalmaneser III records that Aramaeans had taken the cities of Pitru and Mutkinu on the great bend of the Euphrates in the reign of Ashur-rabi II (1012-972) (Luckenbill, *ARAB* 1, §603), and the annals of Ashur-dan II mention Aramean attacks in the reign of the same king (*ARI* 2, §363). Aramaean successes in the upper Habur (*ARI* 2, §427) and the Tur Abdin region (*ARI* 2, §641) in this period are also noted in the annals of later kings.

Ashur-dan II (934-912) claims the first Assyrian success against the Aramaean enemies (*ARI* 2, §363), and greater victories were achieved by Adad-Nirari II (911-891). Adad-Nirari's inscriptions supply the first references to Aramaean political units in northern Mesopotamia; the kingdoms of Bit Bahiani, Laqe, Hindanu, and Bit Halupe were defeated and compelled to pay tribute, and Bit-Adini sent the gift of two apes. Conflicts with 'Ahlamu-Aramaeans' are also recorded (*ARI* 2, §421-434).

THE ARAMAEAN STATES IN SYRIA AND NORTH MESOPOTAMIA

A constellation of states known from Assyrian royal annals or from indigenous royal inscriptions can be designated 'Aramaean' either because the sources specifically refer to the inhabitants or rulers as Aramaeans or because the rulers have Aramaic names. Many of these have the prefix 'Bit' ('House of'), generally taken to imply the 'tribal' nature of the societal organization (Postgate 1974:234). It is in this context that we encounter the first archaeological evidence from excavated sites relevant to the early Aramaeans. The following is a synthesis of the historical and archaeological data available from these states. All of the Aramaean political units were eventually incorporated into the neo-Assyrian provincial system, either in the course of the ninth century, in the case of the northwest Mesopotamian polities (i.e. in the Syrian Jezireh), or by the late eighth century, in the case of the west Syrian states.

Bit Adini was located on the great bend of the Euphrates with its capital at Til-Barsib. Originally under the control of rulers with neo-Hittite names and cultural associations, Bit Adini was later taken over by Aramaean rulers (Malamat 1973:147; Winter 1983:182; cf. Ussishkin 1971 for a reversal of the sequence)[3]. Aramaeans were present in the area by at least the early tenth century, according to the annals of Shalmaneser III. Excavations at Tell Ahmar, ancient Til-Barsib, identified an 'Aramaean' stratum tentatively dated to the ninth century with at least six large-scale multi-room buildings on either side of a street, the most impressive of which had two large courtyards, one paved with tiles (Thureau-Dangin and Dunand 1936:84-96).

Bit Bahiani was located in the eastern Habur triangle, with its capital at Gozan, modern Tell Halaf. The state was first attacked by Adad-Nirari II (911-891), at which time it was ruled by one Abi-salamu (*ARI* 2, §433). The ruler Kapara, attested from finds at Tell Halaf itself (see below), and Had-yiṯ'i (HDYS'Y) of the bilingual Aramaic/Akkadian Tell Fekheriyeh inscription are both dated to the ninth century (Abou Assaf et al. 1982). An altar fragment from Tell Halaf, also dated to the ninth century, bears the name (?) ZDMT in Aramaic (Donner and Röllig *KAI*, 231).

Excavations at Tell Halaf, conducted in the citadel of the first millennium city, discerned two periods before the Assyrian conquest: *Altbau*, with five phases, and *Neubau*, with one phase. *Altbau* buildings included the early *Tempel-Palast*, part of the Northeast Palace, the *Goldgruft*, and tomb installations under a later brick massif. The *Neubau* period was the era of Kapara, whose Akkadian inscriptions celebrated the rebuilding of the bit-hilani style *Tempel-Palast* (Hrouda 1962:117). The duration of the *Altbau* and *Neubau* phases could not have been longer than a century, judging from the architectural remains, and ceramic parallels with the middle to late Amuq O assemblage (which is as yet unpublished) suggest a ninth century B.C. date (Naumann 1950:376ff.; but cf. Canby 1964 on the possibility of earlier occupation). Particularly diagnostic of the ceramic assemblage are a red burnished ware also found in western Syria and a local painted ware with horizontal red stripes. A comparable assemblage was recovered in brief soundings at Tell Fekheriyeh, ancient Sikan, across the river from Tell Halaf, where part of an early first millennium B.C. monumental building of bit-hilani type was excavated (McEwan et al. 1958).

Another state seems to have existed east of Bit-Bahiani, south of the Tur Abdin in the vicinity of Nasibina (Nisibin); Adad-Nirari II mounted a series of campaigns against rulers in this area called 'Temanites' (*ARI* 2, §424-7) and noted that Gidara, one of its cities, had been held by Aramaeans since the reign of Tiglath-pileser II (966-935).

Bit Halupe was located on the lower Habur, with Suru its major center. It too was first mentioned in the campaigns of Adad-Nirari II; later in the ninth century, Ashurnasirpal II overthrew its rebellious ruler Ahi-yababa, who himself had displaced a governor loyal to the Assyrians (*ARI* 2, §547)[4].

Laqe, on the middle Euphrates near its juncture with the Habur, first appears in the campaigns of Adad-Nirari II (*ARI* 2, §434), as does *HINDANU*, downstream from Laqe.

Bit Zamani, in the Tur Abdin region, is first mentioned in a campaign of Tukulti-Ninurta II (890-884) against its ruler Amme-baala (*ARI* 2, §467), but Aramaean military power was operative in the region by at least the late eleventh century, according to Ashurnasirpal II (*ARI* 2, §641). The latter king, contending with a rebellion in the area, deported 1500 'troops of the Ahlamu-Aramaeans' to Assyria.

Bit Agushi, or *Yahan*, in northern Syria, may have been referred to in a text of Ashur-dan II noting the loss of two cities on the great bend of the Euphrates to (Ia?)-ha-nu in the reign of Ashur-rabi II (*ARI* 2, §363). Alternatively, it has been suggested that Yahan was under the control of neo-Hittite rulers until the ninth century, when the Aramaean Agushi established a new kingdom (Landsberger 1948:29). Subjugated by Shalmaneser III in 849, Bit Agushi's capital was changed from Arne to Arpad (Tell Rifa'at?), and the area was finally incorporated into the Assyrian provincial system ca. 740 B.C. The Sefire Treaties of the mid-eighth century, inscribed on steles found southeast of Aleppo, document an alliance between King Matī'-'el of Arpad and Barga'yā of KTK (*KAI* 222-224); associated with this alliance were 'all Aram' and 'all Upper and Lower Aram.'

Tell Rifa'at, possibly ancient Arpad, yielded excavated early first millennium B.C. remains in phase IIc (ca. tenth-ninth century), which had several destruction levels, and phase IIb (ca. ninth-seventh century) (Seton-Williams 1961; 1967; Matthers 1981: 415ff.). The architectural evidence from these phases includes a small sample of domestic structures; a monumental gate with associated orthostats dates from the same period. The ceramic assemblage consists of grit-tempered fruitstands, flat-and ring-base bowls, and platters, and is also characterized by coarse ware. Red-slipped and burnished platters are similar to examples from Hama E (Matthers et al. 1978:144). Excavations in the early part of the century uncovered what was said to be an Iron Age palace with 'two columns in the Hittite style' and a 'fragment of an inscription in Phoenician letters' on the site, but these have not been published (Matthers1981:415). Early first millennium occupation at Abou Danné, southeast of Aleppo, is documented in levels III-IV (Tefnin 1980; LeBeau 1983).

Sam'al, to the north, with its capital at the site of modern Zinjirli, is another putatively neo-Hittite state Aramaized somewhere between the tenth and eighth centuries (Landsberger 1948:21). The Samalian dialect appears on two eighth century inscriptions: *KAI* 214, a stele of Hadad dedicated by King Panammuwa, and *KAI* 215, an inscription of the king's son Barrakib recounting his father's achievements. In contrast, the orthostat of Barrakib relating that ruler's own accomplishments is written in standard Aramaic, not Samalian (*KAI* 216).

Zinjirli was surrounded by a double wall with towers and had a citadel which included an Upper Palace (possibly built in the ninth century by Kilamuwa) and a Lower Palace dating to the reign of Barrakib, both characterized by bit-hilani structures (Von Luschan 1893-1911).

Hamath, first ruled by neo-Hittites, became 'Aramaized' in the late ninth or early eighth century, according to the evidence of the mid-ninth century hieroglyphic Hittite inscriptions of Urhilinas (Irhuleni) and Uratamis and the late ninth/early eighth century Aramaic stele of Zakur (*KAI* 202, see below) (Hawkins 1972:68).

Excavations were conducted on the citadel of modern Hama (Fugmann 1958). Hama level E, primarily representative of the destruction level of ca. 720 B.C., consisted of a group of large-scale monumental buildings associated with epigraphic evidence including hieroglyphic Hittite inscriptions, Aramaic graffiti inscribed on bricks (*KAI* 203-213), and twenty cuneiform texts. The ceramics of Hama E included the red-slipped and burnished platters known from Rifa'at and the Amuq O assemblage.

Aram-Damascus, abundantly attested in the Old Testament as an adversary of Israel, was also a vigorous opponent of Assyria in the ninth century[5]. Made tributary by Adad-Nirari III (810-783), it was eventually reduced to provincial status ca. 732 B.C. A ninth century stele dedicated to Melqart by Bar-Hadad, 'king of Aram' (= Damascus) was discovered north of Aleppo (*KAI* 201), and an inscription found at Afis, southwest of Aleppo, describes the conflict between Zakur of Hamath and Bar-Hadad (III), son of Hazael of Aram (*KAI* 202). A late ninth century B.C. fragment from Arslantash also includes the name Hazael (*KAI* 232).

A precursor to Aram-Damascus may have been Aram-Zobah, an entity controlling much of southern Syria and the Beq'a which David reportedly campaigned against (I Samuel 14:47, II Samuel 8:3) (Malamat 1973:141-142).

SUMMARY

This review has shown that the historical and archaeological evidence for the early Aramaeans is biased in several crucial ways. First, the vast majority of the historical evidence originates, not with the Aramaeans themselves, but with their enemies, the kings of Assyria. Aramaic written evidence surfaces in the ninth century, but it is relatively rare and is only available two centuries after the first Aramaean appearance in the cuneiform record. Second, both the historical and archaeological evidence are biased chronologically. From the eleventh to the late tenth centuries, the relatively sparse material in the Assyrian royal annals is all that is available from the historical sources; only in the ninth century do both Assyrian and Aramaean written sources begin to proliferate. Likewise, the archaeological record is completely restricted to the ninth century and after. Finally, the evidence which we possess on the early Aramaeans is predominantly concerned with the royal, urban sector of Aramaean society. The Assyrian sources concentrate on military interaction with Aramaean rulers and the exaction of tribute from them, and the Aramaic inscriptions are predominantly royal documents concerned with royal affairs. The archaeological record has not filled this gap: excavations have thus far concentrated on monumental architecture in the capitals of the Aramaean states, with the inevitable consequence that no data are available on non-royal social strata or populations outside the large urban centers.

Our evidence, skewed as it is, affirms that the term Aramaean has utility as an ethnic designation for people of northern Mesopotamia and Syria in the late second and early first millennium. The term is employed not only by the Assyrians but by the Aramaeans themselves in the Sefire Treaties and in the title 'king of Aram' found on the Bar-Hadad inscriptions. The group is associated with a specific language, which is identified with their ethnic designation 'Aramaic', and the names of the rulers of the 'Aramaean' states are appropriately Aramaic[6].

While Syrian and northwest Mesopotamian material culture can be clearly distinguished from the material culture of surrounding regions, there is no significant difference between the neo-Hittite centers and the Aramaean centers, demonstrating the familiar point that material culture and ethnicity need not be congruent (Kamp and Yoffee 1980). On the other hand, the ethnic configuration of the general population in the neo-Hittite or Aramaean states is far from clear, and it is possible that the majority of the people of Syria and northwest Mesopotamia considered themselves Aramaeans and spoke Aramaic, whether their rulers were Aramaean or 'neo-Hittite'.

THE ARAMAEANS AND NOMADIC PASTORALISM

The physical environment associated with the Aramaeans in their earliest appearance in the historical record strongly suggests their association with an economic strategy of nomadic pastoralism. The earliest recorded campaign of Tiglath-pileser I against the Aramaeans is localized in the 'desert' (*mudbara*), the arid steppeland along the Euphrates from Suhu to Carchemish. Later attacks by Tiglath-pileser invariably took place across the Euphrates in the Syrian desert[7]. In an Adad-Nirari II inscription, the Ahlamu-Aramaeans are specifically called 'people of the steppe' (*ṣābē*^{MEŠ} *ṣēri*) (*ARI* 2, §421)[8]. The elusive nature and mobility of the Aramaeans - who failed to succumb to Tiglath-pileser's army despite twenty-eight campaigns - and the arid physical environment of the region (ca. 100-200 mm average annual rainfall today) (Wirth 1971) imply the utilization of anomadic pastoral mode of subsistence.

The conventional interpretation of the Aramaean appearance in the neo-Assyrian records is that of an onslaught of nomadic peoples sweeping in from the heart of the desert to overwhelm the sedentary zone (Malamat 1973:135; Klengel 1972:74; Albright 1975:532; Snell 1985:326). But scholars have recently become aware of the complex nature of nomadic-sedentary interaction, the movement back and forth between strategies of nomadic pastoralism and sedentary agriculture, and the degrees of mixture of both strategies by different groups. Research on the Amorites of the Mari archives of the early second millennium B.C. (Kupper 1957; Luke 1965; Rowton 1976; Matthews 1978) has shown that Amorite groups included nomadic pastoralists (who could occasionally engage in agriculture), villagers, and urban dwellers; the risks involved in Mesopotamian agriculture and pastoralism alike resulted in a continual movement back and forth between these two economic strategies and the stages in between (Adams 1974), such that a strict dichotomy of nomad/sedentary is an inadequate representation of the ancient reality. Further, it has been shown that the nomads, rather than keeping to the fringes of sedentary society, moved well within the borders of the settled zone, where nomad and sedentist existed in a mutually dependent symbiotic relationship.

This model of 'enclosed nomadism' (Rowton 1974) is applicable for the late second millennium B.C. as well. The Aramaeans are associated with the Euphrates and the Syrian desert on their first appearance in cuneiform annals, the same areas associated with pastoral nomadic groups known in the Mari archives. These regions are traditionally associated with sheep and goat pastoralists. Migration routes followed by nomadic pastoralists of the Mari archives and by sheep/goat bedouin of recent ethnographic literature extend from winter pastures as far south as the central Syrian desert to summer pastures in the sedentary zones of Mesopotamia or western Syria: the Sutu of the Mari texts, for example, ranged from the inner Syrian desert near Palmyra to the Syrian Jezireh north of the Euphrates (Luke 1965:114-121; Klengel 1972:51; Heltzer 1981) and modern sheep/goat pastoralists such as the Anaiza, Shammar and Tayy winter as far south as the Syrian desert and utilize summer pasture as far north as the dry farming plains of the Habur triangle (Dillemann 1962:74; Muller 1931:26; Von Oppenheim 1939).

Possible evidence for Aramaean mixed pastoral and sedentary economic strategies, characteristic of enclosed nomadism, may be gleaned from the historical record: Tiglath-pileser I claims to have destroyed six settlements (URU.MEŠ) of Aramaeans in the Jebel Bishri region, which could have been assemblies of tents (Briant 1982:140) but may have been communities of sedentary Aramaeans. During his campaigns against the urban-based Aramaean states, Adad-Nirari II also battled Ahlamu-Aramaeans, 'people of the steppe'. Muquru the 'Temanite' is said to trust in his 'fortified city, his strong bow, his extensive troops, and the Aramaeans' (*ARI* 2, §427), who may signify a nomadic Aramaean group employed militarily, analogous to the nomads in the Mari military or to the maintenance of 'reservoirs' of nomadic pastoralists in the steppe by east Asian dynasties of nomadic origin (Lattimore 1940:248). Similarly, Ashurnasirpal II reports that he exiled '1500 troops of the Ahlamu-Aramaeans belonging to Amme-baala', ruler of Bit Zamani, to Assyria (*ARI* 2, §641)[9].

The alternate model for the origins of the Aramaeans involves a 'wave' of people moving in from the desert to overwhelm the agricultural zone. Albright maintained that the Aramaeans were camel nomads, not sheep and goat pastoralists, and that their use of camels was an important variable in their military as well as mercantile success (Albright 1975: 532). This hypothesis was based on analogy with the supposedly contemporary Midianites of the Old Testament who are said to have 'camels without number' (Judges 6:5) (Malamat 1952:5). But the evidence in favor of Aramaeans as camel nomads is slight.

The dromedary was apparently domesticated in Arabia by the end of the second millennium B.C. (Ripinsky 1975:296; Zarins 1978; Lambert 1960). In the early first millennium, the long-distance mobility afforded by the domestication of the dromedary allowed for the participation of Arabian camel nomads in the conduct of the incense trade from south Arabia[10]; it appears that the incense trade and the domestication of the camel were interdependent processes (Eph'al 1982; Liverani 1987:72; Briant 1982:142). But there is no evidence for the involvement of Aramaean nomadic pastoralists in this system, which developed several centuries after the Aramaeans' initial appearance in the cuneiform record.

Nor are there significant military associations of Aramaeans with camels in the Assyrian sources. We first encounter camels in a military context at the battle of Qarqar, where Shalmaneser III fought a confederation which included 1,000 Arabian camel riders (*ARAB* 1, §611). In the eighth century, 30,000 camels were collected as booty from the Arabian queen Samsi, after which the camel was used ever more extensively as a beast of burden in the Assyrian army (Bulliet 1975:84). The Assyrian evidence indicates that the only Assyrian opponents who made extensive military use of camels were the Arabians (*Aribi*).

An orthostat from the *Tempel-Palast* at Tell Halaf depicts a camel with its rider (Moortgat 1955, Plate 27a), but this lone camel amongst 179 small orthostats from the *Tempel-Palast*, with representations of archers, spearmen, slingers, horsemen, and chariots, does not imply a significant role for the camel in the Aramaean army, much less the Aramaeans' original status as camel nomads. Tukulti-Ninurta II received tribute from Amme-alaba of Hindanu which included thirty camels (*ARI* 2, §472; cf. also *ARI* 2, §547; *ARAB* 1, §598), the only instance of tribute in dromedaries collected from a non-Arabian ruler. It has been proposed that Hindanu was a node along the incense trade route from south Arabia, accounting for the presence of camels there and the tribute exacted (Jankowska 1969:269; Bulliet 1975:78). Given the willingness of the Assyrians to collect thousands of camels from the Arabians, one would expect to find larger numbers of camels taken as tribute from the Aramaean states if the animals were as important and numerous as Albright maintained.

I would suggest that the term Aramaean was simply a new ethnic designation for sheep/goat nomadic pastoralists operating in the Euphrates and Habur regions in patterns comparable to the nomadic pastoralists of preceding centuries. The Amorites of the Mari letters are the first historically attested of these groups, followed by the Sutu, who initially appear, like the early Aramaeans, in an area from the Syrian desert to Mesopotamia (Heltzer 1981), followed by the Ahlamu, mentioned with particular frequency in mid-second millennium Assyrian sources (Brinkman 1968: 277, note 1799). References to the Sutians and Ahlamu decline when Aramaean references appear in force, and later references to both groups are probably not ethnolinguistic terms but designations for 'nomads', much as 'Kurd', 'Arab' or 'Turk' have been used in medieval and subsequent periods (Kramer 1977:92).

The Aramaeans appear on the scene in the middle Assyrian sources as the Ahlamu disappear. If the Aramaeans are indeed the latest manifestation of a nomadic pastoralist group engaged in 'enclosed nomadism' in north Mesopotamia and Syria, we must address the questions of why the Assyrians used a new ethnic term to distinguish them and why they spoke a hitherto unknown language.

It is evident that ethnic designations are relevant to the ancient reality, since the Aramaeans identified themselves with the same ethnic term as that used by the Assyrian sources [11]. We need to know, therefore, why and when ethnic affiliations change (Keyes 1979; 1981). In the case of nomadic pastoralists, social organizations are characteristically fluid, tribal reintegration and disintegration are constants, and tribal names are continually changing (Rowton 1977:185). One reason for this is military: nomadic pastoralists typically engage in mutual conflict, and tribal affiliations change as political fortunes change (Marx 1978:59; Lambton 1971). Economic variables can also account for shifting ethnic identities among pastoralists (Haaland 1969). Further, Barth (1969) shows how different political structures of adjacent ethnic groups can result in changing ethnic affiliation. The egalitarian structure of southern Pathan groups in Pakistan makes it difficult to incorporate new group members successfully, while the Baluch authoritarian political structure allows for easy assimilation of new individuals. As a consequence, individuals dislodged from their social contexts due to economic or other factors are drawn into a Baluch political structure rather than a Pathan, and the Pathans who join

Baluch society are subsequently identified as ethnic Baluch. It has often been suggested that the excessive exploitation of the agricultural producers by the royal establishment and the urban elite of the Late Bronze Age resulted in the creation of large numbers of runaway social outcasts, who would have added to the ranks of nomadic groups (Liverani 1987). Such groups that were particularly conducive to the receipt of new members may have consequently increased in power.

The second question hinges on the problem of our considerable ignorance of the linguistic geography of late second millennium Syria and northern Mesopotamia. Aramaic may not have been a new language in the area at all, nor need it have been exclusively associated with nomadic pastoralists: our knowledge of the Semitic languages spoken in Syria and north Mesopotamia in the late second millennium B.C. is meager in the extreme, and one cannot rule out the possibility that Aramaic was spoken in the region by sedentists as well as nomadic pastoral groups[12]. The Amorite names from Mari have sometimes been hypothesized to be 'proto-Aramaic' (Edzard 1964:142; but cf. Albright 1975:530), and the possibility of a development of Amorite into what eventually became Aramaic may not be out of the question. If, on the other hand, Aramaic was a newly important language in the area, its preeminence could be linked to the recently acquired political power of its associated ethnic group. Linguistic diversity in nomadic pastoralist ethnic groups is observable in the Iranian Zagros, for example, where Kurds, Lurs and Bakhtiari each speak their own dialects which are interrelated but not identical.

Explanatory discussion of the Aramaean conflict with Assyria in the late twelfth and eleventh centuries must be tentative, but certain variables deserve mention. The cuneiform sources disclose that the first major Aramaean invasion coincided with two years of crop failure and famine (Brinkman 1968:92). Short term climatic change has been noted as an effective trigger for social change in marginal areas (Parry 1978; Oates 1972:77), and desiccation of winter pastures is a well-known incentive for nomadic pastoral groups to move further into the wetter agricultural zones, a movement often accompanied by military action against the sedentists. A longer-term dry period has also been invoked as a factor in the breakdown of urban systems in the late second millennium and could have been an important variable in the movement of nomadic pastoralists further into the agricultural zones (Neumann and Parpola 1987), though one hesitates to venture too far into the realm of environmental determinism. The military advantages of nomad pastoralists over the sedentary are celebrated, particularly in periods of weak central authority in the urban-based zone (Asad 1973), and the collapse of the Hittite empire in the early twelfth century left a power vacuum in northern Mesopotamia which the Assyrians only fitfully exploited, the campaigns of Tiglath-pileser consisting more of raids for plunder than the establishment of political control.

But it may be a mistake to conceive of this conflict as a struggle solely between nomadic pastoralists and Assyrian troops. Although it is clear that Aramaeans were engaged in nomadic pastoralist subsistence strategies, nomadic pastoralists may not have constituted the entirety of the Aramaean ethnic group; ethnic identity is not necessarily restricted to one economic strategy, and nomads and sedentists may have the same ethnic affiliation (Rosman and Rubel 1976; Dyson-Hudson and Dyson-Hudson 1980), as the Amorites of the Mari archives demonstrate. Our views tend to be Assyro-centric, due to the origin of our sources, but we must remember that urban societies had existed for centuries in the Habur and Euphrates valleys with only occasional political association with the Assyrians, and military action against the Assyrians could have been taken by sedentists as well as by nomadic pastoralists, both of whom may have included ethnic Aramaeans among their numbers. The processes involved in the subsequent formation of Aramaean-ruled states in Syria and north Mesopotamia and the role of nomadic pastoralists in such processes have yet to be clarified, but again, an appropriate analogue for the construction of explanatory models would be the Amorites of the early second millennium and the development of states ruled by Amorite individuals.

CONCLUSIONS

To summarize, the Aramaeans represented in Assyrian sources as nomadic pastoralists of the Syrian desert and northwest Mesopotamia appear to be a late incarnation of the same sheep and goat pastoralists who interacted with the sedentists in the Syrian Jezireh from the Amorites on in a complex process of economic, political, and ethnic symbiosis. This interpretation rests on relatively sparse evidence, however, and much more research needs to be done to better understand this phenomenon. The brunt of new research in this field must be borne by archaeologists, since Aramaic texts are not likely to turn up in great numbers due to the perishability of the materials they were written on, nor is it likely that the bureaucratic systems of the late second millennium maintained extensive cuneiform record-keeping, given the collapse of scribal schools and palace administration at the end of the thirteenth century (Liverani 1987; Tadmor 1979:1).

Regional surveys have begun to contribute information on settlement pattern change on the ancient Syrian landscape, but these have been hampered thus far by inadequate ceramic chronologies for the late second and early first millennium B.C. Monchambert's survey of the middle Habur, for example, suggests considerable population growth from the Late Bronze to Iron Ages. Six Late Bronze sites were recognized, all but one on the east bank of the river, but as many as 28 Iron Age sites were identified (Monchambert 1984). Continuity of settlement is indicated by the presence of Iron Age pottery on all the Late Bronze Age sites. But because there is no intra-period differentiation, one cannot determine if Iron Age occupation dates to the period of Aramaean political autonomy or to the subsequent period of neo-Assyrian control.

A contrary pattern of population decline and discontinuity is exhibited by the results of Meijer's survey of the area surrounding Tell Leilan in the eastern Habur triangle (Meijer 1986): 38 sites had Late Bronze material, 21 sites had Iron Age material, and only six sites yielded sherds from both periods. But again the ceramic uncertainties (Meijer 1986:33) require that these data be viewed with extreme caution. Likewise, eight sites are associated with the Late Bronze period in a survey of the Balikh valley, but only two are dated to the Iron Age (Akkermans 1984).

A survey of the Qoueiq valley in the vicinity of Tell Rifa'at tentatively identified a pattern of growth in the Iron Age, with nine Late Bronze sites and 30 Iron Age sites; six sites had traces of both Late Bronze and Iron Age occupation (Matthers 1981, Fig. 238). General preliminary results have been presented from a survey of the middle and lower Habur, indicating continued occupation from Late Bronze to Iron (Kühne 1974/77:254), and from surveys of the Euphrates between the Tabqa dam and the Tell Ashara vicinity (Kohlmeyer 1984; Simpson 1984).

The inadequate control over ceramic chronology renders these results necessarily tentative: until there is a useful ceramic sequence spanning the late second to early first millennium, material from surface survey cannot be dated with any precision. Most important, it will be necessary to distinguish between late second millenium, early first millennium, and neo-Assyrian conquest material. These gaps in our archaeological knowledge exist in large part because the great majority of relevant excavation was conducted in the early decades of this century, when little attention was paid to stratigraphic precision, ceramic dating, or publication of artifacts of minimal aesthetic or historic interest. A renewal of interest in the problems of this period should generate important new results with speed. Once chronological control is achieved, many important questions on this period can be addressed: data from surface survey can be used to examine issues relevant to the extent of urban collapse and pastoralist growth in the late second millennium, through examination of the changes in the size and number of larger centers, the proportion of settled area occupied by large centers relative to small settlements, the area dominated by individual urban centers, and the extent of land under cultivation. Excavation can contribute data on changes in social systems and economic strategies - the degree of reliance on pastoralism, for example, can be studied via analysis of faunal material (Zeder 1984). With the acquisition of new ceramic sequences, new research agendas will be able to effect significant changes in our understanding of the late second millennium 'dark age' and those ethnic groups associated with it.

ACKNOWLEDGEMENTS

I offer this study as a small token of my esteem for Maurits N. van Loon, whose interests and accomplishments have encompassed a wide variety of disciplines and a broad range of issues from the Neolithic period to the first millennium B.C. and beyond; I concern myself here with the later end of this timescale.

NOTES

1. The term Ahlamu-Aramaean probably means 'nomad Aramaean' and need not signify an identification of the Aramaeans with the Ahlamu nomads of the mid-second millennium B.C. or a confederation of two nomadic groups (Grayson, *ARI* 2, note 70; Albright 1975:529; Brinkman 1968:277).
2. In Babylonia, depredations by Aramaeans and 'Sutians' continued until the ninth century B.C., and references to sedentary Aramaeans in the Babylonian countryside appear by the eighth century (Brinkman 1968:267ff.).
3. While north Syrian art and architecture is distinguishable from contemporary Assyrian styles, there is no obvious distinction between the material culture of the neo-Hittite states and that of the Aramaean kingdoms (Frankfort 1954:166; Winter 1983:180).
4. Assyrian control of the middle Habur during the reign of Ashur-rabi II (1012-972) may be attested by a cylinder from Assur with an inscription of one Bel-erish, who records various building projects in the region of Shadikanni (*ARI* 2, §346). Millard (1970:172-3) dates the inscription to the reign of Ashur-dan I in the twelfth century, however.
5. The patriarchal references to Aram and Aramaeans are of such uncertain historicity as to be better left out of this discussion (van Seters 1975; Thompson 1974).
6. See Kamp and Yoffee (1980) and Narroll (1964) on the definition and utility of the concept of ethnicity. It is noteworthy that the term Aramaean used in a Syrian or north Mesopotamian context virtually disappears from the Assyrian royal annals after the reign of Ashur-dan II, when the names

of individual states are mentioned instead. Exceptions occur with reference to past Aramaean military successes which took place prior to the reign of Ashur-dan II, in the individual references to Ahlamu-Aramaeans by Adad-Nirari II and Ashurnasirpal II, and in the designation of the state in the Damascus region as 'Aram' (*ARAB* 1, §563).

7. The association of the Aramaeans with the Jebel Bishri parallels the Amorites' association with the same region early in their appearance in the cuneiform record.
8. *CAD ṣēru* A 3f, 'the defeat of the peoples of the steppeland was effected' (p. 145); Grayson's translation is 'field troops'.
9. According to evidence of the Aramaeans living in Babylonia, Aramaeans characteristically possessed a 'tribal' organization, each tribe with one or more 'sheikhs' (*nasīkātu*) (Brinkman 1968:273; cf. Marx 1977 on the utility of the concept of tribe in nomadic pastoralist contexts). The term nasīku is used by the Assyrians for the ruler of Laqe as well. Tribal affiliation of the Aramaeans in Syria and northern Mesopotamia may also be indicated by the designation of the rulers Nur-Adad, Muquru, and Mammi as 'Temanites'.
10. Camel-raising nomads' migration routes can cover up to 800 km, while sheep and goat nomads rarely travel more than 200 km (Müller 1931:96ff.; Wirth 1971:255-256).
11. The considerable importance of ethnic identity among contemporary Near Eastern pastoralists is well-attested in the ethnographic literature (Rosman and Rubel 1976).
12. For northwest Semitic at Emar, see Huehnergard 1983:39; see also Sivan 1984 and Huehnergard 1987 on northwest Semitic elsewhere in inland Syria. No useful ethnolinguistic data on the Syrian Jezireh in the pre-Aramaean era is forthcoming from the middle Assyrian sources.

REFERENCES

ABOU ASSAF A. ET AL.
1982 *La Statue de Tell Fekherye et son Inscription bilingue assyro-araméenne.* Paris: Editions Recherche sur les Civilisations.

ADAMS R. McC.
1974 'The Mesopotamian Social Landscape: A View from the Frontier', in: C. Moore (Ed.), *Reconstructing Complex Societies*, Supplement to the *Bulletin of the American Schools of Oriental Research*, 20. Baltimore. Pp. 1-11.
1981 *Heartland of Cities.* Chicago and London: University of Chicago Press.

AKKERMANS P.M.M.G.
1984 'Archäologische Geländebegehung im Balih-Tal', *Archiv für Orientforschung* 31: 188-189.

ALBRIGHT W.F.
1975 'Syria, the Philistines, and Phoenicia', in: *Cambridge Ancient History II, 2.* Cambridge: Cambridge University Press. Pp. 507-536.

ASAD T.
1973 'The Beduin as a Military Force',in: C. Nelson (Ed.), *The Desert and the Sown.* Berkeley: University of California Press. Pp. 59-68.

BARTH F.
1969 'Pathan Identity and its Maintenance', in: F.Barth (Ed.), *Ethnic Groups and Boundaries.* Boston: Little,Brown. Pp. 117-134.

BRIANT P.
1982 *État et Pasteurs au Moyen-Orient ancien.* Cambridge and Paris: Cambridge University Press and Éditions de la Maison des Sciences et de l'Homme.

BRINKMAN J.A.
1968 *A Political History of Post-Kassite Babylonia.* Rome: Pontificium Institutum Biblicum.

BUCCELLATI G.
1966 *The Amorites of the Ur III Period*, Naples: Istituto Orientale.

BULLIET R.W.
1975 *The Camel and the Wheel.* Cambridge: Harvard University Press.

CANBY J.V.

1964 Review of B. Hrouda, *Tell Halaf IV*, *American Journal of Archaeology* 68: 71-72.

DILLEMANN L.

1962 *Haute Mésopotamie Orientale et Pays Adjacents*. Paris: Librairie Orientaliste Paul Geuthner.

DONNER H. AND RÖLLIG W.

1964 *Kanaanäische und Aramäische Inschriften* (*KAI*). Wiesbaden: Otto Harrassowitz.

DYSON-HUDSON R. AND DYSON-HUDSON N.

1980 'Nomadic Pastoralism', *Annual Review of Anthropology* 9:15-61.

EDZARD D.O.

1964 'Mari und Aramäer', *Zeitschrift für Assyriologie* 56: 142-149.

EPH'AL I.

1982 *The Ancient Arabs*. Jerusalem: Magness Press.

FRANKFORT H.

1954 *The Art and Architecture of the Ancient Orient*. Harmondsworth: Penguin.

FUGMANN E.

1958 *Hama, fouilles et recherches 1931-1938. L'architecture des périodes pré-hellénistiques*. Copenhagen: Nationalmuseet.

GELB I.J.

1961 'The Early History of the West Semitic Peoples', *Journal of Cuneiform Studies* 15: 27-47.

GRAYSON A.K.

1976 *Assyrian Royal Inscriptions, Volume 2* (*ARI* 2) Wiesbaden: Otto Harrasowitz.

HAALAND G.

1969 'Economic Determinants in Ethnic Processes', in: F.Barth (Ed.), *Ethnic Groups and Boundaries*, Boston: Little, Brown. Pp. 58-73.

HAWKINS J.D.

1972 'Hamath', *Reallexikon der Assyriologie* IV: 67-70.

HELTZER M.

1981 *The Sutaeans*, Naples: Istituto Universitario Orientale.

HROUDA B.

1962 *Tell Halaf IV: Die Kleinfunde aus historischer Zeit*. Berlin: Walter de Gruyter.

HUEHNERGARD J.

1983 'Five Tablets from the Vicinity of Emar', *Revue d'Assyriologie* 77: 11-43.

1987 'Northwest Semitic Vocabulary in Akkadian Texts', *Journal of the American Oriental Society* 107: 713-726.

JANKOWSKA N.B.

1969 'Some Problems of the Economy of the Assyrian Empire', in: I. Diakonoff (Ed.), *Ancient Mesopotamia*. Moscow: Nauka. Pp. 253-276.

KAMP K. AND YOFFEE N.

1980 'Ethnicity in Ancient Western Asia During the Early Second Millennium B.C.: Archaeological Assessments and Ethnoarchaeological Perspectives', *Bulletin of the American Schools of Oriental Research* 237: 85-104.

KEYES C.F.

1981 'The Dialectics of Ethnic Change', in: C. Keyes (Ed.), *Ethnic Change*. Seattle: University of Washington. Pp. 4-30.

KEYES C.F. (Ed.)

1979 *Ethnic Adaptation and Identity*, Philadelphia: Institute for the Study of Human Issues.

KLENGEL H.

1972 *Zwischen Zelt und Palast*. Vienna: Verlag Anton Schroll.

KOHLMEYER K.

1984 'Archäologische Geländebegehung im Euphrat-Tal zwischen Ṭabqa/aṯ-Ṯaura und Dair az-Zaur', *Archiv für Orientforschung 31: 190-193*.

KRAMER C.

1977 'Pots and Peoples', in: L. Levine and T.C. Young (Eds.), *Mountains and Lowlands*. Malibu: Undena. Pp. 91-112.

KÜHNE H.

1974/7 'Zur historischen Geographie am Unteren Habur', *Archiv für Orientforschung 25: 249-255.*

KUPPER J.-R.

1957 *Les Nomades en Mésopotamie au temps des rois de Mari.* Paris: Societé d'Édition 'Les Belles Lettres'.

LAMBERT W.G.

1960 'The Domesticated Camel in the Second Millennium', *Bulletin of the American Schools of Oriental Research* 160: 42-43.

LAMBTON A.K.S.

1971 'Īlāt', *Encyclopedia of Islam III.* Leiden: E.J. Brill. Pp. 1095-1110.

LANDSBERGER B.

1948 'Sam'al', Ankara: Drückerei der Türkischen Historischen Gesellschaft.

LATTIMORE O.

1940 *Inner Asian Frontiers of China*, Boston: Beacon Press.

LEBEAU M.

1983 *La céramique de l'Age du Fer II-III à Tell Abou Danné.* Paris: Éditions Recherches sur les Civilisations.

LIVERANI M.

1987 'The Collapse of the Near Eastern Regional System at the End of the Bronze Age: the Case of Syria', in: M. Rowlands et al. (Eds.), *Centre and Periphery in the Ancient World*, Cambridge: Cambridge University Press. Pp. 66-73.

LUCKENBILL D.D.

1926 *Ancient Records of Assyria and Babylonia (Vol.1), (ARAB).* Chicago: University of Chicago Press.

LUKE J.T.

1965 *Pastoralism and Politics in the Mari Period*, Unpublished Ph.D. dissertation, University of Michigan.

MALAMAT A.

1952 *The Aramaeans in Aram Naharaim and the Rise of their States* (Hebrew). Jerusalem: Israel Exploration Society.

1973 'The Aramaeans', ,in: D.J. Wiseman (Ed.), *Peoples of Old Testament Times*, Oxford: Clarendon Press. Pp. 134- 155.

MARX E.

1977 'The Tribe as a Unit of Subsistence: Nomadic Pastoralism in the Middle East', *American Anthropologist* 79: 343-363.

1978 'The Ecology and Politics of Nomadic Pastoralists in the Middle East', in: W. Weissleder (Ed.), *The Nomadic Alternative.* The Hague: Mouton. Pp. 41-74.

MATTHERS J. ET AL.

1978 'Tell Rifa'at 1977: Preliminary Report of an Archaeological Survey', *Iraq* 40: 119-162.

MATTHERS J. (ED.)

1981 *The River Qoueiq, North Syria, and its Catchment.* Oxford: British Archaeological Reports, International Series, 98.

MATTHEWS V.H.

1978 *Pastoral Nomadism in the Mari Kingdom (ca. 1830-1760 B.C.).* Cambridge: American Schools of Oriental Research.

MEIJER D.J.W.

1986 *A Survey in Northeast Syria*, Leiden: Nederlands Archeologisch-Historisch Instituut te Istanbul.

MILLARD A.R.

1970 'Fragments of Historical Texts from Nineveh: Middle Assyrian and Later Kings', *Iraq* 32: 167-176.

MONCHAMBERT J.-Y.

1984 'Le futur lac du Moyen Khabour: Rapport sur la prospection archéologique menée en 1983', *Syria* 61: 181-218.

MOORTGAT A.
1955 *Tell Halaf III: Die Bildwerke*, Berlin: Walter de Gruyter.
MÜLLER V.
1931 *En Syrie avec les Bedouins, Paris: Librairie Ernest Leroux.*
NARROLL R.
1964 'On Ethnic Unit Classification', *Current Anthropology* 5: 283-291.
NEUMANN R.
1950 *Tell Halaf II: Die Bauwerke*, Berlin: Walter de Gruyter.
NEUMANN J. AND PARPOLA S.
1987 'Climatic Change and the Eleventh-Tenth Century Eclipse of Assyria and Babylonia', *Journal of Near Eastern Studies* 46: 161-182.
OATES D.
1972 'The Excavations at Tell al Rimah, 1971', *Iraq* 34: 77- 86.
PARRY M.L.
1978 *Climatic Change, Agriculture and Settlement.* Hamden, Ct.: Archer.
POSTGATE J.N.
1974 'Some Remarks on Conditions in the Assyrian Countryside', *Journal of the Economic and Social History of the Orient 17: 225-243.*
RIPINSKI M.
1975 'The Camel in Ancient Arabia', *Antiquity* 49: 295-298.
ROSMAN A. AND RUBEL P.G.
1976 'Nomad-Sedentary Interethnic Relations in Iran and Afghanistan', *International Journal of Middle Eastern Studies* 7: 545-570.
ROWTON, M.
1974 'Enclosed Nomadism', *Journal of the Economic and Social History of the Orient* 17: 1-30.
1976 'Dimorphic Structure and Topology, *Oriens Antiquus* 15: 17-31.
1977 'Dimorphic Structure and the Parasocial Element', *Journal of Near Eastern Studies* 36: 181-198.
SETON-WILLIAMS M.V.
1961 'Preliminary Report on the Excavations at Tell Rifa'at', *Iraq* 23: 68-87.
1967 'The Excavations at Tell Rifa'at, 1964. Second Preliminary Report', *Iraq* 29: 16-33.
SIMPSON K.
1984 'Archaeological Survey in the Vicinity of Tall al-Asharah', *Archiv für Orientforschung* 31: 185-188.
SIVAN D.
1984 *Grammatical Analysis and Glossary of the Northwest Semitic Vocables in Akkadian Texts of the 15th-13th Century B.C. from Canaan and Syria.* Kevelaer: Neukirchener Verlag.
SNELL D.C.
1985 'The Aramaeans', in: H. Weiss (Ed.), *Ebla to Damascus: the Art and Archaeology of Ancient Syria.* Washington D.C.: Smithsonian Institution Traveling Service. Pp. 326-329.
TADMOR H.
1979 'The Decline of Empires in Western Asia ca. 1200 B.C.E.', in: F. Cross (Ed.), *Symposia*, Cambridge: American Schools of Oriental Research. Pp. 1-14.
TEFNIN R.
1980 'Deux campagnes de fouilles au Tell Abou Danné in: J. Margueron (Ed.), *Le Moyen Euphrate.* Leiden. Pp. 179- 199.
THOMPSON TH.L.
1974 *The Historicity of the Patriarchal Narratives.* Berlin and New York: Walter de Gruyter.
THUREAU-DANGIN F. AND DUNAND M.
1936 *Til-Barsib*, Paris: Librairie Orientaliste Paul Geuthner.
USSISHKIN D.
1971 'Was Bit-Adini a Neo-Hittite or Aramaean State?', *Orientalia NS* 40: 431-437.
VAN SETERS J.
1975 *Abraham in History and Tradition.* New Haven: Yale University Press.

VON LUSCHAN F.
1893- 1911 *Ausgrabungen in Sendschirli*, Berlin: G. Reimer.
VON OPPENHEIM M.
1938 *Die Beduinen. Band I.* Leipzig: Otto Harrassowitz.
WINTER I.
1983 'Carchemish *ša kišad Puratti*', *Anatolian Studies* 33: 177-197.
WIRTH E.
1971 *Syrien: eine geographische Landeskunde*. Darmstadt: Wissenschaftliche Buchgesellschaft.
YOFFEE N.
1977 *The Economic Role of the Crown in the Old Babylonian Period*, Malibu: Undena (Bibliotheca Mesopotamica 5).
ZARINS J.
1978 'The Camel in Ancient Arabia: a Further Note', *Antiquity* 52: 44-46.
ZEDER M.A.
1984 'Meat Distribution at the Highland Iranian Urban Center of Tal-e Malyan', in: J. Clutton-Brock and C. Grigson (Eds.), *Animals and Archaeology, 3*. Oxford: BAR International Series 202. Pp. 279-307.

Metal restoration in the field

RON LEENHEER

In this paper some simple methods of treatment will be presented which are especially intended for those who may have to treat metal finds in the field but lack a basic knowledge of cleaning and conservation of corroded objects.

INTRODUCTION

Restoration – anyone involved in archaeology is familiar with the concept; nevertheless, it is not superfluous to examine more closely the constituent aims of this specialist field:[1]

1. to preserve the object and extend its 'life'
2. to restore its historical value
3. to restore aesthetic qualities.

These three objectives involve respectively:

1. achieving optimal retardation of the corrosion process [= stabilization]
2. restoring the object to such a form that easily identifies its function, its constituent material, and the method of manufacture [= cleaning]
3. recreating the object of glueing all cleaned fragments together, and, if so desired, adding missing parts out of (coloured) plastics.

Before technical data are given as to how these tasks could be performed in the field, this may be the place to raise some fundamental questions in order to help the reader to decide whether in his specific case restoration in the field is desirable or not. The first question to be answered is: Will the finds be available to the expedition once the excavation season has ended? Many expeditions going abroad are not allowed to take sherds home with them, let alone complete (metal) finds. In Syria the situation is different, particularly concerning salvage projects, though legislation dealing with the 'export' of archaeological finds is tending to become more restrictive.

It is suggested here that if the finds are available immediately after the excavation, restoration in the field should be discarded as an option. In all other cases restoration in the field is advisable if an experienced restorer is a member of the expedition. Unfortunately, this is normally not the case. One possibility is to *make* a member of the expedition into an experienced restorer. Of course, this is easier said than done. One has not only to be passionately interested in this field, but also not devoid of patience and inventness. The promising volunteer should be provided with dispensible scrap metal finds for

Table 1. Treatment of metal finds suggested in this paper.

	Wet wire brush	Abrading disc	Controlled electrolysis	Epoxy hot wax
Iron				
Clean				*
Almost clean	*			*
Rusted	*		(*)	*
Slightly corroded	*	(*)	(*)	*
Heavily corroded (treatment in the field should not be attempted)	*	*	*	*
Lead tin				
Oxidized (thin white cover)	*			*
Slightly corroded (light-gray cover)	*		(*)	*
Heavily corroded (dark-gray irregular crust)	*		*	*
Bronze, silver				
Clean				*
Almost clean	*		(*)	*
Slightly corroded	*		*	*
Heavily corroded	*		*	*

(*) apply if considered necessary
 * strongly advised

experimentation. In due course experience will grow, and consequently disastrous errors are avoided.

In archaeological circles it is a well-known fact that many excavated metal artefacts are stored in depots without ever having been literatim 'discovered'. It goes without saying that clean, i.e. datable, (metal) finds may help substantially to understand one's stratigraphy; as a result it will surely add to the quality of publications if metal finds (e.g. coins) are cleaned *during* the campaign.

The second question is: Should we add missing parts to the incomplete find? Above all, the museum curator should have the final say in this matter. To avoid disappointments it is advisable to avoid such skilful reconstructions that they can hardly be distinguished from the real thing.

At present it is customary to use soluble, or at least removable, materials for protection and completion, since there is a fair chance that the first restoration will not be the last one: sometimes the object proves instable, requiring more treatment or, more likely, the object breaks spontaneously into pieces. A future restorer would welcome easily removable additions.

Last but not least one may ask: How reliable is the 'final' restoration? Is it possible to restore the object to its 'original' form?

The possibility is admittedly infinitely small. If completion should be required, the best we can do is to give the object a 'modal' shape, derived from a variety of comparable well-documented examples.

THE FIELDWORK

The only task of the restorer at the findspot is ensuring that the finds are salvaged in a proper way; the rest of his work has to be done at his table. Even careful removal of dirt with a trowel or knife is not advisable. If the find is fragile, the object is better lifted together with the surrounding earth and put in a box instead of a plastic bag. No doubt it is less exciting for the square supervisor and his team, but it is better for the object's chances of survival to be cleaned by the restorer with his professional equipment in a relatively quiet domain.

THE CLEANING

It may seem overdone, but the first thing to do when starting to clean a find is to make a sketch of it and add its registration number to the drawing. The reason is simple: gradually you will find yourself cleaning a dozen or more finds simultaneously and confusion is inevitable, without documentation. Metal finds are best cleaned individually in a bowl of water. Loose dirt is removed with a paintbrush, whereas more persistent dirt should be attacked with a nylon brush. It is not exceptional for an object to break into pieces and then, the object, together with the rest of the bowl's contents, should be transferred to a sieve. Thus, all fragments of sufficient size to be glued together again will be cleaned and salvaged.

When the find is stripped of dirt, it is time to ask yourself: Is the artefact important enough to be cleaned further and to be stabilized afterwards? This is not a problem if a metal object is found only occasionally, but dozens of Roman nails may be a real burden to the restorer, the more so when their contribution to the understanding of the excavation is relatively low. Therefore, in my view, the restorer should not waste time and materials on relatively redundant finds that may be worked on at home.

The next aspect to examine is whether the selected artefact is corroded or not. If there are no crusts over the object or differently coloured spots on its surface, the only thing left to do is to protect the find from aggressive influences.

A simple way to achieve this is the saturation-in-hot-wax method, which will be described in a later section.

CORRODED METALS

Corroded finds outnumber clean ones and they require a more elaborate treatment or rather combinations of treatment. One delicate way of removing the crust uses scalpels to chip off the fragments in the hope that they let go at the exact transition of the object's original surface and the overlying crust. This is a tedious job, and to do it accurately a stereo microscope is required.

Other techniques are available, and their proper application depends upon discriminating between the various metals.

Lead and tin are mostly found oxidized, not corroded. The thin white surface layer can be removed, only to return in a few hours' time if the object is not treated with hot wax *immediately* after the cleaning. Corroded lead and tin have irregular gray crusts and may be treated in the same way as corroded bronze.

Silver is usually found corroded and can be treated like corroded bronze.

Iron is notorious for its propensity to rust. Less well known is how deformed corroded iron objects can be: frequently the crust exceeds the object itself in thickness. Professional equipment is indispensable, e.g. pneumatically powered abraders, to clean such heavily corroded finds. In the field such finds should not even be seriously considered, only a well-equipped laboratory can deal with them.

Less corroded iron objects come in two varieties:

- those with a shiny metallic core directly below the crust, and
- those with a black core beneath the brown crust.

To remove the crust a small tool is used, i.e. a 'mini-drill' to which all kinds of miniature accessories can be attached, resembling a dentist's equipment. With the aid of a small grindstone attached to the drill, the crust should be removed in a short time. As soon as metal lustre is contacted the abrading disc should be moved to another part of the crust. In places no metal lustre will appear and it is better left as it is: do not 'dig' too deep!

After the complete removal of all crusts, the object should be:

- glued together/reinforced with epoxy, and/or
- impregnated with a tannin-alcohol solution.

The best suitable epoxy, to my knowledge, is ARALDITE AY 103/HY 956, made by CIBA. This combination gives the restorer sufficient working time, whereas other epoxies harden much faster (so take care that the numbers are correct: CIBA makes fast expoxies, too). The impregnation solution is easily made: one part of 'Tanninum pulv.', as it is known at the chemist's, is dissolved in nine parts of alcohol, e.g. methylated spirits. With a small brush this solution should be generously applied to the object, and then left to dry for some 24 hours. After this treatment the metallic parts can be rendered lustrous again by using fine sandpaper. All non-metallic parts will be black instead of the former brown. The final treatment, advisable for all clean(ed) metal finds, is saturation-in-hot-wax. It will be explained in a later section.

Copper/bronze: Corroded copper (alloys) should not be exposed to abrading discs, because of the likeliness of damage. I prefer to use miniature steel wire brushes, fast rotating. It should be noticed, though, that new brushes are too sturdy for vulnerable surfaces: as a rule I consider a wire brush safe to use on an object as long as it does no harm to my fingertip. It is adviseable to wear the brush down on something non-critical, such as a smooth stone. Sometimes a wet treatment with such a fast rotating brush is sufficient to remove the crust entirely, without leaving any trace. The wet treatment, that is, dipping the object into water regularly while the brushing goes on, is better for both the object and the restorer: most of the ensuing dust is caught in the water.

The final problem/question: How to deal with persistent crusts and different coloured spots on non-ferrous metals in a safe and relatively fast way?

The only method that, to my knowledge, works both relatively fast and safe, is *controlled electrolysis*, alternated with steel wire brush cleaning. The necessary investment is modest when weighed against the possibilities of this fine technique; all that is required is a nickel-cadmium battery charger of a specific kind, a stainless steel strip that should fit inside a plastic bowl, alligator clips, several feet of telephone cable, a mini-drill with its tools and sufficient water from the local tap, well or river. The nickel-cadmium (= NiCad) battery charger has to be of the 'intelligent' variety, that is, it should supply a certain pre-set number of milliamperes at its maximum power, even when a short-circuit is caused in the low-tension part, in other words, when the positive and the negative outlets are connected with one another. The most convenient apparatus has two wires ending in alligator clips: one providing positive tension and the other negative. Normally the negative outlet is coloured blue or black, and the positive one red. The output has to be approximately 75 milliamperes maximum. It should be stressed that voltage is of minor importance for electrolysis: the only decisive factor is the maximum amount of milliamperes available, making voltage automatically vary between 15 and almost 0 volt.

CONTROLLED ELECTROLYSIS IN PRACTICE

The plastic or glass container is filled with sufficient water for the find resting on the bottom to be submerged completely and likewise (most of) the steel strip, placed in an upright position. It is possible to fix the strip to the side of the bowl with a plastic peg, or anything else at hand. The corroded object is connected with the negative outlet, and the steel strip with the positive one. There should be a distance of some 10 centimetres between the object and the strip. Switch the power on and after some time a reaction will be visible: very small gas bubbles will escape and a (not seldom blue-green) haze travels slowly from the object towards the steel strip. To put it simply, soluble salt particles are pushed away from the negative pole (= the object) and are attracted by the positive pole (= the steel strip) where they are neutralized. As a result the crust will become softer and darker, while the water will change into a kind of brew beyond description. Therefore, the water must be changed regularly in order to view the object. The wet wire brush treatment should be applied at intervals. Since progress is normally very slow, it will be safe to leave the contraption switched on with checks every eight hours or so.

It is possible to deal with a hoard of objects simultaneously, provided they fit inside the bowl and are easily individually identifiable. In Syria I worked with separated strands of telephone cable, attaching an alligator clip to each brightly coloured wire. The other end of these wires was connected with the negative outlet of the NiCad charger. Each alligator clip was connected with a metal object and around the wires the respective registration numbers were attached with self-adhesive tags.

Instead of a steel strip a long iron strip (locally available) was used, bent to form a circle all along the inside of the bowl. In this way a circular electric field was created with the metal objects in the centre, each of them individually connected with the negative pole, surrounded by the positive pole, the iron ring. In this way the small quantity of energy available was more or less equally divided over the objects under treatment.

REPARATION AND CORROBORATION

There is a fair change that an object will break during treatment. The glue best suited for metals is the aforementioned epoxy ARALDITE AY 103/ HY 956 by CIBA. For a good result the fragments must be completely dry. Even if the object is still in one piece, this epoxy is useful for strengthening the sometimes spongy core of the find. To make it work, the object has to be heated up to some 50 degrees Celcius (easily reached in the Syrian sun) and the well-stirred epoxy is to be applied to the object with a dispensable plastic or wooden tool. It is possible to make a rough mould from Scotch Magic Tape for small objects, to keep the still liquid glue more or less in the desired position. After the hardening of the epoxy, the tape can be quickly removed, since polyester and epoxy do not bond. Superfluous plastic is rapidly abraded with the mini-drill and a suitable attachment.

SATURATION-IN-HOT-WAX

The final task is to protect the clean object against aggressive influences. A simple and effective method is hot wax treatment. The only complicated aspect of this technique is the heating of the wax itself. One should not heat the wax directly on a gas stove. If temperature should rise far over 80 degrees Celcius, the colourless wax will turn brown and after a while it may spontaneously ignite. A safe way is to heat a container with water in which a smaller container is suspended. The containers should not touch. In the upper one the wax is warmed by the heat of the water. The great advantage of this system is that self-ignition is avoided; in addition, the wax will stay hot for a longer time after removal from the heat source. With a pair of tweezers the object is lowered into the hot, liquid wax, and after a short time tiny air bubbles will escape from crevices and cracks in the object. As soon as the air bubbles stop, each of the following has to be considered:

1. whether the object has to be inverted
2. whether the wax has become too cool
3. whether the object is saturated.

These possibilities have to be checked. Sometimes air bubbles do not appear at all. This may indicate that there are no crevices in the object or that the wax has no access to them.

The saturated object should be lifted from the wax with the pair of tweezers and immediately placed in a fair amount of absorbing paper to remove excess wax. Be careful, use lots of paper, not so much to the benefit of the object, but rather to protect your hands from burning: metal objects stay hot for a surprisingly long time!

The result should be a natural-looking metal object without that artificial lustre so characteristic of lacquered restorations. Since the wax has filled all cavities normally accessible for oxygen and vapours, there is a fair chance that after the (hot wax) treatment no more, or at least, far less chemical reaction will take place on and below the surface of the find. In other words, given the constraints of restoration in the field, we try our best to achieve the optimal retardation of the corrosion process.

CONCLUSION

Restoration-in-the-field is not a viable option for every archaeological expedition. One must acknowledge that restoration in the laboratory is beyond doubt a better choice.

Even so, there are circumstances under which it is defensibly better to treat the excavated finds without delay, during the campaign. It is all too common, unfortunately, to be told that finds, once stored, prove completely untracable at a time when they are wanted for further research. Another point in favour of immediate cleaning and conservation is the likely deterioration of the object in storage.

With this paper is it hoped that more archaeologists will come to realize that restoration is less an alchemistic mystery than a set of feasable techniques and may possibly benefit their own excavations!

WARNING

This paper is intended to give ideas for the successful restoration of metal finds under specific conditions. Therefore, such excellent treatments requiring laboratory facilities as the carcinogenic technique of bronze stabilization with Benzotriazol, have been omitted.

ACKNOWLEDGEMENT

To Prof. M.N. van Loon I owe the great privilege to have participated in some of his Syrian expeditions. While there, I came to understand that metal finds deserve a more gentle way of cleaning than – due to insufficient information – was practised. Prof. Van Loon was amongst the first to show a genuine interest in the techniques presented above, and he generously gave me a free hand. My interest in metals was aroused by Dr J.J. Butler (Biological Archaeological Institute, Groningen) during his interesting lectures on the European Bronze Age, whereas Prof. J.H. Crouwel (Archaeological Historical Institute, Amsterdam), an author and lecturer well-known amongst those interested in the Prae- and Protohistory of the Aegean, gave me the opportunity to become experienced with survey techniques in Laconia, during several campaigns. For the more practical side of the story it is my pleasure to thank my teacher at the R.O.B. (National Service for Archaeological Soil Research) in Amersfoort, Mr Ronnie Meijers, and the director of the Restoration Department, Mr G.J. Rombout: they gave me more than just a course.

Special thanks I owe to Mrs Pamela Armstrong (British School at Athens) who changed the Pidgin English of my first draft into the Queen's Dove.

NOTES

1. Jaro M. 1985 'Chemische und elektrochemische Methoden bei der Restaurierung und Konservierung von Bronsefunden', pp. 144-147, in: *Archaeologische Bronzen - antike Kunst, moderne Technik*, Berlin.

A bibliography of Maurits N. Van Loon up to 1988

ODETTE HAEX

1958 with E. Porada, 'Current Near East Excavations', *Archaeology* 11: 285-286.

1962 'A Lion Bowl from Hasanlu', *Expedition* 4/4: 14-19.

1963 Review of Kurt Bittel et al., *Die Hethitische Grabfunde von Osmankayasi (Boğazköy-Hattusa), Bibliotheca Orientalis* 20: 168-169.

1963 64 'Excavations in Northwest Iran, 1962', *Persica* 1: 18-33.

1964 Review of R.D. Barnett and M.Falkner, *The Sculptures of Aššur-Nasir-Apli II (883-859 BC.), Tiglath-Pileser III (745-727 BC), Esarhaddon (681-669 BC) from the Central and South-West Palaces at Nimrud, American Journal of Archaeology* 68: 72-73.

1966 'Pre-Neolithic Village', *Scientific American* 214: 53-54.

1966 'Syria', in: *Council for Old World Archaeology, Area 15, III*: 23-28.

1966 'Mureybit: An Early Village in Inland Syria', *Archaeology* 19: 215-216.

1966 *Urartian Art, its Distinctive Traits in the Light of New Excavations*, Istanbul: Nederlands Historisch-Archaeologisch Instituut.

1966 'First Results of the 1965 Excavations at Tell Mureybit near Meskene', *Annales Archéologiques Arabes Syriennes* 16:211-217.

1967 Review of Machteld J. Mellink (Ed.), *Dark Ages and Nomads c. 1000 B.C., Bibliotheca Orientalis 24:21-26.*

1967 *The Tabqa Reservoir Survey 1964*, Damas-R.A.S.: Direction Génerale des Antiquités et des Musées.

1967 /68 'The Euphrates Valley Expedition', *Oriental Institute Report*, Chicago. Pp. 8-10.

1968 (with James H. Skinner), 'The Oriental Institute Excavations at Mureybit, Syria: Preliminary Report on the 1965 Campaign', *Journal of Near Eastern Studies* 27: 265-290.

1968 (with Giorgio Buccellati), 'The 1968 Excavation at Korucutepe near Elâzig', *Türk Arkeoloji Dergisi* 17,1: 79-85.

1968 Review of Margarete Riemscheider, *Das Reich am Ararat, Bibliotheca Orientalis* 25: 359-362.

1968 'First Results of the 1967 Excavations at Tell Selenkahiye', *Annales Archéologiques Arabes Syriennes* 18: 21-32.

1969 (with Rudolph H. Dornemann), 'New Sites in the Euphrates Valley', *Archaeology* 22: 65-70.

1969 'New Evidence from Inland Syria for the Chronology of the Middle Bronze Age', *American Journal of Archaeology* 73: 276-277.

1969 'Korucutepe', in: M.J. Mellink, 'Archaeology in Asia Minor', *American Journal of Archaeology* 73: 210-211.

1970 (with Hans G. Güterbock), 'The 1970 Excavation at Korucutepe near Elâzig', *Türk Arkeoloji Dergisi*, 19: 127-131.

1970 (with Giorgio Buccellati), 'The University of Chicago-University of California Excavations at Korucutepe 1968' in: *METU, Keban Project Publications - 1968 Summer Work*, Ankara. Pp. 89-102.

1970 'The 1969 Excavation at Korucutepe near Elâzig', *Türk Arkeoloji Dergisi* 18 (1969): 123-128.

1970 Review of Guitty Azarpay, 'Urartian Art and Artifacts, a Chronological Study', *Oriental Art, New Series* 16: 174-175.

1970 Review of Johannes A.H Potratz, *Die Skythen in Südrussland*, *Journal of Near Eastern Studies* 29: 66-72.

1970 Review of R. Ghirshman, *Tchoga Zanbil (Dur-Untash) I, La Ziggurat* and R. Ghirshman, *Tchoga Zanbil (Dur-Untasch) II, Temenos, Temples, Palais, Tombes*, *Bibliotheca Orientalis* 27: 385-389.

1971 'Korucutepe Excavations, 1969, Architecture and General Finds', in: *METU, Keban Project 1969 Activities*, Ankara. Pp. 59-68.

1971 Review of Vadim Mikhailovich Masson, *Sredniaia Aziia i Drevnii Vostok* ('Central Asia and the Ancient Orient'), *Journal of Near Eastern Studies* 30: 226-232.

1972 Review of L. Vanden Berghe, *Het Archeologisch Onderzoek naar de Bronscultuur van Luristan, Opgravingen in Pusht-i Kuh I: Kalwali en War Kabud (1965 en 1966)*, *Bibliotheca Orientalis* 29: 66-69.

1972 (with Hans G. Güterbock), 'Korucutepe Excavations, 1970', in: *METU, Keban Project 1970 Activities*, Ankara. Pp. 83-85.

1973 'First Results of the 1972 Excavations at Tell Selenkahiye', *Annales Archéologiques Arabes Syriennes* 23: 145-158.

1973 'The Excavations at Korucutepe, Turkey, 1968-70: Preliminary Report', *Journal of Near Eastern Studies* 32: 357-376.

1973 Review of Herman Gasche et al., *Tell ed-Der I. Rapport préliminaire sur la première campagne (Février 1970)*, *Bibliotheca Orientalis* 30: 72-73.

1974 'Oude Lering, Nieuwe Nering: Het Uitzwermen der Noord-Syrische Ambachtslieden in de late 8e eeuw v.Chr.', Rede uitgesproken bij de aanvaarding van het ambt van gewoon hoogleraar in de Prehistorie en Archeologie van West-Azië aan de Universiteit van Amsterdam op 2 december 1974.

1974 'The Euphrates mentioned by Sarduri II of Urartu', in: *Anatolian Studies presented to Hans Gustav Güterbock on the Occasion of his 65th Birthday*, Leiden: Nederlands Historisch-Archaeologisch Instituut in het Nabije Oosten. Pp. 187-194.

1974 'History of Urartu and Armenia', *Encyclopaedia Britannica* (15th ed.): 1039-1044.

1974 Mesopotamia, s.v. 'Dress' *Encyclopaedia Britannica* (15th ed.): 1017-1018.

1975 Ed., Korucutepe. Final Report on the Excavations of the Universities of Chicago, California (Los Angeles) and Amsterdam in the Keban Reservoir, Eastern Anatolia 1968-1970. Vol. 1, Amsterdam, Oxford: North-Holland Publishing Company.

1975 'First results of the 1972 Excavations at Tell Selenkahiye', in: *Le Temple et le Culte, XXe RAI*. Leiden. Pp. 21-25.

1975 'The inscription of Ishpuini and Meinua at Qalatgah, Iran', *Journal of Near Eastern Studies* 34: 201-207.

1975 'Die Kunst von Urartu', in: Winfried Orthmann (Ed.), Der Alte Orient, Propyläen Kunstgeschichte 14. Berlin. Pp. 453-466.

1976 85 *Lemmata* in: G. Bartelink et al. (Eds.), *Woordenboek der Oudheid*. Encyclopedisch overzicht van het Oude Egypte, het oude Nabije Oosten, de Grieks-Romeinse wereld en het vroege Christendom, Weesp, I/III.

1977 'Archaeological Evidence of Trade in Western Asia: Problems and Prospects', in: B.L. van Beek et al., eds., *Ex Horreo*, Amsterdam: Universiteit van Amsterdam. Pp. 1-8.

1977 'The Place of Urartu in First Millennium B.C. Trade', *Iraq* 39: 229-231.

1977 78 'First Results of the 1974 and 1975 Excavations at Selenkahiye near Meskene, Syria', *Annales Archéologiques Arabes Syriennes* 27/28: 165-185.

1978 'Oudheden uit het Heilige land in het Allard Pierson Museum', *Mededelingen Vereniging van Vrienden van het Allard Pierson Museum* 16: 4-6.

1978 Ed., *Korucutepe*. Final Report on the Excavations of the Universities of Chicago, California (Los Angeles) and Amsterdam in the Keban Reservoir, Eastern Anatolia, 1968-1970. Vol. 2, Amsterdam et al.: North-Holland Publishing Company.

1978 (with M.C. Cauvin and D. Stordeur), 'Les Outillages Lithiques et Osseux de Mureybet, Syrie, Fouilles Van Loon 1965', *Cahiers de l'Euphrate* 1.

1978/79 (et al.), 'Tall Buqras, 1978-1979', *Archiv für Orientforschung* 26: 152-155.

1979 '1974 and 1975 Preliminary Results of the Excavations at Selenkahiye near Meskene, Syria', *Annual of the American Schools of Oriental Research* 44: 97-112.

1979 Review of O. Aurenche (Ed.), *Dictionnaire illustré multilingue de l'architecture du Proche Orient Ancien, Bibliotheca Orientalis* 36: 217.

1979 Review of Stephan Kroll, *Keramik urartäischer Festungen in Iran, Bibliotheca Orientalis* 36: 370-372.

1979 Review of W. Kleiss und H. Hauptmann in Verbindung mit R.M Boehmer, P.Calmeyer, A. Erzen, A. v. Gladiss. S. Kroll, M. Salvini, G. Tiracjan, *Topografische Karte von Urartu: Verzeignis der Fundorte und Bibliographie, Bibliotheca Orientalis* 36: 372.

1980 Review of Sylvia A. Matheson, *Persia: an Archaeological Guide*, rev. ed., *Persica* 9: 192-197.

1980 'De Prehistorie van West-Azië, in: M. Chamalaun and H.T. Waterbolk, (Eds.), Voltooid verleden tijd?, Intermediair Bibliotheek. Pp. 167-177.

1980 Ed., *Korucutepe*. Final Report on the Excavation of the Universities of Chicago, California (Los Angeles) and Amsterdam in the Keban Reservoir, Eastern Anatolia 1968-1970. Vol. 3, Amsterdam et al.: North-Holland Publishing Company.

1981 'Zegels uit Syrië': Tentoonstelling in het Allard Pierson Museum, Mededelingen van het Allard Pierson Museum 23: 5-8.

1981 Review of H. Gasche and M.J. Steve, l'Acropole de Suse, Nouvelles fouilles (Rapport préliminaire), and H. Gasche, La poterie élamite du deuxième millénaire, a.C., Bibliotheca Orientalis 38: 707-716.

1981 Review of George M.A. Hanfmann and Jane C. Waldbaum, *A Survey of Sardis and the Major Monuments Outside the City Walls, Bibliotheca Orientalis* 38: 698-701.

1982 (et al.), 'The 1976-1977 Excavations at Tell Bouqras', *Annales Archéologiques Arabes Syriennes* 32: 45-57.

1982 'Mitanni: An Early Link between East and West?', *Persica* 10: 47-85.

1982 'Hammam et-Turkman on the Balikh: Background and First Results of the University of Amsterdam's 1981 Excavation', *Akkadica* 27: 30-45.

1983 (with D.J.W. Meijer), 'Hammam et-Turkman on the Balikh: First Results', *Annales Archéologiques Arabes Syriennes* 33,1: 131-152.

1983 (with D.J.W. Meijer), 'Hammam et-Turkman on the Balikh: First Results of the University of Amsterdam's 1984 Excavation', *Annales Archéologiques Arabes Syriennes* 33,2: 299-313.

1983 'Hammam et-Turkman on the Balikh: First results of the University of Amsterdam's 1982 Excavation', *Akkadica* 35: 1-23.

1983 'Teksten uit Urartu uit de 9e-8e eeuw v.Chr.', in: K.R. Veenhof, (Ed.), *Schrijvend Verleden*, Leiden: Ex Oriente Lux. Pp. 32-45.

1983 (et al.), *Op zoek naar Zalpa*, Exhibition in the Allard Pierson Museum. Amsterdam.

1983 Review of David Stronach, *Pasargadae: A Report on the Excavations conducted by the British Institute of Persian Studies from 1961 to 1963, Bibliotheca Orientalis* 40: 742-745.

1984 'Hammam et-Turkman, een stad uit de Bronstijd in Syrië', *Spiegel Historiael* 19: 552-557.

1984 Review of B. Pritchard, *Recovering Sarepta, A Phoenician City, Bibliotheca Orientalis* 41: 485-486.

1984 Review of Mohammad Rahim Sarraf, *Die Keramik von Tell-i Iblis und ihre zeitliche und räumliche Beziehung zu den anderen iranischen und mesopotamischen Kulturen, Bibliotheca Orientalis* 41: 495-496.

1984 'The Netherlands Excavations at Bouqras', Symposium International 'Histoire de Deir ez-Zur', *Annales Archéologiques Arabes Syriennes* 34: 43-61.

1985 'Tall Hammam et-Turkman', Archiv für Orientforschung 31: 140-150.

1985 'Hammam et-Turkman en Syrie du Nord: Récentes découvertes au pays de Rébecca et Rachel', *Archéologia* 208: 38-47.

1985 'Hammam et-Turkman on the Balikh: First result of the University of Amsterdam's 1984 Excavation', *Akkadica* 44: 21-40.

1985 'Konets perioda Rannei Bronzy v Sirii: problemy khronologii i interpretatsii', in: P. Matthiae and I.M. Diakonov (Eds.), *Drevniaia Ebla*. Pp. 52-60.

1985 Review of R. Ghirshman, *Tombe princière de Ziwiyé et le début de l'art animaler, Scyte, Bibliotheca Orientalis* 42: 199-202.

1985 Review of Wolfram Kleiss, *Bastam I: Ausgrabungen in den Urartäischen Anlagen 1972-1975, Bibliotheca Orientalis* 42: 187-194.

1985 Review of Marguerite Yon (Ed.), *Dictionnaire illustré multilingue de la céramique du Proche Orient Ancient, Bibliotheca Orientalis* 42: 171-174.

1985 *Anatolia in the Second Millennium B.C.* Iconography of Religions XV, 12, Institute of Religious Iconography, State University Groningen. Leiden.

1985 'Recent Discoveries in Syria: The Excavations at Hammam et-Turkman', *Proceedings of the British Academy*.

1986 Review of Asad Mahmoud, *Neo-Assyrian Sculptures from Saddikanni (Tell Ajaja), Bibliotheca Orientalis* 43: 794-795.

1986 Review of Hermann Müller-Karpe (Ed.), *Archäologie und Geschichtsbewusstsein, Bibliotheca Orientalis* 43: 790.

1986 'The Drooping Lotus Flower', In: *Insight through Images. Studies in Honour of Edith Porada, Bibliotheca Mesopotamica* 21: 245-252.

1987 (with D.J.W. Meijer), 'Hammam et-Turkman on the Balikh: First Results of the University of Amsterdam 1986 Excavation', *Akkadica* 52: 1-9.

1987 (with P.M.M.G. Akkermans, H.H. Curvers and D.J.W. Meijer), 'De Balikh-Vallei, Archeologie in Tweestromenland', *Natuur en Techniek* 55: 330-343.

1987 Review of Seyyare Eichler, *Götter, Genien und Mischwesen in der urartäischen Kunst, Bibliotheca Orientalis* 44: 227-230.

1987 Review of Igor M. Djakonoff, *The Pre-History of Armenian People, Bibliotheca Orientalis* 44: 230-233.

1987 Review of *Archäologische Mitteilungen aus Iran, Vol. 16, Bibliotheca Orientalis* 44: 251-252.

1987 Review of Paolo Emilio Pecorella and Mirjo Salvini, eds., *Tra lo Zagros e l'Urmia: Ricerche storiche ed archeologiche nell' Azerbaigan Tranico, Bibliotheca Orientalis* 44: 552-261.

1988 Review of Paul E. Zimansky, *Ecology and Empire: The Structure of the Urartian State, Journal of the American Oriental Society* 108: 163-165.

1988 'New evidence for North Syrian Chronology from Hammam et-Turkman', *American Journal of Archaeology* 92: 582-587.

in press *Hammam et-Turkman I. Report on the University of Amsterdam's 1981-1984 Excavations in Syria*. Uitgaven van het Nederlands Historisch-Archeologische Instituut te Istanbul.

in press (with H.H. Curvers) *The Holmes Expeditions to Luristan*, Oriental Institute Publications. Chicago.

in press (with D.J.W. Meijer and P.M.M.G. Akkermans), 'Hammam et-Turkman and Sabi Abyad on the Balikh, The University of Amsterdam Excavation 1986', *Annales Archéologiques Arabes Syriennes*.

in press *Gids voor de afdeling West Azië van het Allard Pierson Museum Amsterdam*, Leiden: Ex Oriente Lux.